Bacterial Adhesion to Host Tissues
Mechanisms and Consequences

This book is about the adhesion of bacteria to their human hosts. Although adhesion is essential for maintaining members of the normal microflora in/on their host, it is also the crucial first stage in any infectious disease. It is important, therefore, to fully understand the mechanisms underlying bacterial adhesion so that we may be able to develop methods of maintaining our normal (protective) microflora and of preventing pathogenic bacteria from initiating an infectious process. These topics are increasingly important because of the growing prevalence of antibiotic-resistant bacteria and, consequently, the need to develop alternative approaches for the prevention and treatment of infectious diseases. This book describes the bacterial structures responsible for adhesion and the molecular mechanisms underlying the adhesion process. A unique feature is that it also deals with the consequences of adhesion for both the adherent bacterium and the host cell/ tissue to which is has adhered. Researchers and graduate students in microbiology and molecular medicine will find this book to be a valuable overview of current research on this exciting and rapidly developing topic.

MICHAEL WILSON is Professor of Microbiology in the Faculty of Clinical Sciences, University College London, and Head of the Department of Microbiology at the Eastman Dental Institute, University College London. He is the co-editor of *Community Structure and Co-operation in Biofilms* (2000) and co-author of *Bacterial Disease Mechanisms: An Introduction to Cellular Microbiology* (2002).

Over the past decade, the rapid development of an array of techniques in the fields of cellular and molecular biology have transformed whole areas of research across the biological sciences. Microbiology has perhaps been influenced most of all. Our understanding of microbial diversity and evolutionary biology, and of how pathogenic bacteria and viruses interact with their animal and plant hosts at the molecular level, for example, have been revolutionized. Perhaps the most exciting recent advance in microbiology has been the development of the interface discipline of Cellular Microbiology, a fusion of classical microbiology, microbial molecular biology and eukaryotic cellular and molecular biology. Cellular Microbiology is revealing how pathogenic bacteria interact with host cells in what is turning out to be a complex evolutionary battle of competing gene products. Molecular and cellular biology are no longer discrete subject areas but vital tools and an integrated part of current microbiological research. As part of this revolution in molecular biology, the genomes of a growing number of pathogenic and model bacteria have been fully sequenced, with immense implications for our future understanding of microorganisms at the molecular level.

Advances in Molecular and Cellular Microbiology is a series edited by researchers active in these exciting and rapidly expanding fields. Each volume will focus on a particular aspect of cellular or molecular microbiology, and will provide an overview of the area, as well as examining current research. This series will enable graduate students and researchers to keep up with the rapidly diversifying literature in current microbiological research.

CELLULAR MICROBIOLOGY

ADVANCES IN MOLECULAR AND

Series Editors

Professor Brian Henderson
University College London

Professor Michael Wilson
University College London

Professor Sir Anthony Coates
St George's Hospital Medical School, London

Professor Michael Curtis
St Bartholemew's and Royal London Hospital, London

Advances in Molecular and Cellular Microbiology 1

Bacterial Adhesion to Host Tissues

Mechanisms and Consequences

EDITED BY

Michael Wilson

University College London

CAMBRIDGE
UNIVERSITY PRESS

CAMBRIDGE UNIVERSITY PRESS
Cambridge, New York, Melbourne, Madrid, Cape Town, Singapore,
São Paulo, Delhi, Dubai, Tokyo

Cambridge University Press
The Edinburgh Building, Cambridge CB2 8RU, UK

Published in the United States of America by Cambridge University Press, New York

www.cambridge.org
Information on this title: www.cambridge.org/9780521126755

First published 2002
This digitally printed version 2009

A catalogue record for this publication is available from the British Library

ISBN 978-0-521-80107-2 Hardback
ISBN 978-0-521-12675-5 Paperback

Additional resources for this publication at www.cambridge.org/9780521126755

Contents

Plate section is between pp. 240 and 241*

*A colour version of these plates is available for download
from www.cambridge.org/9780521126755

Contributors

Goran Bergsten
Department of Microbiology, Immunology and Glycobiology (MIG)
Institute of Laboratory Medicine
Lund University
Sölvegatan 23
S-223 62 Lund
Sweden

Ian Blomfield
Department of Biosciences
University of Kent at Canterbury
Canterbury
Kent CT2 7NJ
UK

Frank Ebel
Unité de Génétique Moléculaire
Institut Pasteur
25 rue du Dr. Roux
75724 Paris
France

Richard P. Ellen
Faculty of Dentistry
University of Toronto
124 Edward Street
Toronto
Ontario
Canada M5G 1G6

Hans Fischer
Department of Microbiology, Immunology and Glycobiology (MIG)
Institute of Laboratory Medicine
Lund University
Sölvegatan 23
S-223 62 Lund
Sweden

Timothy J. Foster
Department of Microbiology
Moyne Institute of Preventive Medicine
Trinity College
Dublin 2
Ireland

Gad Frankel
Department of Biochemistry and Centre for Molecular Microbiology & Infection
Imperial College of Science Technology and Medicine
London SW7 2AZ
UK

Bjorn Frendéus
Department of Microbiology, Immunology and Glycobiology (MIG)
Institute of Laboratory Medicine
Lund University
Sölvegatan 23
S-223 62 Lund
Sweden

Janet R. Gilsdorf
Department of Pediatrics and Communicable Diseases
University of Michigan Medical School
Ann Arbor, MI 48109
USA

Gabriela Godaly
Department of Microbiology, Immunology and Glycobiology (MIG)
Institute of Laboratory Medicine
Lund University
Sölvegatan 23
S-223 62 Lund
Sweden

Erika Gustafsson
Department of Microbiology, Immunology and Glycobiology (MIG)
Institute of Laboratory Medicine
Lund University
Sölvegatan 23
S-223 62 Lund
Sweden

Carlos A. Guzmán
Vaccine Research Group
Division of Microbiology
GBF-German Research Centre for Biotechnology
Mascheroder Weg. 1
D-38124 Braunschweig
Germany

Pauline S. Handley
1.800 Stopford Building
School of Biological Sciences
University of Manchester
Oxford Road
Manchester M13 9PT
UK

Long Hang
Department of Microbiology, Immunology and Glycobiology (MIG)
Institute of Laboratory Medicine
Lund University
Sölvegatan 23
S-223 62 Lund
Sweden

Elizabeth L. Hartland
Department of Microbiology
Monash University
Clayton 3800
Victoria
Australia

Maria Hedlund
Department of Microbiology, Immunology and Glycobiology (MIG)
Institute of Laboratory Medicine
Lund University
Sölvegatan 23
S-223 62 Lund
Sweden

Howard F. Jenkinson
Department of Oral and Dental Science
University of Bristol Dental School
Bristol BS1 2LY
UK

Stuart Knutton
Institute of Child Health
University of Birmingham
Birmingham B4 6NH
UK

Andreas U. Kresse
Vaccine Research Group
Division of Microbiology
GBF-German Research Centre for Biotechnology
Mascheroder Weg. 1
D-38124 Braunschweig
Germany

Richard J. Lamont
Department of Oral Biology
University of Washington
Seattle, WA 91895-7132
USA

Ann-Charlotte Lundstedt
Department of Microbiology, Immunology and Glycobiology (MIG)
Institute of Laboratory Medicine
Lund University
Sölvegatan 23
S-223 62 Lund
Sweden

Roderick McNab
Department of Microbiology
Eastman Dental Institute
256 Grays Inn Road
University College London
London WC1X 8LD
UK

Dorothy E. Pierson
Department of Microbiology and Immunology
C5181 Veterinary Medical Center
Cornell University College of Veterinary Medicine
Ithaca, NY 14853
USA

Ian S. Roberts
1.800 Stopford Building
School of Biological Sciences
University of Manchester
Oxford Road
Manchester M13 9PT
UK

Martin Samuelsson
Department of Microbiology, Immunology and Glycobiology (MIG)
Institute of Laboratory Medicine
Lund University
Sölvegatan 23
S-223 62 Lund
Sweden

Patrik Samuelsson
Department of Microbiology, Immunology and Glycobiology (MIG)
Institute of Laboratory Medicine
Lund University
Sölvegatan 23
S-223 62 Lund
Sweden

Catharina Svanborg
Department of Microbiology, Immunology and Glycobiology (MIG)
Institute of Laboratory Medicine
Lund University
Sölvegatan 23
S-223 62 Lund
Sweden

Majlis Svensson
Department of Microbiology, Immunology and Glycobiology (MIG)
Institute of Laboratory Medicine
Lund University
Sölvegatan 23
S-223 62 Lund
Sweden

Muhamed-Kheir Taha
Unité des Neisseria and Centre National de Référence des Méningocoques
(CNRM)
Institut Pasteur
28 rue du Dr Roux
75724 Paris cedex 15
France

Clare Taylor
1.800 Stopford Building
School of Biological Sciences
University of Manchester
Oxford Road
Manchester M13 9PT
UK

Marjan van der Woude
Department of Microbiology
University of Pennsylvania
202A Johnson Pavilion
3610 Hamilton Walk
Philadelphia, PA 19104
USA

Mumtaz Virji
Department of Pathology and Microbiology
School of Medical Sciences
University of Bristol
Bristol BS8 1TD
UK

Bjorn Wullt
Department of Microbiology, Immunology and Glycobiology (MIG)
Institute of Laboratory Medicine
Lund University
Sölvegatan 23
S-223 62 Lund
Sweden

CONTRIBUTORS

Preface

Except when *in utero*, every human being is colonized by approximately 10^{14} microbes (mainly bacteria) that constitute the normal microflora. Apart from those organisms that are present in the lumen of the intestinal tract, all of these microbes maintain an association with their host by adhering to some cell, tissue or secretion. It is increasingly being realized that not only is this normal microflora beneficial to its host (e.g. by providing vitamins and protection from exogenous pathogens) but it is, indeed, essential for the host's proper development, for example in the differentiation and maturation of the intestinal tract, immune system, etc. A knowledge of how members of the normal microflora adhere to their host is, therefore, important in understanding the mutualistic association we know as *Homo sapiens*.

Adhesion of a pathogenic organism to its host is also the first stage in any infectious disease and this truism provides another justification for studying bacterial adhesion. Interest in this aspect of bacterial virulence is increasing rapidly as our armamentarium of antibiotics dwindles in effectiveness owing to the development of resistance in major pathogens. It is believed (and strongly hoped) that research into bacterial adhesion mechanisms will identify new targets for therapeutic intervention.

This book is intended to update the reader in key areas of research in the field of bacterial adhesion to host cells and tissues. It is divided into three parts that deal with: (I) the mechanisms underlying the adhesion of bacteria to host structures, (II) the effect that adhesion to host cells has on bacteria and (III) the consequences for the host cell (or tissue) of bacterial adhesion. The first part describes recent advances in our understanding of the adhesive structures and adhesins found in bacteria colonizing (and causing disease in) a variety of habitats in a human being – the oral cavity, respiratory tract, gut, urinary tract, skin and internal tissues. The remaining two parts concentrate on less well understood aspects of adhesion – the effects that this process has on the adherent organism and on the cell/tissue to which it has adhered. While we know very little about the former, our knowledge of what happens to host cells following bacterial adhesion is increasing at a rapid rate. With regard to the latter, one frequent consequence of bacterial adhesion to a host cell is invasion of that cell. Only one chapter dealing with this topic has been included (this focuses on organisms other than the classical invasive pathogens) as this huge subject will be covered in a later volume in this series.

Michael Wilson

PART I   Bacterial adhesins and adhesive structures

CHAPTER 1

Surface protein adhesins of staphylococci

Timothy J. Foster

1.1 INTRODUCTION

Staphylococcus aureus is primarily an extracellular pathogen. In order to initiate infection it adheres to components of the host extracellular matrix (ECM). Adherence is mediated by surface protein adhesins called MSCRAMMs (microbial surface components recognizing adhesive matrix molecules) (Patti *et al.*, 1994a). In most cases the MSCRAMMs are covalently bound to peptidoglycan in the cell wall. However, there are several examples of MSCRAMMs that are non-covalently associated with the wall. Coagulase-negative staphylococci also express MSCRAMMs. This chapter will discuss the mechanisms of attachment of proteins to the cell wall and will review the properties of MSCRAMM proteins that have been characterized at the molecular level.

1.2 ANCHORING OF PROTEINS TO THE CELL WALL

Cell-wall-anchored proteins that are covalently bound to peptidoglycan are recognizable by a motif located at the C-terminus (Navarre and Schneewind, 1999). This comprises the sequence LPXTG (Leu-Pro-X-Thr-Gly) followed by hydrophobic residues that span the cytoplasmic membrane and by several positively charged residues. The positively charged residues are required to hold the protein transiently in the membrane during secretion through the Sec secretome (Schneewind *et al.*, 1993). The LPXTG sequence is recognized by an enzyme called sortase that cleaves LPXTG between the Thr and Gly residues (Navarre and Schneewind, 1994; Ton-That *et al.*, 1999). The carboxyl group of the Thr is joined to the amino group of the branch peptide in nascent peptidoglycan. In the case of *S. aureus*, linkage occurs to the NH_2 group of the fifth Gly residue of the branch peptide, which would otherwise form the interpeptide bridge of cross-linked peptidoglycan

(Ton-That and Schneewind, 1999; Ton-That *et al.*, 1999). The wall-anchored protein becomes joined to the lipid-linked intermediate prior to its incorporation into peptidoglycan (Ton-That *et al.*, 1997). The covalently linked protein can be released from the cell only by enzymatic degradation of peptidoglycan. The glycine endopeptidase lysostaphin releases proteins of homogeneous size whereas muramidases release proteins of heterogeneous size owing to varying amounts of peptidoglycan attached to the C-terminus (Schneewind *et al.*, 1993).

The enzyme that catalyses the sorting reaction is sortase. The protein has an N-terminal hydrophobic domain that provides anchorage to the outer face of the cytoplasmic membrane (Mazmanian *et al.*, 1999; Ton-That *et al.*, 1999). Indeed it is likely that sortase is closely associated with the secretome because it must capture proteins as they are in the process of being secreted through the Sec pathway. Interestingly, a sortase-defective mutant can grow normally *in vitro*, which shows that sortase is not an essential enzyme. The mutant is defective in the expression of several LPXTG-anchored surface proteins and has reduced virulence in murine infection models (Mazmanian *et al.*, 2000). This indicates that sortase could be a novel target for antimicrobial agents.

1.3 CELL-WALL-ASSOCIATED PROTEINS

1.3.1 Protein A

Protein A (Spa) is the archetypal cell-wall-anchored protein of *S. aureus*. It is known primarily for its ability to bind the Fc region of immunoglobulin (Ig) G. Its structural organization is somewhat different from that of other surface proteins in that the N-terminal signal sequence is followed by tandem repeats of five homologous IgG binding domains (Fig. 1.1; Uhlén *et al.*, 1984). Each is composed of an approximately 60 amino acid residue unit that forms three α-helices (Starovasnik *et al.*, 1996). The structure of the subdomain B in complex with the Fc region of IgG subclass 1 has been solved by X-ray analysis of a co-crystal (Deisenhofer, 1981). The binding between the two molecules involves nine amino acid residues in the IgG fragment and 11 amino acid residues in the protein A domain (Gouda *et al.*, 1998). The binding characteristics and specificity of the Spa–IgG interaction have been analysed in great detail (Langone, 1982).

One important role of Spa in staphylococcal infections is that it is antiphagocytic. By binding to protein A on the bacterial surface, the Fc region of IgG is not available for recognition by the Fc receptor on polymorphonuclear leukocytes (PMNLs) (Gemmell *et al.*, 1990). A protein A-defective mutant

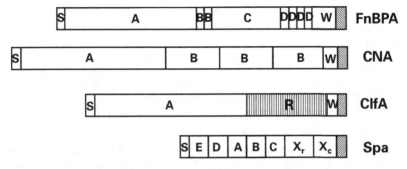

Figure 1.1. Organization of surface proteins of *S. aureus*. The domain organization of the fibronectin-binding protein A (FnBPA), the collagen-binding protein CNA, the fibrinogen-binding protein (ClfA) and protein A (Spa). The signal sequences (S) are removed during secretion across the cytoplasmic membrane. Region X_r of protein A is a proline-rich octapeptide repeat that spans the cell wall. Region X_c is a non-repeated (constant) region. Each protein has common features at the C-terminus indicated by the cross-hatched box (LPXTG motif, hydrophobic region, and positively charged residues). Regions W and R are peptidoglycan-spanning regions. (From Foster and Höök, 2000; with permission from American Society for Microbiology Press.)

was more avidly phagocytosed by PMNLs in the presence of normal serum opsonins than was the wild type, and the mutant was less virulent in murine infection models (Patel *et al.*, 1987). The recent observation that Spa can mediate adherence of bacteria to von Willebrand factor, an extracellular matrix protein important in normal haemostasis, suggests that protein A may have an additional role in the infection process (Hartlieb *et al.*, 2000).

1.3.2 Fibronectin-binding proteins

Fibronectin (Fn) is a dimeric glycoprotein that occurs in a soluble form in body fluids and in a fibrillar form in the ECM (Hynes, 1993). A primary function of insoluble Fn is to act as a substratum for the adhesion of cells mediated by integrin receptors that bind to specific sites in the central part of Fn (Yamada, 1989). The primary binding site for staphylococcal Fn-binding protein is in the 29 kDa N-terminal domain, which is composed of five type I modules (Sottile *et al.*, 1991; Potts and Campbell, 1994).

Most strains of *S. aureus* express two related Fn-binding proteins FnBPA and FnBPB, which are encoded by closely linked genes (Signás *et al.*, 1989; Jönsson *et al.*, 1991). One survey of 163 isolates comprising carriage strains, as well as strains from invasive disease and orthopaedic device-related infection, found that 77% had both *fnbA* and *fnbB* genes, while 23% had only *fnbA*

(Peacock *et al.*, 2000). Strains from orthopaedic infections adhered to Fn at a significantly higher level than did carriage strains or strains from non-device-related infections.

FnBPA and FnBPB have a structural organization similar to that of FnBPs from streptococci (Fig. 1.1; Joh *et al.*, 1994; Patti *et al.*, 1994a; Foster and Höök, 1998). The primary ligand-binding domain (D), which is almost identical in FnBPA and FnBPB, is located very close to the cell-wall-spanning domains (region W) and is composed of three to five repeats of an approximately 40 residue motif. Synthetic peptides mimicking repeated units effectively inhibit Fn binding to bacteria and bacterial attachment to immobilized Fn (Raja *et al.*, 1990).

Studies with peptides and recombinant proteins expressing combinations of different FnBP D repeats indicate that the major interaction occurs between FnBP repeat D3 and Fn modules 4 and 5, but that other discrete sequences within the D region bind to different Fn type I module pairs (Joh *et al.*, 1998). The ligand-binding domain of FnBP reacts simultaneously at multiple sites with Fn (Fig. 1.2). A consensus Fn-binding motif is present with each D repeat (McGavin *et al.*, 1991, 1993). The interaction between the MSCRAMM and Fn involves structural rearrangements in the D repeat region. The ligand-binding D repeat region has an unordered structure and acquires a defined conformation upon binding to the rigid type I modules of Fn (House-Pompeo *et al.*, 1996). This conformational change is accompanied by the formation of neo-epitopes called ligand-induced binding site (LIBS) epitopes that can be demonstrated by monoclonal antibodies and by antibodies isolated from patients recovering from staphylococcal infection (Speziale *et al.*, 1996). The antibodies that recognize the neo-epitopes do not interfere with the MSCRAMM–Fn interaction but rather stabilize the FnBP–ligand complex and appear to promote Fn binding. The immunodominant non-LIBS epitopes in D1–D3 are confined to repeats D1 and D2 (Sun *et al.*, 1997). They are very similar to each other and bind Fn with lower affinity than does D3. The region in D3 corresponding to epitopes in D1–D2 contains the Fn-binding consensus but is otherwise divergent (8/21 different residues). Antibodies in polyclonal sera preferentially react with the low affinity D1–D2 domains and block Fn binding to bacteria by no more than 50%.

Staphylococcus aureus can invade cultured fibroblasts, endothelial and epithelial cells, and Fn plays a major role (Dziewanowska *et al.*, 1999; Lammers *et al.*, 1999; Peacock *et al.*, 1999a; Sinha *et al.*, 1999; Fowler *et al.*, 2000). Bacteria either recruit soluble Fn or bind to Fn bound to the surface of host cells. Bacteria bind Fn via the type I modules at the N-terminus. Fn

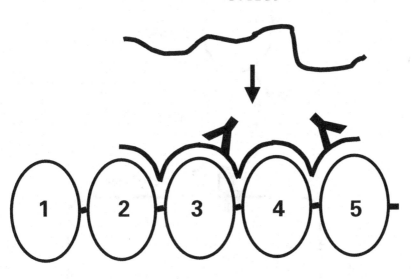

D1-D2-D3

Fibronectin type I modules

Figure 1.2. Interaction of the ligand-binding region of Fn-binding proteins with Fn. The wavy line represents the ligand-binding D1–D2–D3 repeats of FnBPs, which do not have secondary structure. The protein interacts with the type I modules of Fn and takes on a discernible secondary structure with the formation of neo-epitopes (ligand-induced binding site epitopes). (From Foster and Höök, 2000; with permission from American Society for Microbiology Press.)

is bound to the $\alpha_5\beta_1$ integrin on the surface of the host cell at the central Fn RGD (Arg-Gly-Asp) motif-bearing module. Thus Fn forms a bridge between the bacterial FnBP adhesin and the mammalian cell integrin (Fig. 1.3). This results in stimulation of phagocytosis and bacteria become internalized. FnBP-defective mutants of *S. aureus* are not taken up, non-invasive bacteria that acquire FnBP expression become invasive, and bacterial internalization is blocked by soluble recombinant D repeat regions of FnBP and by anti-integrin function-blocking antibodies. The importance of internalization *in vivo* is unclear, but it could be involved in bacterial escape from the bloodstream and invasion of internal organs, in the initiation of invasive endocarditis, and in bacterial persistence.

FnBPs are considered to be important virulence factors in the initiation

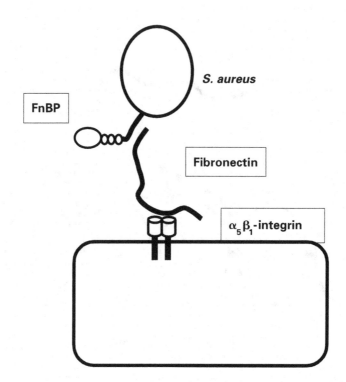

Figure 1.3. Role of Fn in promoting bacterial attachment to mammalian cells. The N-terminal type I modules of Fn are bound to the D1–D2–D3 region of FnBP attached to the cell surface of *S. aureus*. The same molecule of Fn binds to the $\alpha_5\beta_1$ integrin via the centrally located RGD motif. This stimulates actin rearrangement and bacterial internalization.

of foreign body infection. They promote bacterial adherence to immobilized Fn *in vitro*, and to implanted biomaterial that has been in long-term contact with the host such as plastic coverslips implanted subcutaneously in guinea pigs (Greene *et al.*, 1995) and titanium screws implanted in the iliac bone of guinea pigs (Fischer *et al.*, 1996). In contrast, fibrinogen appears to be the major adhesion-promoting factor in the conditioning layer on biomaterial that had been in short-term contact with the host (Vaudaux *et al.*, 1995; François *et al.*, 2000). There are conflicting data concerning the ability of FnBPs to promote infection in experimental animals. One study with a FnBP-defective mutant of strain 879 indicated that the bacterial MSCRAMM is important in promoting bacterial adhesion to damaged heart valve tissue in the rat model for endocarditis (Kuypers and Proctor, 1989), while another report with mutants of strain 8325–4 found no effect (Flock *et al.*, 1996).

These contradictory data may reflect different bacterial strains (8325–4 is known to express FnBPs poorly *in vitro*) or differences in performing the infection models.

1.3.3 Fibrinogen-binding proteins

Fibrinogen is a large protein of M_r 340 000. It is composed of three polypeptide chains (α, β, γ) that are extensively linked by disulphide bonds to form an elongated dimeric structure (Ruggeri, 1993). It is the most abundant ligand for the integrin $\alpha_{IIb}/\beta3$ (glycoprotein gpIIb/IIIa) on the surface of platelets. The binding of Fg to the integrin receptor on activated platelets results in platelet aggregation and the formation of platelet–fibrin thrombi (Hawiger, 1995). The C-terminal sequences of the α-, β- and γ-chains form independently folded globular domains. The three dimensional structure of the γ-chain module is known (Spraggon *et al.*, 1997; Doolittle *et al.*, 1998)

Until recently it was thought that the ability of *S. aureus* to adhere to Fg-coated substrates and to form clumps in a solution containing Fg (e.g. plasma) was solely due to the clumping factor ClfA (McDevitt *et al.*, 1994, 1995). It is now known that *S. aureus* can express other Fg-binding adhesins: the ClfB protein, which is related to ClfA (Ní Eidhin *et al.*, 1998); and the Fg-binding proteins, which can also interact with Fg via their A domains (Wann *et al.*, 2000). The *clfA* and *clfB* genes are not allelic variants but are distinct genes. They are not closely linked, in contrast to the *fnbA* and *fnbB* genes.

The structural organization of ClfA and ClfB is very similar (Fig. 1.4). The surface-exposed approximately 500 residue ligand-binding A domains are linked to the cell wall via the R domain, which comprises mainly the Ser-Asp dipeptide repeats. The R domain appears to serve as a flexible stalk, allowing the presentation of the A domain on the surface for ligand interactions (Hartford *et al.*, 1997). FnBPA and FnBPB also possess N-terminal A domains that have sequence similarity with the A domains of ClfA and ClfB, and in the case of FnBPA promote binding to Fg (Wann *et al.*, 2000). Otherwise, the FnBP and Clf proteins have a completely different structural organization, apart from the typical cell-wall-anchoring signals at the extreme C-terminus.

Although the structural organization of ClfA and ClfB is similar, the amino acid sequences of the ligand-binding A domains are only 27% identical. ClfA and ClfB bind to different sites in Fg. ClfA recognizes the flexible peptide that extends from the γ-module at the C-terminus of the γ-chain (Fig. 1.5; McDevitt *et al.*, 1997) whereas ClfB binds to the α-chain (Ní Eidhin *et al.*, 1998). The A domain of FnBPA is approximately 25% identical with that of ClfA and binds to the same site in the Fg γ-chain (Wann *et al.*, 2000).

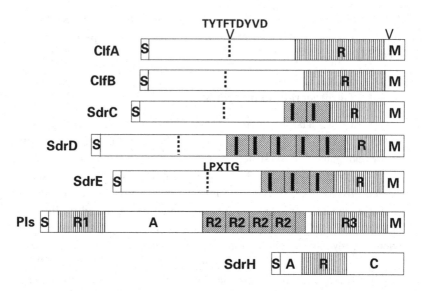

Figure 1.4. The Sdr multigene protein family in staphylococci. The domain organization of different members of the Sdr protein family is shown. The approximately 500 residue A domain of ClfA, ClfB, SdrC, SdrD, SdrE, SdrF and SdrG have a conserved sequence TYTFTDYVD. SdrF and SdrG of *S. epidermidis* have an organization similar to that of the SdrC, SdrD and SdrE proteins of *S. aureus* but are not shown. The A domain of Pls has no sequence similarity to the A domains of the Clf–Sdr proteins, which are related by about 25–30%.

The ClfA domain recognizes a site located at the extreme C-terminus of the γ-chain of Fg (residues 399–411). These residues form a flexible extension from the globular γ-module (Fig. 1.5; Spraggon *et al.*, 1997: Doolittle *et al.*, 1998). A 17 amino acid residue synthetic peptide corresponding to the Fg γ-chain residues 399–411 binds to a recombinant form of the A domain in an interaction that is inhibited by Ca^{2+} (O'Connell *et al.*, 1998). The A domain contains a motif that is reminiscent of a Ca^{2+}-binding EF-hand, and site-specific mutations in this motif resulted in a protein with lower affinity for the γ-chain peptide and less sensitivity to Ca^{2+}. Thus ClfA exhibits fibrinogen-binding characteristics similar to those of the platelet integrin α_{IIb}/β_3 (O'Connell *et al.*, 1998). Both bind to the same site in Fg in interactions that are affected by Ca^{2+}. The recombinant form of the ClfA A domain is a potent inhibitor of Fg-dependent platelet aggregation (McDevitt *et al.*, 1997). This could be a bacterial defence mechanism to prevent release of antimicrobial peptides during degranulation (Yeaman *et al.*, 1992) that might occur during platelet aggregation in the vicinity of colonizing bacteria.

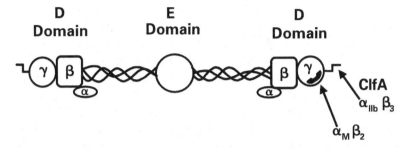

Figure 1.5. Structure of fibrinogen. Schematic diagram showing the structural organization of fibrinogen. The globular D domains comprise the C-terminal residues of the α-, β- and γ-chains. The C-terminus of the γ-chain protrudes from the globular γ-module. Binding sites for ClfA and integrins are shown. The E domain contains the N-terminal residues of α-, β- and γ-chains cross-linked by disulphide bridges. (From Foster and Höök, 2000; with permission from American Society for Microbiology Press.)

The binding of ClfA to fibrinogen is progressively inhibited by Ca^{2+} in the range 1–10 mM (O'Connell *et al.*, 1998). The concentration of free Ca^{2+} in blood plasma is 1.3 mM and is closely regulated at the threshold of the inhibitory range, although concentrations can vary more widely in extracellular spaces (Brown *et al.*, 1995). However, at platelet-rich thrombi, and possibly on the surface of freshly implanted biomaterial, the Ca^{2+} concentration appears to be considerably lower and may allow ClfA to bind fibrinogen. Thus, as bacteria circulate in plasma, they will tend to adhere to Fg/platelet-containing coagulation sites.

The ClfB region A binds to the Fg α-chain (Ní Eidhin *et al.*, 1998) but the precise binding site has not been defined. ClfB-promoted binding to immobilized Fg is also inhibited by millimolar concentrations of Ca^{2+}. The ClfB protein is expressed maximally during the early part of the exponential phase of growth (Ní Eidhin *et al.*, 1998; McAleese *et al.*, 2001). Transcription (and hence translation) terminates before the culture reaches stationary phase, so ClfB molecules become diluted amongst the progeny cells during the remaining cell divisions and some are released into the culture supernatant by cell wall turnover. Also, some of the ClfB protein is cleaved by the metalloprotease aureolysin (McAleese *et al.*, 2001). Protease cleavage results in loss of an N-terminal domain and the protein loses its ability to bind Fg. In contrast, ClfA is expressed abundantly on cells from the stationary phase of growth. It is also cleaved at a site similar to that of ClfA but does not lose ligand-binding activity (McDevitt *et al.*, 1995).

The ClfA protein is the primary adhesin of *S. aureus* for promoting bacterial interactions with Fg. This is particularly the case for cells in the stationary phase of growth where ClfB is absent. ClfA promotes bacterial attachment to immobilized Fg (McDevitt *et al.*, 1994), to plasma clots formed *in vitro* (Moreillon *et al.*, 1995) and to plastic biomaterial that had been exposed to blood for short-term conditioning (Vaudaux *et al.*, 1995; François *et al.*, 2000). In addition, a ClfA⁻ mutant had reduced ability to bind to damaged heart tissue in the rat endocarditis model (Moreillon *et al.*, 1995). The infection rate was lower in the ClfA⁻ mutant but was restored in the mutant complemented with the wild-type *clfA* gene. However, reduced virulence was not manifested at higher infection doses, indicating that other adhesins can compensate for lack of ClfA. Similarly, when exponential phase cells that lacked ClfA were used, the ClfB protein was shown to act as a bacterial adhesin and virulence factor (Ní Eidhin *et al.*, 1998; Entenza *et al.*, 2000).

ClfA has recently been shown to be an important virulence factor in the mouse model of septic arthritis (Josefsson *et al.*, 2001). The ClfA⁻ mutant was significantly less virulent in the bacteraemia phase of the infection, causing fewer mortalities and less weight loss than did the wild type. The mutant also caused much reduced tissue damage in infected joints as compared with wild-type infection. The effect in joints was observed both in mice that were injected intravenously, where the bacteria may have been eliminated in the bloodstream by more efficient phagocytosis, and also when bacteria were injected directly into the joint. It is possible that reduced virulence is linked to the observation that ClfA⁻ mutants are more susceptible to phagocyotsis by PMNLs.

1.3.4 Proteins of the Sdr family in *S. aureus*

ClfA and ClfB are members of a larger family of structurally related surface proteins characterized by the R domain containing Ser-Asp dipeptide repeats. Strains of S. aureus contain a tandem array of related genes, sdrC, sdrD and sdrE. The proteins are predicted to have a structural organization similar to that of ClfA and ClfB, except for an additional B repeat comprising 110–113 residues located between the A domain and the R domain (Fig. 1.4; Josefsson *et al.*, 1998a). The function of the B domain is not known. Each repeat has three high affinity Ca^{2+}-binding sites (Josefsson *et al.*, 1998b). Bound Ca^{2+} is required to promote the rigid rod-like structure of the B repeat array. It is possible that the B region acts as a non-flexible stalk, which, in combination with the more flexible R region, is required for surface display of the (putative) ligand-binding A domain. The A domains of the Sdr proteins are of a size similar to that of the A domains of ClfA and ClfB and have about 30% sequence identity where any pairwise combination is compared. However, unlike ClfA and ClfB, the ligands of SdrC, SdrD and SdrE have not been elucidated, with the exception of a variant of SdrD (Bbp) that binds to bone sialoprotein (Tung *et al.*, 2000). In contrast, SdrD, the A domain of which is about 75% identical with Bbp, does not bind bone sialoprotein. We have recently shown that SdrE, when expressed on the surface of *Lactococcus lactis*, can promote platelet aggregation, most likely mediated by binding to a plasma protein that acts as a bridge between the bacteria and a platelet receptor (O'Brien *et al.*, 2001).

Some strains of methicillin-resistant *S. aureus* (MRSA) express a high molecular weight protein Pls, which masks different surface proteins such as Spa, Clf and FbBPs and prevents ligand binding (Hildén *et al.*, 1996; Savolainen *et al.*, 2001). The protein is susceptible to degradation by host plasmin. It appears that the *pls* gene is closely linked to the *mec* genes and may be part of the *mec* element in these strains. Protein Pls has a Ser-Asp dipeptide region R but is otherwise quite distinct from the typical ClfA-Sdr proteins. It carries two other repeated regions, R1 at the extreme N-terminus and R2 towards the C-terminus, specifying repeats of 12–14 and 128–129 residues, respectively. They flank an approximately 400 residue A domain. It is not known whether Pls has any ligand-binding activity.

1.3.5 Collagen-binding protein

Some strains of *S. aureus* express a collagen-binding protein called CNA. The presence of CNA is necessary and sufficient for bacteria to adhere to collagenous tissues such as cartilage (Switalski *et al.*, 1993). The collagen-binding activity has been located in a 190 amino acid residue segment within the N-terminal A domain. The crystal structure of the subdomain has been solved at 1.8 Å (1 Å = 0.1 nm) resolution (Symersky *et al.*, 1997). The polypeptide folds like a jelly roll in two β-sheets connected by a short α-helix. A trench traverses one of the β-sheets, and molecular modelling suggests that this trench can accommodate a collagen triple helix. The collagen-binding function of the trench was further demonstrated by site-directed mutagenesis, where changes in single residues forming the walls of the trench resulted in proteins with reduced collagen-binding activity.

Biophysical analysis of the recombinant A domain of CNA suggest that it is folded into an elongated structure rather than a sphere (Rich *et al.*, 1998). This might suggests that the A domain is a mosaic protein composed of several subdomains that are independently folded and have different functions. The A regions of ClfA and ClfB have similar physical properties (S. Perkins, E. Walsh, T. Foster and M. Höök, unpublished data). .

The collagen-binding protein is an important virulence factor in the mouse septic arthritis model (Patti *et al.*, 1994b). Fewer mice developed arthritis when injected with the CNA⁻ mutant, as compared with those infected with the wild type. While CNA contributes to the development of arthritis, other bacterial components can compensate if it is lacking. This is clearly demonstrated in the experiments described above, demonstrating the role of ClfA in the same model. The strain used for those experiments did not express CNA.

1.3.6 Other LPXTG-anchored proteins

The genome sequence of strain N315 was published recently (Kuroda *et al.*, 2001) and the sequencing of four other strains of *S. aureus* is nearing completion. Bioinformatic analysis has revealed the existence of 10 previously unknown genes capable of expressing proteins with the typical LPXTG motif, followed by hydrophobic residues and positively charged residues typical of the sorted/covalently linked family of surface proteins (M. Pallen and T.J. Foster, unpublished data). A challenge for investigators is to identify the ligands bound by these proteins and their role in colonization and pathogenesis.

1.3.7 Non-covalently anchored proteins in *Staphylococcus aureus*

The Map protein of *S. aureus* is associated with the cell wall and surface of the organism but it does not have a cell-wall-spanning domain or a membrane anchor and LPXTG motif (McGavin *et al.*, 1993; Jönsson *et al.*, 1995). It contains a Sec-dependent signal sequence at the N-terminus. It can be quantitatively released from cells by treatment with LiCl and is thus not covalently anchored. The Map protein comprises six repeats of a 110 amino acid residue motif with a central portion composed of a subdomain with high homology to the peptide-binding groove of mammalian major histocompatibility complex class II (MHCII) molecules. Map is capable of interacting with a variety of proteins and peptides, including many ECM proteins of the host. Map is related to the extracellular adherence protein Eap, which can act as a transplantable substrate for promoting bacterial attachment to mammalian cells, to immobilized ECM components and possibly to implanted biomaterial (Palma *et al.*, 1999).

Elastic fibres are components of the mammalian ECM and are present in abundance in tissues that require elasticity, such as skin, the lungs and the aorta. Mature elastin is a polymer of tropoelastin monomers that are secreted from mammalian cells prior to deposition in tissue and cross-linking via modified lysine side chains (Mecham and Davis, 1994). *Staphylococcus aureus* expresses a surface-associated protein that promotes bacterial interactions with elastin at a region that is distinct from the binding site for the mammalian elastin receptor (Wrenn *et al.*, 1986; Park *et al.*, 1991). The initial description of the elastin-binding protein (EbpS) reported a 25 kDa protein encoded by a 606 bp gene (Park *et al.*, 1996). We have recently shown that EbpS in fact comprises 486 residues and is an integral membrane protein (F. Roche, R. Downer, P. Park, R. Mecham and T. J. Foster, unpublished data). The elastin-binding domain is located at the N-terminus, as reported by Park *et al.* (1999) and is exposed at the surface. The C-terminus contains a motif that is implicated in binding to peptidoglycan. Topological analysis of EbpS using PhoA and LacZ fusions supports a model where both the N-terminus and the C-terminus of EbpS are located on the outer face of the cytoplasmic membrane.

1.4 SURFACE PROTEIN ADHESINS OF COAGULASE-NEGATIVE STAPHYLOCOCCI

1.4.1 *Staphylococcus epidermidis*

Staphylococcus epidermidis has a propensity to form a biofilm on implanted medical devices such as intravenous catheters (Peters *et al.*, 1981). Biofilms comprise multiple layers of cells embedded in an amorphous extracellular glycocalyx. The majority of cells in the biofilm have no contact with the biomaterial surface. Many strains can produce a visible adherent biofilm in test tubes or tissue culture plates. The hypothesis that biofilm formation is essential for pathogenicity is supported by several studies that correlated biofilm formation with ability to cause infection (Davenport *et al.*, 1987; Diaz-Mitoma *et al.*, 1987). There is debate as to whether adherence to naked polymer surfaces or to surfaces conditioned by deposition of host proteins is most relevant in a clinical setting (discussed by Mack, 1999). It can be argued that colonization of a catheter prior to implantation requires bacteria to adhere to naked polymer surfaces, whereas colonization of an already implanted device requires the ability to interact with host proteins.

Biofilm formation *in vitro* can be separated into two phases: (i) primary attachment to the polymer surface, and (ii) biofilm accumulation in multi-layered cell clusters, paralleled by glycocalyx production. Primary attachment is a complex process. It can vary due to the type of polymer and to the hydro-phobicity of the bacterial cell surface (for a review, see Mack, 1999). Several bacterial factors have been implicated in attachment: (i) a capsular polysac-charide adhesin PS/A (Tojo *et al.*, 1988), (ii) a 220 kDa cell-surface-associated fibrillar protein (Timmerman *et al.*, 1991), and (iii) the major autolysin AtlE (Heilmann *et al.*, 1997). Autolysin mutants failed to adhere to polymer sur-faces *in vitro*, but this could be due to pleiotropic alterations to the cell surface caused by the defect in cell wall turnover rather than AtlE itself being an adhesin. In addition, purified AltE bound to vitronectin in ligand affinity blots, but AtlE was not shown directly to be an MSCRAMM.

The accumulation phase of biofilm formation requires the expression of a polysaccharide intercellular adhesin (PIA). Transposon insertion mutants defective in PIA adhered to polymer surfaces but could not form multilay-ered aggregates (Mack *et al.*, 1994; Heilmann *et al.*, 1996). The polysaccha-ride is synthesized by the *ica* genes. Indeed, the cloned *ica* genes can be expressed in another staphylococcal species, *S. carnosus*, and promote expression of PIA and cell aggregation.

Staphylococcus epidermidis can adhere to different host proteins that have been immobilized on polymer surfaces (Herrmann *et al.*, 1988; Vaudaux *et al.*, 1989), but only binding to Fg has been characterized at the molecular level. *Staphylococcus epidermidis* strains can express a Fg-binding surface protein called Fbe (M. Nilsson *et al.*, 1998; Pei *et al.*, 1999;; Hartford *et al.*, 2001) or SdrG (McCrea *et al.*, 2000). SdrG/Fbe has the same sequence organ-ization as the Sdr proteins of *S. aureus*, comprising a 548 residue A domain, two B repeats, an R region containing the dipeptide repeat Ser-Asp followed by a cell-wall-anchoring domain at the C-terminus (Fig. 1.4). The A domain of SdrG/Fbe binds to the β-chain of Fg (Pei *et al.*, 1999).

Staphylococcus epidermidis can express two other Sdr proteins (McCrea *et al.*, 2000). SdrF, like SdrG, is a close relative of the *S. aureus* SdrCDE pro-teins. SdrH, however, is quite distinct. It contains a short A domain at the N-terminus, followed by an R domain comprising the dipeptide Ser-Asp and a 277 residue domain C that contains a hydrophobic stretch at the C-terminus. As it lacks a typical LPXTG motif, it is not clear whether SdrH is anchored to peptidoglycan. The functions of SdrF and SdrH are unknown.

1.4.2 *Staphylococcus saprophiticus*

This bacterium causes urinary tract infections in females. The Fn-binding activity and haemagglutinin were attributed to the autolysin Aas (Hell *et al.*, 1998), which has sequence similarity to Alt of *S. aureus* and AltE of *S. epidermidis.* An Aas-defective mutant of *S. saprophiticus* lacked the ability to bind Fn and to cause haemagglutination. The binding activity was localized in recombinant proteins to the repeat region R1–R3.

1.4.3 *Staphylococcus schleiferi*

Staphylococcus schleiferi subsp. *schleiferi* is an emerging nosocomial pathogen. The ability to bind Fn is a common trait of this organism and could be an important virulence factor. Western ligand blotting identified a FnBP of between 180 and 200 kDa (Peacock *et al.*, 1999b). It is likely that the Fn-binding region is very similar to that of the *S. aureus* FnBPA and FnBPB because polymerase chain reaction primers specific for region D1–D3, the major Fn-binding region, amplified a fragment of the same size as in the *S. aureus* genes and binding of *S. schleiferi* was blocked by recombinant D1–D3 protein

1.5 VACCINATION

There is accumulating experimental evidence that *S. aureus* infections can be prevented by vaccination targeted at single surface protein antigens. Thus protection against mastitis and endocarditis was achieved by immunizing animals with FnBPs (Nelson *et al.*, 1992; Rozalska and Wadström, 1993; Schennings *et al.*, 1993). Immunization with the recombinant A domain of CNA protected against septic death in mice (I.M. Nilsson *et al.*, 1998). Protection was shown to be antibody mediated because mice were also protected by passive immunization. Significant protection in the same model was also obtained by active immunization with the A domain of ClfA and by passive transfer of rodent or human serum with high titre of anti-ClfA antibodies (Josefsson *et al.*, 2001). It is not known whether protection is due to adhesion-blocking activity or to opsonization (or both). There is optimism, therefore, that human nosocomial *S. aureus* infections, particularly those caused by MRSA, can be combatted by passive immunization with intravenous immunoglobulin and ultimately with humanized monoclonal antibodies.

ACKNOWLEDGEMENTS

I acknowledge the Wellcome Trust for a project grant (50558), as well as support from BioResearch Ireland, The Health Research Board, Enterprise Ireland and Inhibitex Inc. I also thank members of my group who have contributed unpublished data.

REFERENCES

Brown, E.M., Vassilev, P.M. and Hebert, S.C. (1995). Calcium ions as extracellular messengers. *Cell* **83**, 679–682.

Davenport, D.S., Massanari, R.M., Pfaller, M.A., Bale, M.J., Streed, S.A. and Hierholzer, W.J. (1987). Usefulness of a test for slime production as a marker for clinically significant infections with coagulase-negative staphylococci. *Journal of Infectious Diseases* **153**, 332–339.

Deisenhofer, J. (1981). Crystallographic refinement and atomic models of a human Fc fragment and its complex with fragment B of protein A from *Staphylococcus aureus* at 2.9 and 2.8 Å resolution. *Biochemistry* **20**, 2361–2370.

Diaz-Mitoma, F., Harding, G.K.M., Hoban, D.J., Roberts, R.S. and Low, D.E. (1987). Clinical significance of a test for slime production and ventriculoperitoneal shunt infections caused by coagulase-negative staphylococci. *Journal of Infectious Diseases* **156**, 555–560.

Doolittle, R.F., Spraggon, G. and Everse, S.J. (1998). Three-dimensional structural studies on fragments of fibrinogen and fibrin. *Current Opinion in Structural Biology* **8**, 792–798.

Dziewanowska, K., Patti, J.M., Deobald, C.F., Bayles, K.W., Trumble, W.R. and Bohach, G.A. (1999). Fibronectin binding protein and host cell tyrosine kinase are required for internalization of *Staphylococcus aureus* by epithelial cells. *Infection and Immunity* **67**, 4673–4678.

Entenza, J.M., Foster, T.J., Vaudaux, P., Francioli, P. and Moreillon, P. (2000). Contribution of clumping factor B to the pathogenesis of experimental endocarditis due to *Staphylococcus aureus*. *Infection and Immunity* **68**, 5443–5446.

Fischer, B., Vaudaux, P., Magnin, M., El Mestikawy, Y., Proctor., R.A., Lew, D.P. and Vasey, H. (1996). Novel animal model for studying the molecular mechanisms of bacterial adhesion to bone-implanted metallic devices: role of fibronectin in *Staphylococcus aureus* adhesion. *Journal of Orthopaedic Research* **14**, 914–920.

Flock, J.I., Hienz, S.A., Heimdahl, A. and Schennings, T. (1996). Reconsideration of the role of fibronectin binding in endocarditis caused by *Staphylococcus aureus*. *Infection and Immunity* **64**, 1876–1878.

Foster, T.J. and Höök, M. (1998). Surface protein adhesins of *Staphylococcus aureus*. *Trends in Microbiology* **6**, 484–488.

Foster, T.J. and Höök, M. (2000). Molecular basis of adherence of *Staphylococcus aureus* to biomaterials. In *Infections Associated with Indwelling Medical Devices*, 3rd edn, ed. F.A. Waldvogel and A.L. Bisno, pp. 27–39. Washington, DC: American Society for Microbiology Press.

Fowler, T., Wann, E.R., Joh, D., Johansson, S., Foster, T.J. and Höök, M. (2000). Cellular invasion by *Staphylococcus aureus* involves a fibronectin bridge between the bacterial fibronectin-binding MSCRAMMs and host cell α1 integrins. *European Journal of Cell Biology* **79**, 672–679.

François, P., Schrenzel, J., Stoerman-Chopard, C., Favre, H., Herrmann, M., Foster, T.J., Lew, D.P. and Vaudaux, P.E. (2000). Identification of plasma proteins adsorbed on hemodialysis tubing that promote *Staphylococus aureus* adhesion. *Journal of Laboratory and Clinical Medicine* **135**, 32–42.

Gemmell, C.G., Tree, R., Patel, A., O'Reilly, M. and Foster T.J. (1990). Susceptibility to opsonophagocytosis of protein A, alpha-haemolysin and beta-toxin deficient mutants of *Staphylococcus aureus* isolated by allele-replacement. *Zentralblatt für Bakteriologie Supplement* **21**, 273–277.

Gouda, H., Shiraiski, M., Takahashi, H., Kalo, K., Torigoe, H., Arala, Y. and Shimada, I. (1998). NMR study of the interaction between the B domain of staphylococcal Protein A and the Fc portion of immunoglobulin G. *Biochemistry* **37**, 129–136.

Greene, C., McDevitt, D., François, P., Vaudaux, P.E., Lew, D.P. and Foster, T.J. (1995). Adhesion properties of mutants of *Staphylococcus aureus* defective in fibronectin-binding proteins and studies on the expression of the *fnp* genes. *Molecular Microbiology* **17**, 1143–1152.

Hartford, O., François, P., Vaudaux, P. and Foster, T.J. (1997). The dipeptide repeat region of the fibrinogen-binding protein (clumping factor) is required for functional expression of the fibrinogen-binding domain on the *Staphylococcus aureus* cell surface. *Molecular Microbiology* **25**, 1065–1107.

Hartford, O., O'Brien, L., Schofield, K., Wells, J. and Foster, T.J. (2001). The SdrG protein of *Staphylococcus epidermidis* HB promotes bacterial adherence to fibrinogen. *Microbiology*, in press.

Hartleib, J., Kohler, N., Dickinson, R.B., Chhatwal, G.S., Sixma, J.J., Foster, T.J., Peters, G., Kehrel, B.E. and Herrmann, M. (2000). Protein A is the von Willebrand factor binding protein on *Staphylococcus aureus*. Evidence for a novel function characterizing protein A as an MSCRAMM adhesin. *Blood* **96**, 2149–2156.

Hawiger, J. (1995). Adhesive ends of fibrinogen and its anti-adhesive peptides: the end of a saga? *Seminars in Haematology* **32**, 99–109.

Heilmann, C., Schweitzer, O., Gerke, C., Vanittanakom, N., Mack, D. and Gotz, F. (1996). Molecular basis of intercellular adhesion in the biofilm-forming *Staphylococcus epidermidis*. *Molecular Microbiology* **20**, 1083–1091.

Heilmann, C., Hussain, M., Peters, G. and Gotz, F. (1997). Evidence for autolysin-mediated attachment of *Staphylococcus epidermidis* to a polystyrene surface. *Molecular Microbiology* **24**, 1013–1024.

Hell, W., Meyer, H.-G. W. and Gatermann, S.G. (1998). Cloning of *aas*, a gene encoding a *Staphylococcus saprophiticus* surface protein with adhesive and autolytic properties. *Molecular Microbiology* **29**, 871–881.

Herrmann, M., Vaudaux, P.E., Pittet, D., Auckenthaler, R., Lew, D.P., Schumacher-Perdreau, F., Peters, G. and Waldvogel, F.A. (1988). Fibronectin, fibrinogen, and laminin act as mediators of adherence of clinical staphylococcal isolates to foreign material. *Journal of Infectious Diseases* **158**, 693–701.

Hildén, P., Savolainen, K., Tyynelä, J., Vuento, M. and Kuusela, P. (1996). Purification and characterisation of a plasmin-sensitive surface protein of *Staphylococcus aureus*. *European Journal of Biochemistry* **236**, 904–910.

House-Pompeo, K., Xu, Y., Joh, D., Speziale, P. and Höök, M. (1996). Conformational changes in the fibronectin binding MSCRAMMs are induced by ligand binding. *Journal of Biological Chemistry* **271**, 1379–1384.

Hynes, R. (1993) Fibronectins. In *Guidebook to the Extracellular Matrix and Adhesion Proteins*, ed. T. Kreis and R.Vale, pp. 56–58. Oxford: Oxford University Press.

Joh, D., Speziale, P., Gurusiddappa, S., Manor, J. and Höök, M. (1998). Multiple specificities of the staphylococcal and streptococcal fibronectin-binding microbial surface components recognizing adhesive matrix molecules. *European Journal of Biochemistry* **258**, 897–905.

Joh, H.J., House-Pompeo, K., Patti, J., Gurusiddappa, S. and Höök, M. (1994). Fibronectin receptors from Gram-positive bacteria: comparison of sites. *Biochemistry* **33**, 6086–6092.

Jönsson, K., Signäs, C., Müller, H.P. and Lindberg, M (1991). Two different genes encode fibronectin binding proteins in *Staphylococcus aureus*. The complete nucleotide sequence and characterization of the second gene. *European Journal of Biochemistry* **202**, 1041–1048.

Jönsson, K., McDevitt, D., Homonylo McGavin, M., Patti, J.M. and Höök, M. (1995). *Staphylococcus aureus* expresses a major histocompatibility complex class II analog. *Journal of Biological Chemistry* **270**, 21457–21460.

Josefsson, E., McCrea, K.W., Ní Eidhin, D., O'Connell, D., Cox, J., Höök, M. and Foster, T.J. (1998a). Three new members of the serine-aspartate repeat protein multigene family of *Staphylococcus aureus*. *Microbiology* **144**, 3387–3395.

Josefsson, E., O'Connell, D., Foster, T.J., Durussel, I. and Cox, J.A. (1998b). The binding of calcium to the B-repeat segment of SdrD, a cell surface protein of *Staphylococcus aureus*. *Journal of Biological Chemistry* **273**, 31145–31152.

Josefsson, E., Hartford, O., Patti, J. and Foster T.J. (2001) Protection against *Staphylococcus aureus* arthritis by vaccination with clumping factor A, a novel virulence determinant. *Journal of Infectious Diseases*, in press.

Kuroda, M., Ohta, T., Uchiyama, I., Baba, T., Yuzawa, H., Kobayashi, I. *et al.* (2001). Whole genome sequencing of methicillin-resistant *Staphylococcus aureus*. *Lancet* **357**, 1225–1239.

Kuypers, J.M. and Proctor, R.A. (1989). Reduced adherence to traumatized rat heart valves by a low-fibronectin-binding mutant of *Staphylococcus aureus*. *Infection and Immunity* **57**, 2306–2312.

Lammers, A., Nuijten, P.J.M. and Smith, H.E. (1999). The fibronectin binding proteins of *Staphylococcus aureus* are required for adhesion to and invasion of bovine mammary gland epithelial cells. *FEMS Microbiology Letters* **180**, 103–109.

Langone, J.J. (1982). Protein A of *Staphylococcus aureus* and related immunoglobulin receptors produced by streptococci and pneumococci. *Advances in Immunology* **32**, 157–252.

Mack, D. (1999). Molecular mechanisms of *Staphylococcus epidermidis* biofilm formation. *Journal of Hospital Infection* **43**, Supplement S113–S125.

Mack, D., Nedelmann, M., Krokotsch, A., Schwartzkopf, A., Heesemann, J. and Laufs, R. (1994). Characterization of transposon mutants of biofilm-producing *Staphylococcus epidermidis* impaired in accumulation phase of biofilm production: genetic identification of a hexosamine containing polysaccharide intercellular adhesin. *Infection and Immunity* **62**, 3244–3253.

Mazmanian, S.K., Liu, G., Ton-That, H. and Schneewind, O. (1999). *Staphylococcus aureus* sortase, an enzyme that anchors surface proteins to the cell wall. *Science* **285**, 760–763.

Mazmanian, S.K., Liu, G., Jensen, E.R., Lenoy, E. and Schneewind, O. (2000). *Staphylococcus aureus* sortase mutants defective in the display of surface proteins and in the pathogenesis of animal infections. *Proceedings of the National Academy of Sciences, USA* **97**, 5510–5515.

McAleese, F.M., Walsh, E.J., Sieprawska, M., Potempa, J. and Foster, T.J. (2001). Loss of clumping factor B fibrinogen binding activity by *Staphylococcus aureus* involves cessation of transcription, shedding and cleavage by metalloprotease. *Journal of Biological Chemistry*, **276**, 29969–29978.

McCrea, K.W., Hartford, O., Davis, S., Ní Eidhin, D., Lina, G., Speziale, P., Foster, T.J. and Höök, M. (2000). The serine-asparate repeat (Sdr) protein family in *Staphylococcus epidermidis*. *Microbiology* **146**, 1535–1546.

McDevitt, D., François, P., Vaudaux, P. and Foster, T.J. (1994). Molecular characterization of the fibrinogen receptor (clumping factor) of *Staphylococcus aureus*. *Molecular Microbiology* **11**, 237–248.

McDevitt, D., François, P., Vaudaux, P. and Foster, T.J. (1995). Identification of the ligand-binding domain of the surface-located fibrinogen receptor (clumping factor) of *Staphylococcus aureus*. *Molecular Microbiology* **16**, 895–907.

McDevitt, D., Nanavaty, T., House-Pompeo, K., Bell, E.C., Turner, N., McIntire, L., Foster, T.J. and Höök, M. (1997). Characterization of the interaction between the *Staphylococcus aureus* fibrinogen-binding MSCRAMM clumping factor (ClfA) and fibrinogen. *European Journal of Biochemistry* **247**, 416–424.

McGavin, M.J., Raucci, G., Gurusiddappa, S. and Hook, M. (1991). Fibronectin binding determinants of the *Staphylococcus aureus* fibronectin receptor. *Journal of Biological Chemistry* **266**, 8343–8347.

McGavin, M.J., Gurusiddappa, S., Lindgren, P.E., Lindberg, M., Raucci, G. and Höök, M. (1993). Fibronectin receptors from *Streptococcus dysgalactiae* and *Staphylococcus aureus*. Involvement of conserved residues in ligand binding. *Journal of Biological Chemistry* **268**, 23946–23953.

Mecham, R.P. and Davis, E.C. (1994). Elastic fiber structure and assembly. In *Extracellular Matrix Assembly and Structure*, ed. P.D. Yurchenco, D.E. Birk and R.P. Mecham, pp. 281–314. San Diego: Academic Press.

Moreillon, P., Entenza, J.M., Francioli, P., McDevitt, D., Foster, T.J., François, P. and Vaudaux, P. (1995). Role of *Staphylococcus aureus* coagulase and clumping factor in the pathogenesis of experimental endocarditis. *Infection and Immunity* **63**, 4738–4743.

Navarre, W.W. and Schneewind, O. (1994). Proteolytic cleavage and cell wall anchoring at the LPXTG motif of surface proteins in Gram-positive bacteria. *Molecular Microbiology* **14**, 115–121.

Navarre, W.W. and Schneewind, O. (1999). Surface proteins of Gram-positive bacteria and mechanisms of their targeting to the cell wall envelope. *Microbiology and Molecular Microbiology Reviews* **63**, 174–229.

Nelson, L., Flock, J.-I., Höök, M., Lindberg, M., Müller, H.P. and Wadström, T. (1992). Adhesins in staphylococcal mastitis as vaccine components. *Flemish Veterinary Journal* **62** (Supplement 1), 111–125.

Ní Eidhin, D., Perkins, S., François, P., Vaudaux, P., Höök, M. and Foster, T.J. (1998). Clumping factor B (ClfB) a new surface-located fibrinogen-binding adhesin of *Staphylococcus aureus*. *Molecular Microbiology* **30**, 245–257.

Nilsson, I.M., Patti, J.M. Bremell, T., Höök, M. and Tarkowski, A. (1998). Vaccination with a recombinant fragment of the collagen adhesin provides protection against *Staphylococcus aureus*-mediated septic death. *Journal of Clinical Investigation* **101**, 2640–2649.

Nilsson, M., Frykberg, L., Flock, J.-I., Pei, L., Lindberg, M. and Guss, B. (1998). A fibrinogen-binding protein from *Staphylococcus epidermidis*. *Infection and Immunity* **66**, 2666–2673.

O'Brien, L.M., Kerrigan, S., Foster, T.J. and Cox, D. (2001). Multiple mechanisms for *Staphylococcus aureus*-induced platelet aggregation: roles for clumping factors A and B and serine-aspartate repeat protein E. *Journal of Clinical Investigation*, in press.

O'Connell, D.P., Nanavaty, T., McDevitt, D., Gurusiddappa, S., Höök, M. and Foster. T.J. (1998). The fibronectin-binding MSCRAMM (clumping factor) of *Staphylococcus aureus* has an integrin-like Ca^{2+}-dependent inhibitory site. *Journal of Biological Chemistry* **273**, 6821–6829.

Palma, M., Haggar, A. and Flock, J.-I. (1999). Adherence of *Staphylococcus aureus* is enhanced by an endogenous secreted protein with broad binding activity. *Journal of Bacteriology* **181**, 2840–2845.

Park, P.W., Roberts, D.D., Grosso, L.E., Parks, W.C., Rosenbloom, J., Abrams, W.R. and Mecham, R.P. (1991). Binding of elastin to *Staphylococcus aureus*. *Journal of Biological Chemistry* **266**, 23399–23406.

Park, P.W., Rosenbloom, J., Abrams, W.R., Rosenbloom, J. and Mecham, R.P. (1996). Molecular cloning and expression of the gene for elastin binding protein (EbpS) in *Staphylococcus aureus*. *Journal of Biological Chemistry* **271**, 15803–15809.

Park, P.W., Broekelmann, T.J., Mecham, B.R. and Mecham, R.P. (1999). Characterization of the elastin binding domain in the cell-surface 25-kDa elastin-binding protein of *Staphylococcus aureus*. *Journal of Biological Chemistry* **274**, 2845–2850.

Patel, A.H., Nowlan, P., Weavers, E.D. and Foster, T.J. (1987). Virulence of protein A-deficient and alpha-toxin-deficient mutants of *Staphylococcus aureus* isolated by allele replacement. *Infection and Immunity* **55**, 3103–3110.

Patti, J.M., Allen, B.A., McGavin, M.J. and Höök, M. (1994a). MSCRAMM-mediated adherence of microorganisms to host tissues. *Annual Reviews of Microbiology* **45**, 585–617.

Patti, J.M., Bremell, T., Krajewska-Pietrasik, D., Abdelnour, A., Tarkowski, A., Ryden, C. and Höök (1994b). The *Staphylococcus aureus* collagen adhesin is a virulence determinant in experimental septic arthritis. *Infection and Immunity* **62**, 152–161.

Peacock, S.J., Foster, T.J., Cameron, B. and Berendt, A.R. (1999a). Bacterial fibronectin-binding proteins and endothelial cell surface fibronectin mediate adherence of *Staphylococcus aureus* to resting human endothelial cells. *Microbiology* **145**, 3477–3486.

Peacock, S.J., Lina, G., Etienne, J. and Foster, T.J. (1999b). *Staphylococcus schleiferi* subsp. *schleiferi* expresses a fibronectin-binding protein. *Journal of Clinical Microbiology* **67**, 4272–4275.

Peacock, S.J., Day, N.P.J., Thomas, M.G., Berendt, A.R. and Foster, T.J. (2000). Clinical isolates of *Staphylococcus aureus* exhibit diversity in *fnb* genes and adhesion to human fibronectin. *Journal of Hospital Infection* **41**, 23–31.

Pei, L., Palma, M., Nilsson, M., Guss, B. and Flock, J.-I. (1999). Functional studies of a fibrinogen binding protein from *Staphylococcus epidermidis*. *Infection and Immunity* **67**, 4525–4530.

Peters, G., Locci, R. and Pulverer, G. (1981). Microbial colonization of prosthetic devices, II. Scanning electron microscopy of naturally infected intravenous catheters. *Zentralblatt für Bakteriologie, Mikrobiologie und Hygiene B* **173**, 293–299.

Potts, J. R. and Campbell, I.D. (1994). Fibronectin structure and assembly. *Current Opinion in Cell Biology* **6**, 648–655.

Raja, R.H., Raucci, G. and Höök, M. (1990). Peptide analogs to a fibronectin receptor inhibit attachment of *Staphylococcus aureus* to fibronectin-coated substrates. *Infection and Immunity* **58**, 2593–2598.

Rich, R.L., Demeler, B., Ashby, K., Deivanayagam, C.C.S., Petrich, J.W., Patti, J.M., Sthanam, V.L.N. and Höök, M. (1998). Domain structure of the *Staphylococcus aureus* collagen adhesin. *Biochemistry* **37**, 15423–15433.

Rozalska, B. and Wadström, T. (1993). Protective opsonic activity of antibodies against fibronectin-binding proteins (FnBPs) of *Staphylococcus aureus*. *Scandinavian Journal of Immunology* **37**, 575–580.

Ruggeri, Z.M. (1993). Fibrinogen/fibrin. In *Guidebook to the Extracellular Matrix and Adhesion Proteins*. ed. T. Kreis and R. Vale, pp. 52–53. Oxford: Oxford University Press.

Savolainen, K,. Paulin, L., Westerlund-Wikström, B., Foster, T.J., Korhonen, T.K. and Kuusela, P. (2001). Expression of *pls*, a game closely associated with the *mecA* gene of methicillin-resistant *Staphylococcus aureus* prevents bacterial adhesion *in vitro*. *Infection and Immunity* **69**, 3013–3020.

Schennings, T., Heimdahl, A., Coster, K. and Flock, J.I. (1993). Immunization with fibronectin binding protein from *Staphylococcus aureus* protects against experimental endocarditis in rats. *Microbial Pathogenesis* **15**, 227–236.

Schneewind, O., Mihaylova-Petkov, D. and Model, P. (1993). Cell wall sorting signals in surface proteins of Gram-positive bacteria. *EMBO Journal* **12**, 4803–4811.

Signás, C., Raucci, G., Jönsson, K., Lindgren, P.E., Anantharamaiah, G.M., Höök, M. and Lindberg, M. (1989). Nucleotide sequence of the gene for a fibronectin-binding protein from *Staphylococcus aureus*: use of this peptide sequence in the synthesis of biologically active peptides. *Proceedings of the National Academy of Sciences, USA* **86**, 699–703.

Sinha, B., François, P.P., Nüße, O., Foti, M., Hartford, O.M., Vaudaux, P., Foster, T.J., Lew, D.P., Herrmann, M. and Krausse, K.H. (1999). Fibronectin-binding protein acts as *Staphylococcus aureus* invasin *via* fibronectin bridging to integrin $\alpha_5\beta_1$. *Cellular Microbiology* **1**, 101–107.

Sottile, J., Schwarzbauer, J., Selegue, J. and Mosher, D.F. (1991). Five type I modules of fibronectin form a functional unit that binds to fibroblasts and to *Staphylococcus aureus*. *Journal of Biological Chemistry* **266**, 12840–12843.

Speziale, P., Joh, D., Visai, L., Bozzini, S., House-Pompeo, K., Lindberg, M. and Höök, M. (1996). A monoclonal antibody enhances ligand binding of fibronectin MSCRAMM (adhesin) from *Streptococcus dysgalactiae*. *Journal of Biological Chemistry* **271**, 1371–1378.

Spraggon, G., Everse, S.J. and Doolittle, R.F. (1997). Crystal structures of fragment D from human fibrinogen and its crosslinked counterpart from fibrin. *Nature* **389**, 455–462.

Starovasnik, M.A., Skelton, N.J., O'Connell, M.P., Kelly, R.F., Reilly, D. and Fairbrother, W.J. (1996). Solution structure of the E-domain of staphylococcal protein A. *Biochemistry* **35**, 1558–1569.

Sun, Q., Smith, G.M., Zahradka, C. and McGavin, M. (1997). Identification of D motif epitopes in *Staphylococcus aureus* fibronectin-binding protein for the production of antibody inhibitors of fibronectin binding. *Infection and Immunity* **65**, 537–543.

Switalski, L.M., Patti, J.M., Butcher, W., Gristina, A.G., Speziale, P. and Höök, M. (1993). A collagen receptor on *Staphylococcus aureus* strains isolated from patients with septic arthritis mediates adhesion to cartilage. *Molecular Microbiology* **7**, 99–107.

Symersky, J., Patti, J.M., Carson, M., House-Pompeo, K., Teale, M., Moore, D., Jin, L., Schneider, A., DeLucas, L.J., Höök, M. and Narayana, S.V.L. (1997). Structure of the collagen-binding domain from a *Staphylococcus aureus* adhesin. *Nature Structural Biology* **4**, 833–838.

Timmerman, C.P., Fleer, A., Besnier, J.M., deGraaf, L., Cremers, F. and Verhoef, J. (1991). Characterization of a proteinaceous adhesin of *Staphylococcus epidermidis* which mediates attachment to polystyrene. *Infection and Immunity* **59**, 4187–4197.

Tojo, M., Yamashita, N., Goldmann, D.A. and Pier, G.B. (1988). Isolation and characterization of a capsular polysaccharide adhesin from *Staphylococcus epidermidis*. *Journal of Infectious Diseases* **157**, 713–722.

Ton-That, H. and Schneewind, O. (1999). Anchor structure of staphylococcal surface proteins. IV. Inhibitors of the cell wall sorting reaction. *Journal of Biological Chemistry* **34**, 23416–24320.

Ton-That, H., Faull, K.F. and Schneewind, O. (1997). Anchor structure of staphylococcal surface proteins. I. A branched peptide that links the carboxyl terminus of proteins to the cell wall. *Journal of Biological Chemistry* **272**, 22285–22292.

Ton-That, H., Lui, G., Mazmanian, S.K., Faull, K.F. and Schneewind, O. (1999).

Purification and characterization of sortase, the transpeptidase that cleaves surface proteins of *Staphylococcus aureus* at the LPXTG motif. *Proceedings of the National Academy of Sciences ,USA* **96**, 12424–12429.

Tung, H.-S., Guss, B., Hellman, U., Persson, L., Rubin, K. and Rydén, C. (2000). A bone sialoprotein-binding protein from *Staphylococcus aureus*: a member of the staphylococcal Sdr family. *Biochemical Journal* **345**, 611–619.

Uhlén, M., Guss, B., Nilssön, B., Gatenbeck, S., Philipson, L. and Lindberg, M. (1984). Complete sequence of the staphylococcal gene encoding protein A. A gene evolved through multiple duplications. *Journal of Biological Chemistry* **259**,1695–1702.

Vaudaux, P., Pittet, D., Haeberli, A., Huggler, E., Nydegger, U.E., Lew, D.P. and Waldvogel, F.A. (1989). Host factors selectively increase staphylococcal adherence on inserted catheters: a role for fibronectin and fibrinogen or fibrin. *Journal of Infectious Diseases* **160**, 865–875.

Vaudaux, P.E., François, P., Proctor, R.A., McDevitt, D., Foster, T.J., Albrecht, R.M., Lew, D.P., Wabers, H. and Cooper, S.L. (1995). Use of adhesion-defective mutants of *Staphylococcus aureus* to define the role of specific plasma proteins in promoting bacterial adhesion to canine arterio-venous shunts. *Infection and Immunity* **63**, 585–590.

Wann, E.R., Gurusiddappa, S. and Höök, M. (2000). A fibronectin-binding MSCRAMM FnbpA of *Staphylococcus aureus* is a bifunctional protein that also binds to fibrinogen. *Journal of Biological Chemistry* **275**, 13863–13871.

Wrenn, D.S., Griffin, G.L., Senior, R.M. and Mecham, R.P. (1986). Characterization of biologically active domains on elastin: identification of a monoclonal antibody to a cell recognition site. *Biochemistry* **25**, 5172–5176.

Yamada, K.M. (1989). Fibronectins: structure, function and receptors. *Current Opinion in Cell Biology* **1**, 956–963.

Yeaman, M.R., Puentes, S.M., Norman, D.C. and Bayer, A.S. (1992). Partial purification and staphylocidal activity of thrombin-induced platelet microbicidal protein. *Infection and Immunity* **60**, 1202–1209.

CHAPTER 2

Mechanisms of utilization of host signalling molecules by respiratory mucosal pathogens

Mumtaz Virji

2.1 INTRODUCTION

Microbes such as *Neisseria meningitidis* (meningococcus) and *Haemophilus influenzae* that reside in a single niche, the human upper respiratory tract, are particularly adept at genotypic plasticity and generate phenotypic variants at high frequency. They elaborate multiple surface-expressed adhesive ligands, which undergo antigenic and phase variation with remarkable rapidity. Antigenic variation represents primary structural alteration, often arising from genetic rearrangements, and phase variation represents 'on' or 'off' mode of expression. The evolution of such variation as well as the redundancy in adhesins reflects the polymorphic nature of the host's immune response. The microbial counter-strategies accomplish not only immune evasion but also tissue tropism.

Multiple microbial adhesins may act individually or in concert to interact with target cell molecules, enabling the bacterium primarily to achieve the fundamental requirement of colonization, i.e. adherence to mucosa. Many of the target molecules are involved in normal cell–cell communications or in the reception of hormonal and other signals. As a result, targeting of these signalling molecules additionally allows bacteria to manipulate host cell functions, which may lead to intracellular location, transcytosis or paracytosis across the epithelial barrier. This chapter will outline some general mechanisms of host cell targeting that determine between adhesion and invasion. Further, it will describe several major surface structures of meningococci and their interplay in host cell recognition or evasion. Two specific receptor-targeting mechanisms are explored in detail, since they provide a paradigm to explain epidemiological evidence that implicates host-associated factors in increased microbial invasion.

The chapter is primarily set around meningococcal adhesion mechanisms but, since *N. meningitidis* and *H. influenzae* share a family of host receptors, a comparative analysis of the mechanisms of targeting of these molecules and consequent events will be discussed at the end.

2.2 AN OVERVIEW OF PATHOGENESIS AND EPIDEMIOLOGY OF *NEISSERIA MENINGITIDIS* AND *HAEMOPHILUS INFLUENZAE*

2.2.1 *Neisseria meningitidis*

Neisseria meningitidis strains are isolated from the nasopharynx of up to 30% of healthy individuals and may be classified as commensals of the human respiratory tract. Individual strains may produce one of several capsular chemotypes (serogroups), designated: A, B, C, 29E, H, I, K, L, W135, X, Y, and Z (Cartwright, 1995, p. 24). Of these A, B, and C are most commonly encountered during disease. Serogroup A predominates in Africa and is responsible for epidemic spread whereas serogroups B and C prevail in the West and are associated with sporadic outbreaks. Capsules constitute 'shielding' molecules of meningococci that enhance bacterial survival in distinct environments. Capsule and lipopolysaccharides (LPS) sialylation and perhaps IgA protease also aid in evasion of the host immune mechanisms. However, acapsulate (non-groupable) meningococci are frequently isolated from the nasopharynx of carriers and several genetic mechanisms are responsible for phase variation of the capsule (Hammerschmidt *et al.*, 1996a, b). Under some situations, meningococci cause serious disease conditions that can be fatal, demonstrating their considerable pathogenic potential. This distinguishes meningococci from the other neisseriae that colonize the human nasopharynx but are rarely associated with disease. The precise bacterial factors responsible for the nature of outbreak and the reasons for geographical differences are not clear. Within *N. meningitidis* isolates, some clones have been identified that are more often associated with disease than others. Acquisition of such 'more pathogenic' clones during endemic situations still induces illness only in 1% of individuals harbouring them (van Deuren *et al.*, 2000). Invasive disease may arise as a result of increased host susceptibility, which in turn may be governed by several factors. It has long been recognized that bactericidal antibodies are important in defence against meningococci. Epidemiological studies also suggest that other factors that damage mucosa such as smoking, prior infection of the host (e.g. respiratory viral infections in the winter months in the UK) or very dry atmospheric conditions (in dry seasons in Africa) may also predispose the host to meningococcal infection (Achtman, 1995; van Deuren *et al.*, 2000).

2.2.2 *Haemophilus influenzae*

Several species of the genus *Haemophilus* colonize the human respiratory tract. Up to 80% of healthy individuals may carry strains belonging to the species *H. influenzae* (Turk, 1984). Six serotypes (a–f) of capsulate *H. influenzae* have been described, depending on the composition of the capsular polysaccharide elaborated. These strains may become acapsulate *in vivo* and *in vitro* by genetic rearrangement (Hoiseth and Gilsdorf, 1988). However, in addition, many strains of the organisms lack the genetic information for capsulation and are truly acapsulate (non-typeable *H. influenzae*, NTHi) *in vivo* and *in vitro* and are more commonly isolated from the nasopharynx than typeable strains (THi) (Gyorkey *et al.*, 1984; St Geme *et al.*, 1994). Capsulate strains, especially those expressing the type b capsule, are capable of producing serious conditions in infants, such as bacteraemia and meningitis. Within NTHi strains, those belonging to the clonal biogroup aegyptius (Hi-aeg) are an important cause of purulent conjunctivitis and have been associated with Brazilian purpuric fever. This syndrome is characterized by a rapid development of fever and of petechiae, purpura and vascular collapse, the characteristics normally assigned to *N. meningitidis* (Brenner *et al.*, 1988). Other carriage and disease isolates of NTHi are genetically more diverse than Hi-aeg, are frequent colonizers of the human nasopharynx and are opportunistic pathogens. NTHi strains cause localized as well as disseminated infections, including otitis media, pneumonia, endocarditis, bacteraemia and meningitis and also with acute recurrent and persistent infections in patients with chronic obstructive pulmonary disease (COPD) and cystic fibrosis (CF). What determines recurrent infections by NTHi in patients with COPD or CF, and multiple episodes of otitis media, which occur in children, is unclear. Interestingly, the frequency of colonization of the upper respiratory tract by NTHi increases during respiratory viral infections, and bacteria occur within tissues and have been seen in tissue macrophages (Henderson *et al.*, 1982; van Alphen *et al.*, 1995; Foxwell *et al.*, 1998).

Thus *N. meningitidis* and *H. influenzae* may colonize human respiratory mucosa without disease but have a pathogenic potential, which is realized infrequently and is dependent on increased susceptibility of individuals. One hypothesis is that the underlying mechanisms of increased host susceptibility and bacterial persistence may involve host receptor molecules targeted by bacteria and associated cell signalling cascades. Under normal situations, the microbe achieves successful colonization and survival in the niche but in certain circumstances it leads to an undesired outcome for the host. Investigations of the mechanisms of mucosal colonization and host cell targeting that may help to explain epidemiological observations will be discussed below.

2.3 POTENTIAL MOLECULAR TARGETS AND MECHANISMS OF LIGATION

Available targets on host cell surfaces for microbial adhesins include cell–cell or cell–matrix interaction molecules, for example adhesion receptors belonging to the integrin and immunoglobulin superfamilies, the selectins and the cadherins, which are subverted by many pathogenic organisms whose ligands often mimic the natural ligands. The interactions with the receptor may occur at the normal host ligand recognition site and may utilize the ligand recognition motif, for example the sequence RGD (Arg-Gly-Asp). This motif is present on a wide variety of microbial structures ranging from viral proteins such as TAT of human immunodeficiency virus (Barillari *et al.*, 1993) to bacterial proteins such as *Bordetella pertussis* filamentous haemagglutinin (FHA) and pertactin (Sandros and Tuomanen, 1993), SpeB2 of streptococci (Stockbauer *et al.*, 1999) and many others (Virji, 1996b). Since RGD is a recognition sequence for several receptors (β_2, β_3/β_5 and some β_1 integrin groups), the possession of this sequence has the potential to mediate interactions with multiple integrin receptors. This mechanism can be described as true ligand mimicry. Meningococci also possess ligands that mimic structures of the host. It is becoming apparent from our recent investigations that sialic acid structures present on meningococcal LPSs are recognized specifically by certain host sialic acid-binding immunoglobulin-like lectins (Siglecs; Crocker *et al.* 1998) expressed on distinct subsets of haemopoietic cells and involved in specific functions in haemopoietic cell biology (P.R. Crocker and M. Virji, unpublished data).

In an indirect manner, via a mechanism described as 'pseudo ligand mimicry', microbial adhesion to host ligands and to their natural receptors occurs by a bridging or sandwich mechanism. This mode of interaction has several advantages in that adhesion to host ligands (which in many cases are extracellular matrix (ECM) proteins) allows an organism to adhere not only to the ECM, which may become exposed on the mucosae during damage resulting from mechanical injury or from infection by other microbes, but also to the host cells via these proteins.

Complement deposition on the microbial surface may opsonise many microbes via CR3 ($\alpha_m\beta_2$-integrin), in addition to the non-integrin receptor CR1. Interactions via CR3 that ensue on the coating of microbes by the natural ligand iC3b apparently render the microbe resistant to intracellular killing and some microbes specifically recruit CR3 ligand deposition. For example, *Legionella pneumophila*, which invades and grows within alveolar macrophages and causes pneumonia, localizes complement components on its major outer membrane proteins (Bellinger-Kawahara and Horwitz, 1990). More recently it has been demonstrated that engagement of anti-CR3, iC3b or microbes with the CR3 receptor leads to negative signalling. In monocytes, suppression of interferon (IFN)-γ induced tyrosine phosphorylation and decreased interleukin (IL)-12 production was observed. It is suggested that such signalling on interactions with CR3 may lead to the inhibition of T_H1-dependent cell-mediated immunity and also to reduced nitric oxide and respiratory burst in monocytes, thereby allowing microbes to survive within intracellular compartments (Marth and Kelsall, 1997). Not all sandwich adhesion results in ligand/microbe internalization. It is reported that covering of the bacterial surface with fibronectin results in some cases in the organism having an extracellular location only (Isberg, 1991; see section 2.3.1). Meningococcal Opc protein is another example of a ligand that interacts with several target integrins via a sandwich mechanism (see section 2.5.1.2).

Other mechanisms exist in meningococci and *H. influenzae* that allow interactions at receptor sites normally involved in the recognition of natural ligands. Presumably, these sites are exposed and readily accessible on the receptors and/or because ligation at these sites results in effective manipulation of host cellular signalling mechanisms by the microbe (see section 2.5.2; Plates 2.1 and 2.2).

2.3.1 Factors that may determine between adhesion and invasion

Studies on the mechanisms of microbial invasion of host cells have suggested that low affinity interactions result in an extracellular location whereas high affinity engagement of the microbe with its receptor leads to cellular invasion. High affinity interactions may be achieved by distinct means. The *Yersinia* protein invasin possesses features that enable it to interact with several β_1 integrins with affinity that is higher than that of the natural ligand, allowing efficient competition with extracellular matrix proteins to which β_1 integrins may be engaged. The result is 'zippering' – a process by which the host cell forms a series of contacts over the surface of the microbe leading to uptake (Isberg, 1991; Isberg and Tran Van Nhieu, 1994). High affinity of interaction may be achieved also by engagement of multiple microbial ligands with a single receptor, as described for *Leishmania* ligands gp63 and lipophosphoglycan, which interact with CR3, leading to cellular invasion (Talamas-Rohana *et al.*, 1990). Additionally, receptor clustering achieved by cross-linking of ligands was also a requirement for the signal leading to uptake (Isberg and Tran Van Nhieu, 1994). It is also suggested that high receptor density may achieve the same final end. Thus up-regulation of receptors during many viral and other (e.g. malarial) diseases may help to augment widespread invasion. The Opc protein of *N. meningitidis*, which interacts with integrins via a sandwich mechanism, leads to efficient invasion. It is possible that, in this case, high affinity of interaction is achieved via a secondary ligand leading to multiple receptor occupancy (section 2.5.1.2).

In the case of meningococcal targeting of carcinoembryonic-antigen-related cell adhesion molecules (CEACAMs), the outer membrane Opa proteins may bind to several distinct members of the family of the cell adhesion molecules (Plates 2.1 and 2.2). Some of these members are tethered to the membrane via glycosyl-phosphatidylinositol (GPI) anchors and, interestingly, interactions with these receptors result in high levels of adhesion but inefficient invasion. Other members of CEACAMs are transmembrane signalling molecules and some contain immunoreceptor tyrosine-based inhibitory motifs (ITIMs) in their cytoplasmic tails. Such interaction results in cellular invasion and signalling, which has the potential to down-modulate effector functions of phagocytes (section 2.5.2).

2.3.2 Receptor modulation

Modulation of signalling molecules occurs in response to circulating cytokines and has been recorded for many receptors targeted by microbes. In addition, certain microbes induce up-regulation of their own receptors by supplementary mechanisms. *Bordetella pertussis* appears to up-regulate CR3 via FHA interactions involving leucocyte signal transduction complex (comprising a β_3 integrin and CD47, section 2.5.1.1) as well as via a selectin-like function of pertussis toxin. Up-regulation of receptors has also been observed during malaria infection. Brain endothelium from patients dying of malaria expressed several of the implicated malarial receptors (CD36, intercellular adhesion molecule (ICAM) 1, endothelial leucocyte adhesion molecule (ELAM) 1, vascular cell adhesion molecule (VCAM) 1) not observed in uninfected brain tissue and may result from increased levels of tumour necrosis factor (TNF) α found in malaria patients (for a review, see Virji, 1996a).

2.4 *NEISSERIA MENINGITIDIS* INTERACTIONS WITH HUMAN CELLS – INTERPLAY BETWEEN SURFACE-LOCATED STRUCTURES

Clinical observations of *N. meningitidis* pathogenesis suggest that dissemination to the central nervous system (CNS) occurs via blood. Therefore bacteria must traverse the epithelial barrier of the nasopharynx and endothelial barriers of the vasculature to reach the CNS and other tissues. The possible routes include direct intra- or intercellular translocation (transcytosis vs. paracytosis) in addition to carriage via phagocytic cells. How surface ligands of meningococci participate in cellular interactions with various host cells encountered by bacteria, leading to passage across cellular barriers, has engaged many investigators and our current understanding of the participation of some of the major surface structures is described below.

2.4.1 Capsule and lipopolysaccharides as masking agents

Since meningococci are typically capsulate organisms, with the capacity to alter their capsule expression, in considering their interactions with target cells it is important to address the modulatory effects of the capsule on the functions of surface ligands. In addition, LPS undergoes phase variation in sialic acid expression on its terminal lacto-N-neotetraose (LNnT) structure (Jennings et al., 1995). Both the capsule and LPS affect interactions mediated via the major outer membrane adhesive proteins.

Capsule and sialylated LPS are invariably expressed in disseminated isolates and are believed to protect the organism against antibody/complement and phagocytosis (Vogel and Frosch, 1999). They are also expressed by a number of carrier isolates and may have functions that allow the organism to exist in the nasopharynx (enabling avoidance of mucosal immunity) or are physically protective against external environment during transmission between hosts (antidesiccation property of capsular polysaccharide). However, acapsulate meningococci are isolated frequently from the nasopharynx (Cartwright, 1995, p. 127). *In vitro* studies show that adhesion, and particularly invasion, of epithelial cells is enhanced (aided by some opacity proteins) in the absence of a capsule (Virji et al., 1992b, 1993a). This invites the hypothesis that loss of capsulation may help to establish long-term nasopharyngeal carriage where the intracellular state would potentially provide protection from host defences. Intracellular meningococci have been observed within tonsillar tissue (Sim et al., 2000). Whether factors in the nasopharynx trigger down-modulation of capsulation is not known, but one study suggests that environmental factors may regulate capsule expression (Masson and Holbein, 1985). Recently, a contact-regulated gene, *crg*, has been described, which is reported to be up-regulated following meningococcal contact with host cells. *Crg* may act to control the expression of several genes, including down-modulation of capsule gene expression (see Chapter 7), although this remains to be demonstrated. It follows that dissemination from the site of colonization would require up-regulation of capsulation, since

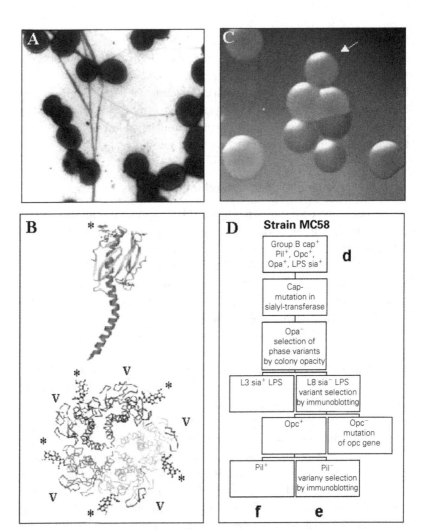

Figure 2.1. Phase- and structurally variable components of *N. meningitidis*. (A) Negative stain transmission electron micrograph of a piliated *N. meningitidis* isolate showing long filamentous pili forming rope-like bundles. (B) *Top:* A three-dimensional molecular model of a pilin of strain C311 based on that of *N. gonorrhoeae* MS11 pilin, which was determined by X-ray crystallography (Parge *et. al.*, 1995). The model was built with the help of structural databases, and minimized using the program X-plor (A. Hadfield and M. Virji, unpublished data).

Below: A diagram showing a cross-section through the pilus fibre based on the proposed pilus assembly model of Parge *et al.* (1995). Virtually all the modifications are located on the outside of the fibre. The positions of the glycans are indicated by asterisks. Thus pilus fibre contains few exposed protein epitopes. Those that are exposed are the variable domains of pili (V). (C)A light micrograph of agar-grown colonies of acapsulate non-

acapsulate bacteria are unlikely to survive in the blood. Alternatively, since blood provides an environment in which meningococci can grow rapidly, it is possible that a small number of capsulate organisms, arising as a result of natural phase variation, will be selected for in the blood. In the case of *H. influenzae*, studies on the infant rat model of haemophilus bacteraemia and meningitis have shown that bacteraemia may arise as a result of the survival of a single organism in the bloodstream (Moxon and Murphy, 1978).

Meningococci from the nasopharynx often express the L8 LPS immunotype that resists sialylation due to the absence of the LNnT structure (Cartwright, 1995, pp. 24, 128). In many *in vitro* studies it has been shown that sialylation of LPS, like capsule, not only imparts resistance to immune mechanisms of the host (Vogel and Frosch, 1999) but also masks the functions of many outer membrane proteins (Virji *et al.*, 1993a, 1995b; Virji, 2000a). The interplay between surface polysaccharides and various adhesins and invasins is a complex area of investigation with antigenically and phase-varying components adding to the complexity. In addition, as mentioned above (section 2.3), LPS itself may act as a ligand for host receptors. Both sialylated (P. Crocker and M. Virji, unpublished data) and asialylated (Harvey *et al.*, 2000) LPSs may interact with specific receptors.

2.4.2 Adhesive proteins of meningococci

Amongst the major adhesive proteins elaborated by *N. meningitidis* are pili (fimbriae) and the outer membrane opacity proteins, Opa and Opc. The pili are long filamentous protein structures composed of multiple pilin subunits (Fig. 2.1). They are generally regarded as the most important

piliated, Opa⁻ meningococci. Opaque granular colonies (arrow) are those arising from bacteria expressing the opacity protein Opc, in this case. Bacteria giving rise to 'transparent' colonies are Opc phase-off variants. (D) Derivation of phenotypic variants of the *N. meningitidis* serogroup B strain MC58. MC58 was isolated from the blood of a patient with meningococcal infection during the Stroud (UK) outbreak in 1983–1985. The parental phenotype contains the full attribute of the major virulence factors as shown. By a combination of genetic manipulation and isolation of phase variants a family of related derivatives was created for studies of the interplay between surface ligands (section 2.4). Pil,pili; LPS sia⁺ represents LPS containing terminal lacto-*N*-neotetraose (LNnT) that is sialylated in this strain, giving rise to L3 LPS immunotype. L8 immunotype that lacks LNnT and sialylation can arise spontaneously. (NB: *N. meningitidis* and *H. influenzae* express short oligosaccharide chains on their 'LPS' and are also termed lipooligosaccharides: LOSs.) For d, e, f, see caption to Fig. 2.2.

adhesins in capsulate bacteria, owing to the fact that, whereas the capsule partly or totally masks outer membrane ligands resulting in their reduced functional efficacy, pili traverse the capsule and remain functional in fully capsulate bacteria. Opa are structurally variable proteins and are encoded by a family of genes of which there are three to four in *N. meningitidis* isolates (Aho *et al.*, 1991; Malorny *et al.*, 1998). In the closely related *N. gonorrhoeae* (gonococci), up to 12 homologous *opa* genes may be present. Distinct Opa proteins are characterized by variant extracellular domains. Of the four domains (loops 1–4) predicted to be exposed on the bacterial surface, only loop 4 (the most proximal to the C-terminus) is conserved. Loop 1 maintains some structural similarities between Opa proteins and is termed semivariable (SV), whereas loops 2 and 3 are hypervariable (HV1 and HV2) (Malorny *et al.*, 1998; section 2.5). Opa proteins undergo antigenic and phase variation at a high frequency. The two events are linked since the random on/off switching of each *opa* locus generates the variation of expression from none to several Opa proteins (Stern *et al.*, 1986). Opc protein is expressed widely by many clinical isolates of meningococci and shares some physicochemical properties with the Opa proteins (Olyhoek *et al.*, 1991). Opc is phase variable but structurally it is largely invariant.

2.4.3 Pili and their importance in multiple cellular targeting and in potentiation of cellular damage

Both carrier and disease isolates are usually piliated; however, pili are lost on non-selective subculture (Virji *et al.*, 1995a), which suggests that pili are selected for *in vivo*. Pili have been implicated in mediating epithelial interactions (Stephens and Farley, 1991) and were shown to mediate haemagglutination (Trust *et al.*, 1983). Pili also mediate adhesion both to human umbilical vein and microvascular endothelial cells (Virji *et al.*, 1991, 1993b, 1995a; Merz and So, 2000), and cells of the human meninges (Hardy *et al.*, 2000). One consequence of pilus-mediated adhesion to endothelial cells is increased cellular damage, which is mediated primarily by LPS (Fig. 2.2) and is dependent on the presence of serum CD14 (Dunn *et al.*, 1995; Dunn and Virji; 1996). Thus multiple ligation of receptors via LPS and pili augment cellular signalling events that result in cell death. These *in vitro* toxic effects reflect the acute toxicity of meningococci for endothelial cells observed during vascular dissemination.

2.4.4 Structure/function relationships of meningococcal pili, pilus-associated adhesins

Meningococci elaborate two structural classes of pili (I and II). However, no discernible functional difference has been assigned to either class. Both undergo antigenic variations, which alters their tissue tropism. Studies using pilus adhesion variants (derived by single colony isolation, with or without prior selection on host cells) implied that structural variations in pilin affect epithelial interactions significantly, but have less effect on endothelial interactions (Virji *et al.*, 1992a, 1993b; Merz and So, 2000; Virji, 2000). Thus the pilin subunit may contain a human cellular binding domain, or at least has influence on adhesion if mediated by an accessory protein such as PilC, which has been implicated in cellular adhesion and in biogenesis in both meningococci and gonococci (Merz and So, 2000). At present, how pilin structural variations modulate PilC or other pilus-associated adhesion functions is not clear. Further analysis of PilC interactions is described in Chapter 7 in this volume.

Studies on adhesion variants of meningococci have revealed that meningococcal pili are subject to post-translational modifications (Virji *et al.*, 1993b; Virji, 2000) and they contain unusual substitutions. A trisaccharide structure (Galβ1–4, Galα1–3-2,4-diacetamido-2,4,6-trideoxyhexose) is present on all variant pili of strain C311 (Stimson *et al.*, 1995; Fig. 2.1). Further studies have shown that, at a distinct site, meningococcal pili contain a, perhaps, unique substitution, α-glycerophosphate (Stimson *et al.*, 1996). The pili of *N. meningitidis* also contain epitopes that bind specifically to anti-phosphorylcholine antibodies (Weiser *et al.*, 1998). Interestingly, this epitope is not found on the homologous pili of commensal neisserial strains (*N. lactamica, N. flavescens, N. subflava*), instead the moiety is present on their LPS. Thus the manner of phosphorylcholine (ChoP) expression distinguishes commensal strains from the pathogenic strains of *Neisseria* (Serino and Virji, 2000). It has been reported that ChoP expressed on *Streptococcus pneumoniae* may interact with platelet-activating factor receptor (Cundell *et al.*, 1995) and signal for cellular invasion. It is possible that its presence on neisserial surface structures could modulate cellular interactions. The functional consequences of pilin modifications are not understood at present.

2.4.5 Phenotypic requirements for interactions mediated via outer membrane proteins

As in fully capsulate bacteria, only pili appear to be effective in mediating cellular adhesion to human epithelial and endothelial cells, acapsulate

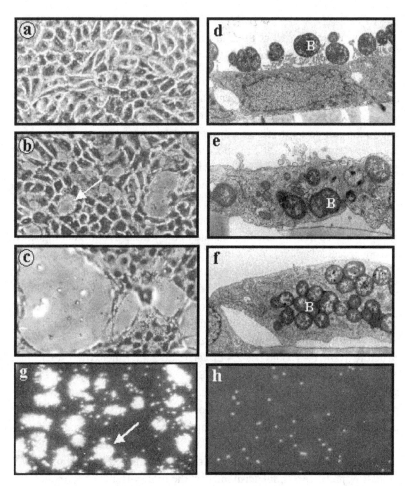

Figure 2.2. Some mechanisms and consequences of receptor ligation by multiple meningococcal ligands. (a–c) Light micrographs of Giemsa-stained human umbilical vein endothelial cell monolayers demonstrating the synergistic effect of LPS and pili in mediating cytopathic damage. Confluent endothelial cell monolayers were inoculated with piliated or non-piliated variants of a capsulate meningococcal strain or a capsulated clinical isolate of *H. influenzae*. No toxicity was observed in the latter case (a). Under the same condition, non-piliated meningococci caused some damage (arrow, b). However, piliated variants of the same meningococcal strain produced a much increased cytotoxic effect (c). The primary factor responsible is LPS since the toxic effect is dependent on serum CD14 (Dunn and Virji, 1996). (d–f) Distinct modes of interaction of variants of *N. meningitidis* strain MC58 with human endothelial cells observed by transmission electron microscopy. The derivatives used here are shown in Fig. 2.1D. (d) The parental isolate ('d' Fig. 2.1D), despite expressing the opacity proteins, is not invasive in this case, due to the presence of capsule. Adherence is mediated by pili that traverse the capsule. (e) Cellular

non-piliated derivatives of a serogroup A strain (C751) were used to demonstrate the invasive potential of the meningococcal proteins Opa and Opc (Virji *et al.*, 1992b, 1993a). In addition, a library of variants and mutants (varying in expression of capsule, LPS, pili, Opa and Opc) created in a serogroup B strain (MC58) was also used to study the roles of outer membrane proteins (Fig 2.1). These studies demonstrated that Opc can act as an invasin in distinct serogroups, and that surface polysaccharides (capsule, LPS) inhibit Opc-mediated invasion. Interestingly, pili may potentiate Opc-mediated invasion of some cells (Virji *et al.*, 1995b; Fig. 2.2e and f). As acapsulate/asialylated phenotypes occur in the nasopharynx, and Opc and Opa are expressed in many nasopharyngeal isolates, these proteins may be important, particularly in interactions with nasopharyngeal epithelial cells. In more recent studies we have also demonstrated that particular ligand–receptor pairs may bind with such high affinity as to overcome the inhibitory effect of the capsule on outer membrane protein function (see section 2.5.2).

2.5 INTERACTIONS OF MENINGOCOCCI WITH MULTIPLE SIGNALLING MOLECULES

Several host cell receptors have been identified that may be targeted by distinct meningococcal ligands, including the membrane cofactor protein CD46 (by pili/pilC), heparan sulphate proteoglycans (by Opa and Opc), asialoglycoprotein receptor (asialylated LPS), Siglecs (sialylated LPS), extracellular matrix proteins (Opc and Opa), integrins (Opc and Opa), and CEACAMs (Opa). Of these, the last three will be discussed in greater detail below to explore how targeting of these receptors may increase bacterial pathogenic potential. Several recent reviews have described some of the other receptor interactions (Merz and So, 2000), although many of the details remain to be investigated. Engagement of multiple receptors by independent adhesins, or ligation of individual receptors at multiple sites,

invasion by acapsulate, asialylated, non-piliated, Opa⁻ bacteria expressing the Opc protein ('e' Fig. 2.ID). (f) Increased cellular invasion primarily mediated by Opc, in a phenotype also expressing pili ('f' Fig. 2.1D). B, bacteria. (g–h) Immunofluorescence micrographs demonstrating Opc-mediated interactions supported entirely by purified vitronectin. Adherence of *N. meningitidis* inoculated in the absence of serum but in media supplemented with vitronectin alone (g), or together with RGDS peptide (h). Bacteria in excess of 100 can be seen covering the entire surface of a human endothelial cell (arrow, g). Meningococci were labelled with anti-LPS antibody and detected with rhodamine-conjugated secondary antibody.

often leads to augmentation or modification of signalling, leading to increased invasion or toxicity (Fig. 2.2).

2.5.1 Targeting of integrins

2.5.1.1 Integrin structure and function

Integrins are a superfamily of heterodimeric transmembrane molecules consisting of diverse α-chains (more than 18 have been described) and relatively more conserved β-chains (8 have been described). The nature of the β-chain defines the family of integrins. β_1, β_3 and β_5 are found on many cell types and interact with matrix and other proteins and may mediate cell–cell adhesion. The β_2 family are leukocyte-associated integrins and include complement receptors. As discussed in section 2.3, the sequence RGD (Arg-Gly-Asp) is a recognition sequence on many serum and ECM ligands of integrins. In addition, receptor specificity may be modified by RGD flanking sequences. Within the RGD motif, aspartic acid is important for recognition by integrins (Ruoslahti and Pierschbacher, 1987) and some integrins that do not recognize the RGD sequence nevertheless require an acidic amino acid (Asp or Glu) for binding to ligands. Interestingly, in the case of the Yersinia invasin that binds multiple β_1-integrins (the majority do not recognize RGD), aspartic acid has been shown to be essential in adhesion (Isberg and Tran Van Nhieu, 1994).

Integrins have short cytoplasmic tails that associate with cytoskeletal proteins such as α-actinin, talin, vinculin, paxillin and tensin in complexes (focal adhesions) and affect cytoskeletal arrangement via actin microfilaments. Thus integrins have the capacity to form a physical link between the ECM and the cytoskeleton. Two principal mechanisms by which integrins are activated to transduce signals involve conformational change and receptor clustering. Conformation-dependent activation is exemplified by thrombin-mediated activation of gpIIb/IIIa ($\alpha_{IIb}\beta_3$) integrin expressed on platelets. Thrombin induces intracellular signalling, leading to modulation of the conformation of gpIIb/IIIa, which is then able to bind fibrinogen, leading to platelet aggregation. This has been called 'inside out' signalling (Hughes and Pfaff, 1998).

Signal transduction via integrins may reside, at least in part, in cytoskeletal rearrangement, which may physically bring together molecular complexes by forming focal adhesions. Inhibition of clustering, achieved by drugs that prevent cytoskeletal rearrangement, also inhibits tyrosine kinases, suggesting that kinase activation may be mediated by receptor clustering. A tyrosine kinase, focal adhesion kinase (FAK) pp125[FAK], is believed to play an important role in integrin-mediated signal transduction. This kinase localizes to focal adhesion complexes and is stimulated to autophosphorylate at Tyr-397 upon integrin binding to the ECM. Studies using cytochalasin D (which inhibits cytoskeletal rearrangement) have shown that signal transduction via integrins requires cytoskeletal reorganization and involves mitogen-activated protein (MAP) kinase activation in addition to pp125[FAK] phosphorylation. FAK may interact with Src and recruit Src to focal adhesions, causing hyperphosphorylation of focal adhesion structures. There is also evidence for association of other tyrosine kinases with integrins; for example, in monocytes, Syk (Src family) may respond to integrin signalling. Certain integrins are associated with CD47, an unusual member of the immunoglobulin superfamily, also known as IAP (integrin-associated protein). CD47 has unique signal-transduction properties, initiating heterotrimeric G-protein signalling that augments the function of integrins of the β_1, β_2 and β_3 families (Clark and Brugge, 1995; Richardson and Parsons, 1995; Rosales et al., 1995; Brown and Frazier, 2001)

Stimulation of integrins may be translated into a variety of intracellular signals that may lead to transcription of inflammatory mediators such as IL-1β and IL-8. Integrins may also act in concert with other receptor pathways to enhance or dampen signals. In particular, activation via growth factor receptors requires adherence of cells to ECM via integrins. The small G-protein Rho appears to be critical in integrating signals induced by integrins and growth factor receptors (Clark and Brugge, 1995).

A recent study has demonstrated that *Borrelia hermsii*, a spirochaete responsible for relapsing fever in humans, is able to alter the conformation of $\alpha_{IIb}\beta_3$ (gpIIb/IIIa) integrin on resting platelets via a contact-dependent platelet activation, thereby promoting its binding to activated integrins (Alugupalli et al., 2001).

2.5.1.2 Interactions of the outer membrane protein Opc with integrins: an example of pseudo ligand mimicry

2.5.1.2.1 RGD-dependent cellular invasion mediated by Opc

Opc, a basic protein, appears to have the capacity to bind to multiple ECM components and serum proteins (Virji et al., 1994a). This, together with the fact that interactions of Opc-expressing phenotype with the apical surface of polarized human endothelial cells require serum-derived factors (Virji et al., 1994b) and that pre-coating of bacteria with serum was sufficient to mediate interactions, suggested a sandwich mechanism of adhesion. The serum factors were shown to be RGD-containing proteins and RGDS but not RGES (Arg-Gly-Glu-Ser) peptides inhibited bacterial invasion of human endothelial cells. Moreover, antibodies against the vitronectin receptor ($\alpha_v\beta_3$, VNR) and also, to some extent, the fibronectin receptor ($\alpha_5\beta_1$, FNR) inhibited adherence and invasion. Bacterial interactions could also be supported by purified vitronectin in the absence of serum (Fig. 2.2g), supporting the notion that Opc interacts with the natural ligands of RGD-recognizing integrins and, via the ligand, indirectly interacts at the ligand-binding site of the receptor. This interaction is extremely effective in mediating cellular invasion (Fig. 2.2e) (Virji *et al.*, 1994b, 1995b).

In addition to utilization of the integrin ligands, further factors may be involved in the interactions via the vitronectin receptor. This was suggested from the observations that cloned Opc does not confer invasive properties on *Escherichia coli*, even though the protein is surface expressed and is immunologically similar to that of *N. meningitidis*. By analogy with the complement receptor CR3, which has been shown to interact simultaneously with C_3bi-coated particles and with microbial glycolipids at distinct sites (Wright *et al.*, 1989), VNR may require multiple ligand engagement. Indeed, the VNR also exhibits binding sites for ganglioside GD2 (Cheresh *et al.*, 1987). Gangliosides and LPS share structural similarities in that both are amphipathic with strongly anionic hydrophilic groups and it is tempting to speculate that some manner of LPS interaction with the VNR may be an additional factor required. However, it is also possible that the level of Opc expressed by *E. coli* is not optimum, since efficient interactions of *N. meningitidis* via Opc require the protein to be expressed at a high density on the bacterial surface (Virji *et al.*, 1995b).

An interesting feature of Opc interaction was the requirement for host cytoskeletal function. Attempts to inhibit host cell invasion by the use of cytochalasin D resulted in inhibition not only of invasion but also of total cell association. This observation is in contrast to cell adhesion mediated by *N. meningitidis* Opa proteins, which apparently increases in the presence of cytochalasin D (Virji *et al.*, 1993b).

2.5.2 Targeting of CEACAMs

2.5.2.1 CEACAMs: structure and biology

CEACAMs belong to the CEA family, a member of the immunoglobulin superfamily that includes clinically important tumour markers such as CEA (Plate 2.1). The antigen was first identified in human colonic carcinoma tissue extracts, but since then it has been demonstrated in numerous normal human tissues. CEACAM1 (previously CD66a, biliary glycoprotein (BGP); Beauchemin *et al.*, 1999) has the broadest tissue distribution and is expressed on the apical surfaces of epithelial cells of human mucosa, cells of myeloid lineage as well as some endothelial cells. CEACAM1 belongs to a transmembrane subgroup and may contain either a short (CEACAM1-S) or a long (CEACAM1-L) cytoplasmic tail. In studies on respiratory pathogens, it is important to note that CEACAM expression on normal epithelial cells in oral, tonsillar and lung tissues has been reported. CEACAM1, especially the isoform with a long cytoplasmic tail (CEACAM1-L) that contains immuno-receptor tyrosine-based activation motif (ITAM)/ITIM-like motifs, has been implicated in signal transduction. In addition, CEACAM3 isoforms may also contain similar motifs whose tyrosine phosphorylation leads to protein tyro-sine kinase or phosphatase activation and stimulation or termination of sig-nalling (Prall *et al* 1996; Obrink 1997; Hammarström 1999). Beauchemin *et al.* (1997) have reported the association of CEACAM1 (BGP) with tyrosine phosphatases SHP-1 (Src-homology containing Tyr phosphatase-1) in colonic epithelial cells and CEACAM1-mediated signalling in these cells results in the inhibition of cell growth.

2.5.2.2 Opa protein targeting of CEACAMs

The identity of the receptor for Opa protein was obtained from studies that examined target cells (COS, African Green monkey kidney cells, lacking Opa receptors) that were transfected with cloned cDNA encoding several dis-tinct human cell surface molecules. These included constitutively expressed or inducible adhesion molecules. The studies led to several interesting obser-vations. First, they identified that CEACAM1 expression on transfected COS cells resulted in increased adherence by meningococcal isolates. Second, they showed that some capsulate meningococci adhered to transfected cells that expressed CEACAM1 at high levels (Virji *et al.*, 1996a; Plate 2.1). The binding epitope (adhesiotope) for Opa proteins was shown to be located at the N-terminal IgV-like domain of the receptor (Plates 2.1 and 2.2). Polymorphonuclear phagocytes (PMNL) and some epithelial cells express several members of the CEACAM family including CEACAM1 (Stanners *et al.*, 1995; Teixeira *et al.*, 1994). These have highly homologous N-terminal domains, and Opa proteins of *N. meningitidis* were shown to mediate inter-actions with PMNL and epithelial cells via CEACAMs (Virji *et al.*, 1996b). Studies on *N. gonorrhoeae* also confirmed that Opa proteins engage with CEACAMs on distinct target cells (Chen and Gotschlich, 1996; Virji *et al.* 1996a,b; Gray-Owen *et al.*, 1997).

2.5.2.2.1 Opa/CEACAM interactions in commensal neisseriae

In addition to pathogenic neisseriae, in several commensal species of *Neisseria* (Cn), genes similar to pathogenic *opa* genes have been reported. Recently, surface expression of Opa-like proteins was shown in many Cn strains, for example *N. lactamica*, *N. subflava*, *N. flavescens* (Toleman *et al.*, 2001). The Cn Opa proteins are structurally similar but are generally (but not always) smaller and contain conserved regions that differ from the pathogenic Opa at many sites. Further, their SV and HV domains are often of different dimensions as compared with the equivalent domains of the pathogenic strains, and may produce distinct surface conformations. This evolutionary distance reflected in the Opa structure raised the question of whether commensal Opa proteins possess the capacity to target CEACAM molecules. Interestingly, within Opa-expressing commensal isolates, >75% had the capacity to target CEACAM1. Therefore the receptor-targeting function is evolutionarily conserved and suggests a primary role of this interaction in neisserial colonization of the human mucosae. The subtleties of receptor targeting and signalling that may differentiate between commensal and pathogenic Opa interactions with their common receptors need to be explored. One observation that may differentiate between commensal and pathogenic Opa/CEACAM interaction is that, in general, the commensals bind with lower affinity to CEACAMs as compared with *N. meningitidis* isolates (Toleman *et al.*, 2001).

2.5.2.2.2 Opa proteins: tissue specificity vs. tropism

Within the neisserial gene pool, a large repertoire of variable surface-exposed HV and SV alleles are found. To investigate the extent to which CEACAM could serve as a target for variant Opa proteins of *Neisseria*, we carried out a large survey of clinical isolates. Numerous human mucosal and disease isolates (including 50 strains each of gonococci and meningococci) were examined for their interactions with soluble CEACAM1-Fc. Specific adherence of the N-terminal domain of the receptor to c. 95% of Opa-expressing pathogenic *Neisseria* spp. was observed (Table 2.1). These studies imply that the receptor is a target for a conserved domain present on the majority of Opa proteins of meningococci and gonococci, although the domain remains to be identified. Alternatively, distinct Opa variable regions contain complementary sequences with the capacity to target the receptor.

To investigate the effect of variable domains of Opa proteins on receptor recognition, we used variants of strain C751, since they share some variable sequences. They form two pairs (OpaA/B and OpaB/D) with only a single difference either in their HV1 or HV2 regions (Plate 2.2c). Using HeLa cells

Table 2.1. *Reactivity of human mucosal isolates with CEACAM1*

Species	No. of strains tested	No. of CEACAM1–binding strains
Neisseria meningitidis	50	47
N. gonorrhoeae	50	48
Commensal *Neisseria* spp.	10	8
Typeable *Haemophilus influenzae*	24	18
Non-typeable *H. influenzae*	15	11
H. influenzae biogroup aegyptius	12	12
H. parainfluenzae	7	0
Pseudomonas aeruginosa	6	0
Streptococcus pneumoniae	18	0

Note: Data obtained from reactivity of whole cell lysates of bacteria in an immuno-dot blot assay that used bacteria in solid phase and soluble chimeric receptors (CEACAM1-human Fc) in liquid phase (Virji *et al.*, 1999, 2000).

transfected with cDNA encoding CEA, CEACAM3 or CEACAM6, we observed that all three Opa of C751 bound equally well to CEA but showed distinct tropism for CEACAM3- and CEACAM6-expressing cells. Since C751 Opa proteins make up distinct structures via combinations of their HV domains, HV1 and HV2 regions appeared to be involved in tropism for the distinct CEACAMs. In further studies, site-directed mutants of surface-exposed residues of CEACAM1 were used for identification of adhesiotopes on the receptor (Plates 2.1 and 2.2). The studies identified a number of critical amino acid residues required for Opa attachment. The residues were scattered on the C/F/G strands and loops of the N-domain but formed a continuous adhesiotope in the three-dimensional model (Plate 2.2). Binding of all Opa tested required Tyr-34 and Ile-91. Further, efficient interaction of distinct Opa proteins were dependent on other residues located in close proximity to Tyr-34 on the CFG face in the three-dimensional model. Studies from three laboratories have confirmed the critical binding regions of Opa proteins on CEACAM receptors (Billker *et al.*, 2000; Virji, 2000b). Using the N-domain mutants, we also observed a striking influence of HV1 and HV2 regions of C751 Opas on receptor targeting. For example, the mutant receptor with a single mutation at Ser-32 on the beta strand 'C' was not recognized by an OpaA variant with a distinct HV2, but OpaB and OpaD bound to the mutant receptor efficiently. In another case, a Gln-89 mutation on the F strand resulted in greatly reduced binding of OpaD with a new HV1 region (Virji *et al.*, 1999).

Overall, the studies identified a binding region in the N-domains of members of the CEACAM family arranged around Tyr34. They indicated that at least two amino acid residues determine the primary receptor specificity of the majority of Opa proteins. The tropism appears to be determined by one to several amino acid residues located on the edges of this region and these are differently recognized by distinct Opa proteins. Thus there are distinct Opa adhesiotopes on the N-domains of CEACAM molecules. The data suggest that Opa proteins have evolved to generate a variety of hypervariable domains that may not only help to avoid the host immune response but retain receptor binding capacity in such a way that they compensate for the heterogeneity in their target receptor family, thus increasing their host tissue range. It remains to be seen whether commensal strains with Opa proteins of greater structural and conformational repertoire provide a further insight into the features that determine receptor tropism vs. specificity as well as colonization vs. pathogenesis.

2.5.2.2.3 Opa signalling via CEACAMs

In addition to receptor tropism, studies of the interactions of meningococci with HeLa transfectants revealed distinct outcomes that depended on the receptor expressed. Expression of the transmembrane receptor CEACAM3 (CGM1) correlated with invasion whereas that of GPI-anchored CEA, although supporting substantial adhesion, did not result in significant internalization. A similar situation was apparent with CEACAM6 (NCA) (Virji et al., 1999). The observations are interpreted in terms of the capacity of the receptor to transduce signals, although it should be noted that, in transfected cells, individual CEA molecules may not have complementary coupling partners for signal transduction and the situation in host cells with simultaneous expression of both signalling and GPI-anchored members may be distinct. Interestingly, it has been suggested that molecules such as CEA may have evolved to reduce the burden of microbes on the mucosa. Late evolutionary development of the GPI-anchored members, apical location and shedding/secretion of large amounts of CEA in the gut suggest the possibility that CEA, which is targeted by enteric pathogens, may have developed as an arm of innate immunity (Hammarström, 1999).

Signalling via Opa/CEACAM interactions has been studied mainly in professional phagocytes, and appears to involve Src kinases, Rac1, p21-activated kinase (PAK) and Jun terminal kinase. The pathway differs from the opsonic pathway of internalization via FcγR, which results in activation of cellular killing mechanisms. It is suggested that signalling via the ITIMs present in the CEACAM cytoplasmic domains may improve the chances of gonococcal survival within the phagocyte (Hauck et al., 1998). Precisely what happens in interactions at the epithelial surface is not clear; our current investigations on meningococci and H. influenzae are discussed below.

2.5.2.3 *Haemophilus influenzae* targeting of CEACAMs

In THi, NTHi as well as Hi-aeg, several adhesins have been identified (St Geme, 1999). Many of these have been shown to target carbohydrate receptors in the host environment (Foxwell et al., 1998). Our recent studies demonstrated that H. influenzae targets CEACAMs and this binding appears to involve protein–protein interactions.

On screening of multiple mucosal isolates to assess their potential for interactions with CEACAMs, we observed that several strains of THi and NTHi were able to bind to the receptor. *Haemophilus parainfluenzae* and several other mucosal isolates did not recognize the receptor (Table 2.1). The studies indicated that, like *N. meningitidis*, *H. influenzae* primarily bound to the N-terminal domain (Virji et al., 2000); however, unlike *N. meningitidis*, *H. influenzae* requires other extracellular domains of CEACAM1 for efficient interactions (D.J. Hill and M. Virji, unpublished data). The precise roles of the other domains in *H. influenzae* interactions are not clear but they may help to determine a particular conformation of the N-domain favoured by this organism for binding. Further analysis, using receptor molecules mutated at distinct residues in the N-domain, indicated that the primary binding sites of *H. influenzae* and *N. meningitidis* were closely positioned on the protein (Plate 2.2). Indeed, the two species may compete for these receptors (Virji et al., 2000).

One of the ligands of *H. influenzae* that is able to recognize CEACAM receptor constructs appears to be the outer membrane P5 protein (Hill et al., 2001). Interestingly, the P5 protein exhibits many structural features that are similar to Opa proteins, for example variable surface-exposed loops. However, there is little sequence homology and no conserved stretches of amino acid residues are present. It will be interesting to compare the features of Opa and P5 that confer the receptor-binding properties. A recent study has indicated that infection with the respiratory syncytial virus significantly enhances NTHi attachment to an A549 lung epithelial cell line and P5 proteins ('P5-homologous fimbriae') play a critical role (Jiang et al., 1999). Indeed, our studies have shown that A549 accommodate the *H. influenzae* strains that we tested via their CEACAM receptors, suggesting a potential role of CEACAMs in increasing susceptibility to infection by this organism.

2.5.2.4 Potential outcomes of CEACAM targeting

2.5.2.4.1 Phagocytic interactions

Interactions with phagocytic cells, which could occur readily between acapsulate phenotypes and PMNL, may appear to be counter-productive. However, as mentioned above, it has been suggested that non-opsonic interactions of *N. gonorrhoeae* Opa/PMNL involve host cell signalling pathways that lead to bacterial localization in an intracellular niche that may allow prolonged survival (Hauck *et al.*, 1998). Although this remains to be demonstrated, it is possible that bacteria able to survive for a relatively short period within the cells may be carried from site to site in a 'Trojan horse' manner. Macrophage trafficking and the spread of mousepox virus during experimental respiratory infection has been ascribed to this mechanism. *In situ* hybridization studies have identified *H. influenzae* within macrophages in adenoid tissues (Forsgren *et al.*, 1994). Since CEA molecules have also been identified on tissue macrophages (Obrink, 1997; Hammarström, 1999), their involvement in *H. influenzae*–macrophage interactions is entirely possible. Whether PMNL or macrophages with engulfed *H. influenzae* help to disseminate the bacteria remains to be investigated.

2.5.2.4.2 Epithelial interactions

Targeting of transmembrane CEACAMs would also be expected to lead to the manipulation of host signalling mechanisms that could result in either transcytosis or paracytosis and, by either mechanism, bacteria may traverse the epithelial barriers. CEACAM-mediated traversal of colonic epithelial cells by a gonococcal strain has been reported (Wang *et al.*, 1998). We have investigated the mechanisms by which meningococci and haemophilus strains utilize their common receptors on target cells to traverse the cellular barriers. The studies have shown that both bacteria are able to manipulate the host cytoskeleton following ligation with CEACAMs (Plate 2.2). Polymerized actin can be seen localized beneath bacterially induced caps in transfected cells as well as in human epithelial cell lines. However, the studies have also indicated that, further downstream, signalling diverges and *H. influenzae* and *N. meningitidis* traverse polarized monolayers in distinct manners with distinct rates and modes of transmigration (Plate 2.2; D.J. Hill, J. Griffith and M. Virji, unpublished data). The molecular details remain to be defined.

2.6 CONCLUSION

Of the critical determinants of receptor targeting that may lead to cellular invasion, the strength of ligand–receptor ligation that may be achieved when receptors are expressed at high levels stands out as a key event that may determine increased host susceptibility to some bacterial infections. An important observation in favour of this argument is that, even in capsulate bacteria, certain Opa proteins with high affinity for CEACAM1 may effectively bind to host cells expressing large numbers of these receptors. This demonstrated that even the inhibitory effects of surface sialic acids may be overcome when appropriate ligand/receptor pairs are present at the required density. Thus targeting of cell adhesion molecules, which are up-regulated during inflammation (Dansky-Ullman *et al.*, 1995) may be critical to bacterial pathogenesis and may shift the balance from a carrier state to dissemination. A low level constitutive expression of the receptor, for example on epithelial cells, may favour attachment without invasion. Viral infections or other conditions during which cytokines may be up-regulated, could result in increased expression of those signalling molecules targeted by bacteria, thereby increasing their potential to enter phagocytic cells as well as mucosal epithelial cells. Massive invasion of epithelial cells could be injurious to the host, while that of phagocytic cells could result in incomplete elimination of bacteria and the possibility of transmission within them.

In summary, recent advances in the molecular mechanisms of meningococcal interactions with human target cells are beginning to provide certain clues that may explain epidemiological observations. Several potential routes of invasion may exist. Acapsulate bacteria may target ECM exposed by epithelial damage following viral or other infections or they may invade epithelial cells. Further dissemination of such a phenotype requires selection of capsulate bacteria arising from natural phase variation. However, epidemiological evidence suggests that disease in susceptible individuals occurs soon after acquisition, with no prolonged carriage. The observations that some receptor–ligand interactions may occur in capsulate bacteria provide a feasible rationale for an alternative, perhaps not exclusive, mechanism. In this scenario, targeting of cell adhesion molecules that are up-regulated by inflammatory cytokines may be the critical determinants of meningococcal invasion. Viral infections, or other conditions leading to inflammation, could result in increased expression of receptors that recruit meningococci via one or more ligands. In a host with inadequate immunological protection against meningococci, invasion would result in rapid growth and dissemination. This may constitute one of a number of mechanisms responsible in distinct circumstances for disease outbreaks.

In the case of *H. influenzae* also, the frequency of colonization of the upper respiratory tract by NTHi increases during respiratory viral infections that stimulate the production of cytokines. Cytokine-stimulated increased receptor expression may also be responsible for the observed tendency of *H. influenzae* to invade tissue macrophages or migrate between cells. These processes, in turn, may be responsible for their escape from antimicrobials, leading to persistence in spite of antibiotic therapy. Thus, as with *Neisseria*, studies on *Haemophilus* interactions with signalling molecules that are upregulated by inflammatory cytokines may help to define the basis of increased susceptibility to infections.

ACKNOWLEDGEMENTS

M.V. is an MRC Senior Fellow. The studies in the laboratory of M.V. were supported by grants from the MRC, the National Meningitis Trust, the Meningitis Research Foundation and the Spencer Dayman Meningitis Research Laboratories. I am grateful to Andrea Hadfield and Mark Jepson for their help with molecular modelling and confocal imaging, respectively.

REFERENCES

Achtman, M. (1995). Epidemic spread and antigenic variability of *Neisseria meningitidis*. *Trends in Microbiology* **3**, 186–192.

Aho, E.L., Dempsey, J.A., Hobbs, M.M., Klapper, D.G. and Cannon, J.G. (1991). Characterization of the *Opa* (class-5) gene family of *Neisseria meningitidis*. *Molecular Microbiology* **5**, 1429–1437.

Alugupalli, K.R., Michelson, A.D., Barnard, M.R., Robbins, D., Coburn, J., Baker, E.K., Ginsberg, M.H., Schwan, T.G. and Leong, J.M. (2001). Platelet activation by a relapsing fever spirochaete results in enhanced bacterium-platelet interaction via integrin alphaIIbbeta3 activation. *Molecular Microbiology* **39**, 330–341.

Barillari, G., Gendelman, R., Gallo, R.C. and Ensoli, B. (1993). The Tat protein of human immunodeficiency virus type 1, a growth factor for AIDS, Kaposi sarcoma and cytokine-activated vascular cells, induces adhesion of the same cell types by using integrin receptors recognizing the RGD amino acid sequence. *Proceedings of the National Academy of Sciences, USA* **90**, 7941–7945.

Beauchemin, N., Kunath, T., Robitaille, J., Chow, B., Turbide, C., Daniels, E. and Veillette, A. (1997). Association of biliary glycoprotein with protein tyrosine phosphatase SHP-1 in malignant colon epithelial cells. *Oncogene* **14**, 783–790.

Beauchemin, N., Draber, P., Dveksler, G., Gold, P., Gray-Owen, S., Grunert, F., Hammarstrom S., Holmes, K.V., Karlsson, A. *et al.* (1999). Redefined nomenclature for members of the carcinoembryonic antigen family. *Experimental Cell Research* **252**, 243–249.

Bellinger-Kawahara, C. and Horwitz, M.A. (1990). Complement component C3 fixes selectively to the major outer membrane protein (MOMP) of *Legionella pneumophila* and mediates phagocytosis of liposome–MOMP complexes by human monocytes. *Journal of Experimental Medicine* **172**, 1201–1210.

Billker, O., Popp, A., Gray-Owen, S.D. and Meyer, T.F. (2000). The structural basis of CEACAM-receptor targeting by neisserial Opa proteins. *Trends in Microbiology* **8**, 258–260.

Brenner, D.J., Mayer, L.W., Carlone, G.M., Harrison, L.H., Bibb, W.F., Brandileone, M.C., Sottnek, F.O., Irino, K., Reeves, M.W., Swenson, J.M., *et al.* (1988). Biochemical, genetic, and epidemiologic characterization of *Haemophilus influenzae* biogroup aegyptius (*Haemophilus aegyptius*) strains associated with Brazilian purpuric fever. *Journal of Clinical Microbiology* **26**, 1524–1534.

Brown, E.J. and Frazier, W.A. (2001). Integrin-associated protein (CD47) and its ligands. *Trends in Cell Biology* **11**, 130–135.

Cartwright K. (ed.) (1995). *Meningococcal Disease*. Chichester: John Wiley & Sons.

Chen, T. and Gotschlich, E.C. (1996) CGM1a antigen of neutrophils, a receptor of gonococcal opacity proteins. *Proceedings of the National Academy of Sciences*, **93**, 14851–14856.

Cheresh, D.A., Pytela, R., Pierschbacher, M.D., Klier, F.G., Ruoslahti, E. and Reisfeld, R.A. (1987). An Arg-Gly-Asp-directed receptor on the surface of human melanoma cells exists in an divalent cation-dependent functional complex with the disialoganglioside GD2. *Journal of Cell Biology* **105**, 1163–1173.

Clark, E.A., and Brugge, J.S. (1995). Integrins and signal transduction pathways: the road taken. *Science* **268**, 233–239.

Crocker, P.R., Clark, E.A., Filbin M., Gordon, S., Jones, Y., Kehrl, J.H., Kelm, S., Le Douarin, N., Powell, L. *et al.* (1998). Siglecs: a family of sialic-acid binding lectins. *Glycobiology* **8**, v.

Cundell D.R., Gerard N.P., Gerard C., Idanpaan-Heikkila I. and Tuomanen E.I. (1995). *Streptococcus pneumoniae* anchor to activated human cells by the receptor for platelet-activating factor. *Nature* **377**, 435–438.

Dansky-Ullman, C., Salgaller, M., Adams, S., Schlom, J. and Greiner, J.W. (1995). Synergistic effects of IL-6 and IFN-gamma on carcinoembryonic antigen

(CEA) and HLA expression by human colorectal carcinoma cells: role for endogenous IFN-beta. *Cytokine* **7**, 118–129.

Dunn, K.L.R. and Virji, M. (1996). *Neisseria meningitidis* toxicity for cultured human endothelial cells requires soluble CD14. In *Pathogenic Neisseria*, ed. W.D. Zollinger, C.E. Frasch, and C.D. Deal, pp. 275–276. Bethesda, MD, NIH.

Dunn, K.L.R., Virji, M. and Moxon, E.R. (1995). Investigations into the molecular basis of meningococcal toxicity for human endothelial and epithelial cells: the synergistic effect of LPS and pili. *Microbial Pathogenesis* **18**, 81–96.

Forsgren, J., Samuelson, A., Ahlin, A., Jonasson, J., Rynnel-Dagoo, B. and Lindberg, A. (1994). *Haemophilus influenzae* resides and multiplies intracellularly in human adenoid tissue as demonstrated by in situ hybridization and bacterial viability assay. *Infection and Immunity* **62**, 673–679.

Foxwell, A.R., Kyd, J.M. and Cripps, A.W. (1998). Nontypeable *Haemophilus influenzae*: pathogenesis and prevention. *Microbiological Molecular Biology Review* **62**, 294–308.

Gray-Owen, S.D., Dehio, C., Haude, A., Grunert, F. and Meyer, T.F. (1997). CD66 carcinoembryonic antigens mediate interactions between Opa-expressing *Neisseria gonorrhoeae* and human polymorphonuclear phagocytes, *EMBO Journal* **16**, 3435–3445.

Gyorkey, F., Musher, D., Gyorkey, P., Goree, A. and Baughn, R. (1984). Nontypable *Haemophilus influenzae* are unencapsulated both in vivo and in vitro. *Journal of Infectious Diseases* **149**, 518–522.

Hammarström, S. (1999). The carcinoembryonic antigen (CEA) family: structures, suggested functions and expression in normal and malignant tissues. *Seminars in Cancer Biology* **9**, 67–81.

Hammerschmidt, S., Hilse, R., van Putten, J.P., Gerardy-Schahn, R., Unkmeir, A. and Frosch, M. (1996a). Modulation of cell surface sialic acid expression in *Neisseria meningitidis* via a transposable genetic element. *EMBO Journal* **15**, 192–198.

Hammerschmidt, S., Muller, A., Sillmann, H., Muhlenhoff, M., Borrow, R., Fox, A., van Putten, J., Zollinger, W.D., Gerardy-Schahn, R., and Frosch, M. (1996b). Capsule phase variation in *Neisseria meningitidis* serogroup B by slipped-strand mispairing in the polysialyltransferase gene (SiaD): correlation with bacterial invasion and the outbreak of meningococcal disease. *Molecular Microbiology* **20**, 1211–1220.

Hardy, S.J., Christodoulides, M., Weller, R.O. and Heckels, J.E. (2000). Interactions of *Neisseria meningitidis* with cells of the human meninges. *Molecular Microbiology* **36**, 817–829.

Harvey, H.A., Porat, N., Campbell, C.A., Jennings, M., Gibson, B.W., Phillips,

N.J., Apicella, M.A. and Blake, M.S. (2000). Gonococcal lipooligosaccharide is a ligand for the asialoglycoprotein receptor on human sperm. *Molecular Microbiology* **36**, 1059–1070.

Hauck, C.R., Meyer, T.F., Lang, F. and Gulbins, E. (1998). CD66-mediated phagocytosis of Opa(52) *Neisseria gonorrhoeae* requires a Src-like tyrosine kinase- and Rac1-dependent signalling pathway. *EMBO Journal*, **17**, 443–454.

Henderson, F.W., Collier, A.M., Sanyal, M.A., Watkins, J.M., Fairclough, D.L., Clyde, W.A. Jr and Denny, F.W. (1982). A longitudinal study of respiratory viruses and bacteria in the etiology of acute otitis media with effusion. *New England Journal of Medicine* **306**, 1377–1383.

Hill, D.J., Toleman, M.A., Evans, D.J., Villullas, S., Van Alphen, L. and Virji, M. (2001). The variable P5 proteins of typable and non-typable *Haemophilus influenzae* target human CEACAM1. *Molecular Microbiology* **39**, 850–862.

Hoiseth, S.K. and Gilsdorf, J.R. (1988). The relationship between type b and non-typable *Haemophilus influenzae* isolated from the same patient. *Journal of Infectious Diseases* **158**, 643–645.

Hughes, P.E. and Pfaff, M. (1998). Integrin affinity modulation. *Trends in Cell Biology* **8**, 359–364.

Isberg, R.R. (1991). Discrimination between intracellular uptake and surface adhesion of bacterial pathogens. *Science* **252**, 934–938.

Isberg, R.R. and Tran Van Nhieu, G. (1994). Binding and internalization of microorganisms by integrin receptors. *Trends in Microbiology* **2**, 10–14.

Jennings, M.P., Hood, D.W., Peak, I.R.A., Virji, M. and Moxon, E.R. (1995). Molecular analysis of a locus for the biosynthesis and phase variable expression of the lacto-N-neotetraose terminal LPS structure in *Neisseria meningitidis*. *Molecular Microbiology* **18**, 729–740.

Jiang, Z., Nagata, N., Molina, E., Bakaletz, L.O., Hawkins, H. and Patel, J.A. (1999). Fimbria-mediated enhanced attachment of nontypeable *Haemophilus influenzae* to respiratory syncytial virus-infected respiratory epithelial cells. *Infection and Immunity* **67**, 187–192.

Malorny, B., Morelli, G., Kusecek, B., Kolberg, J. and Achtman, M. (1998). Sequence diversity, predicted two-dimensional protein structure, and epitope mapping of neisserial Opa proteins. *Journal of Bacteriology* **180**, 1323–1330.

Marth, T. and Kelsall, B.L. (1997). Regulation of interleukin-12 by complement receptor β signaling. *Journal of Experimental Medicine* **185**, 1987–1995.

Masson, L. and Holbein, B.E. (1985). Influence of environmental conditions on serogroup B *Neisseria meningitidis* capsular polysaccharide levels. In *The Pathogenic Neisseriae*, ed. G.K. Schoolnik, G.F. Brooks, S. Falkow, C.E. Frasch, J.S. Knapp, J.A. McCutchan and S.A. Morse, pp. 571–578. Washington, DC: ASM Press.

Merz, A.J. and So, M. (2000). Interactions of pathogenic neisseriae with epithelial cell membranes. *Annual Review of Cell and Developmental Biology* **16**, 423–457.

Moxon, E.R. and Murphy, P.A. (1978). *Haemophilus influenzae* bacteraemia and meningitis resulting from survival of a single organism. *Proceedings of the National Academy of Sciences, USA* **75**, 1534–1536.

Obrink, B. (1997). CEA adhesion molecules: multifunctional proteins with signal-regulatory properties. *Current Opinion in Cell Biology* **9**, 616–626.

Olyhoek, A.J.M., Sarkari, J., Bopp, M., Morelli, G. and Achtman, M. (1991). Cloning and expression in *Escherichia coli* of *opc*, the gene for an unusual class 5 outer-membrane protein from *Neisseria meningitidis*. *Microbial Pathogenesis* **11**, 249–257.

Parge, H.E., Forest, K.T., Hickey, M.J., Christensen, D.A., Getzoff, E.D. and Tainer, J.A. (1995). Structure of the fibre-forming protein pilin at 2.6 Å resolution. *Nature* **378**, 32–38.

Prall, F., Nollau, P., Neumaier, M., Haubeck, H.D., Drzeniek, Z., Helmchen, U., Loning, T. and Wagener, C. (1996). CD66a (BGP), an adhesion molecule of the carcinoembryonic antigen family is expressed in epithelium, endothelium and myeloid cells in a wide range of normal human tissues. *Journal of Histochemistry and Cytochemistry* **44**, 35–41.

Richardson, A. and Parsons, J.T. (1995). Signal transduction through integrins: a central role for focal adhesion kinase? *BioEssays* **17**, 229–236.

Rosales, C., O'Brien, V., Kornberg, L. and Juliano, R. (1995). Signal transduction by cell adhesion receptors. *Biochimica et Biophysica Acta* **1142**, 77–98.

Ruoslahti, E. and Pierschbacher, M.D. (1987). New perspectives in cell adhesion: RGD and integrins. *Science* **238**, 491–497.

Sandros, J. and Tuomanen, E. (1993). Attachment factors of *Bordetella pertussis*: mimicry of eukaryotic cell recognition molecules. *Trends in Microbiology* **1**, 192–196.

Serino, L. and Virji, M. (2000). Phosphorylcholine decoration of LPS differentiates commensal neisseriae from pathogenic strains: identification of *licA*-type genes in commensal neisseriae. *Molecular Microbiology* **35**, 1550–1559.

Sim, R.J., Harrison, M.M., Moxon, E.R. and Tang, C.M. (2000). Underestimation of meningococci in tonsillar tissue by nasopharyngeal swabbing. *Lancet* **11**, 1653–1654.

St Geme J.W. III (1999). Molecular determinants of the interactions between *Haemophilus influenzae* and human cells. *American Journal of Respiratory and Critical Care Medicine* **154**, S192–S196.

St Geme, J.W. III, Takala, A., Esko, E. and Falkow, S. (1994). Evidence for capsule gene sequences among pharyngeal isolates of nontypeable *Haemophilus influenzae*. *Journal of Infectious Diseases* **169**, 337–342.

Stanners, C.P., DeMarte, L., Rojas, M., Gold, P. and Fuks, A. (1995). Opposite functions for 2 classes of genes of the human carcinoembryonic antigen family. *Tumor Biology* **16**, 23–31.

Stephens, D.S. and Farley, M.M. (1991). Pathogenic events during infection of the human nasopharynx with *Neisseria meningitidis* and *Haemophilus influenzae*. *Reviews of Infectious Diseases* **13**, 22–33.

Stern, A., Brown, M., Nickel, P. and Meyer, T.F. (1986). Opacity genes in *Neisseria gonorrhoeae* – control of phase and antigenic variation. *Cell* **47**, 61–71.

Stimson, E., Virji, M., Makepeace, K., Dell, A., Morris, H.R., Payne, G., Saunders, J.R., Jennings, M.P., Barker, S., Panico, M., Blench, I. and Moxon, E.R. (1995). Meningococcal pilin: a glycoprotein substituted with digalactosyl 2,4-diacetamido-2,4,6-trideoxyhexose. *Molecular Microbiology* **17**, 1201–1214.

Stimson, E., Virji, M., Panico, M., Blench, I., Barker, S., Moxon, E.R., Dell, A. and Morris, H.R. (1996). Discovery of a novel protein modification: α-glycerophosphate is a substituent of meningococcal pilin. *Biochemical Journal* **316**, 29–33.

Stockbauer, K,E., Magoun, L., Liu, M., Burns, E.H. Jr, Gubba, S., Renish, S., Pan, X., Bodary, S.C., Baker, E., Coburn, J., Leong, J.M. and Musser, J.M. (1999). A natural variant of the cysteine protease virulence factor of group A *Streptococcus* with an arginine-glycine-aspartic acid (RGD) motif preferentially binds human integrins $\alpha_v\beta_3$ and $\alpha_{IIb}\beta_3$. *Proceedings of the National Academy of Sciences, USA* **96**, 242–247.

Talamas-Rohana, P., Wright, S.D., Lennartz, M.R. and Russell, D.G. (1990). Lipophosphoglycan from *Leishmania mexicana* promastigotes binds to members of the CR3, p150,95 and LFA-1 family of leukocyte integrins. *Journal of Immunology* **144**, 4817–4824.

Teixeira, A.M., Fawcett, J., Simmons, D.L. and Watt, S.M. (1994). The N-domain of the biliary glycoprotein (BGP) adhesion molecule mediates homotypic binding–domain interactions and epitope analysis of BGPc. *Blood* **84**, 211–219.

Toleman, M., Aho, E. and Virji, M. (2001). Expression of pathogen-like Opa adhesins in commensal *Neisseria*: genetic and functional analysis. *Cellular Microbiology* **3**, 33–44.

Trust T.J., Gillespie R.M., Bhatti A.R. and White L.A. (1983). Differences in the adhesive properties of *Neissseria meningitidis* for human buccal epithilial cells and erythrocytes. *Infection and Immunity* **41**, 106–113.

Turk, D.C. (1984). The pathogenicity of *Haemophilus influenzae*. *Journal of Medical Microbiology* **18**, 1–16.

van Alphen, L., Jansen, H.M. and Dankert, J. (1995). Virulence factors in the colonization and persistence of bacteria in the airways. *American Journal of Respiratory and Critical Care Medicine* **151**, 2094–2099.

van Deuren, M., Brandtzaeg, P. and van der Meer, J.W.M. (2000). Update on meningococcal disease with emphasis on pathogenesis and clinical management. *Clinical Microbiology Review* **13**, 144–166.

Virji, M. (1996a). Adhesion receptors in microbial pathogenesis. In *Molecular Mechanisms of Adhesion Molecules*, ed. M. Horton, pp. 99–129. Chichester: John Wiley and Sons.

Virji, M. (1996b). Microbial utilisation of human signalling molecules. *Microbiology* **142**. 3319–3336.

Virji, M. (2000a). Glycans in meningococcal pathogenesis and the enigma of the molecular decorations of neisserial pili. In *Glycomicrobiogy*, ed. R. Boyle, pp. 31–66. London: Plenum Press.

Virji, M. (2000b). The structural basis of CEACAM-receptor targeting by neisserial Opa proteins. *Trends in Microbiology*, **8**, 260–261.

Virji, M., Kayhty, H., Ferguson, D.J.P., Alexandrescu, C., Heckels, J.E. and Moxon, E.R. (1991). The role of pili in the interactions of pathogenic *Neisseria* with cultured human endothelial cells. *Molecular Microbiology* **5**, 1831–1841.

Virji, M., Alexandrescu, C., Ferguson, D.J.P., Saunders, J.R. and Moxon, E.R. (1992a). Variations in the expression of pili: the effect on adherence of *Neisseria meningitidis* to human epithelial and endothelial cells. *Molecular Microbiology* **6**, 1271–1279.

Virji, M., Makepeace, K., Ferguson, D.J.P., Achtman, M., Sarkari, J. and Moxon, E.R. (1992b). Expression of the Opc protein correlates with invasion of epithelial and endothelial cells by *Neisseria meningitidis*. *Molecular Microbiology* **6**, 2785–2795.

Virji, M., Makepeace, K., Ferguson, D.J.P., Achtman, M. and Moxon, E.R. (1993a). Meningococcal Opa and Opc proteins: role in colonisation and invasion of human epithelial and endothelial cells. *Molecular Microbiology* **10**, 499–510.

Virji, M., Saunders, J.R., Sims, G., Makepeace, K., Maskell, D. and Ferguson, D.J.P. (1993b). Pilus-facilitated adherence of *Neisseria meningitidis* to human epithelial and endothelial cells: modulation of adherence phenotype occurs concurrently with changes in amino acid sequence and the glycosylation status of pilin. *Molecular Microbiology* **10**, 1013–1028.

Virji, M., Makepeace, K. and Moxon, E.R. (1994a). Meningococcal outer membrane protein Opc mediates interactions with multiple extracellular matrix components. In *Neisseria 94*, ed. J.S. Evans, S.E. Yost, M.C.J. Maiden and I.M. Feavers, pp. 263–264. Potters Bar, England: NIBSC.

Virji, M., Makepeace, K. and Moxon, E.R. (1994b). Distinct mechanisms of interaction of Opc-expressing meningococci at apical and basolateral surfaces of human endothelial cells; the role of integrins in apical interactions. *Molecular Microbiology* **14**, 173–184.

Virji, M., Makepeace, K., Peak, I., Payne, G., Saunders, J.R., Ferguson, D.J.P. and Moxon, E.R. (1995a). Functional implications of the expression of PilC proteins in meningococci. *Molecular Microbiology* **16**, 1087–1097.

Virji, M., Makepeace, K., Peak, I.R.A., Ferguson, D.J.P., Jennings, M.P. and Moxon, E.R. (1995b). Opc- and pilus- dependent interactions of meningococci with human endothelial cells: molecular mechanisms and modulation by surface polysaccharides. *Molecular Microbiology* **18**, 741–754.

Virji, M., Watt, S.M., Barker, S., Makepeace, K. and Doyonnas, R. (1996a). The N-domain of the human CD66a adhesion molecule is a target for Opa proteins of *Neisseria meningitidis* and *Neisseria gonorrhoeae*. *Molecular Microbiology* **22**, 929–939.

Virji, M., Makepeace, K., Ferguson, D.J.P. and Watt, S.M. (1996b). Carcinoembryonic antigens (CD66) on epithelial cells and neutrophils are receptors for Opa proteins of pathogenic neisseriae. *Molecular Microbiology* **22**, 941–950.

Virji, M., Evans, D., Hadfield, A., Grunert, F., Teixeira, A. M. and Watt, S. M. (1999). Critical determinants of host receptor targeting by *Neisseria meningitidis* and *Neisseria gonorrhoeae*: identification of Opa adhesiotopes on the N-domain of CD66 molecules. *Molecular Microbiology* **34**, 538–551.

Virji, M., Evans, D., Griffith, J., Hill, D., Serino, L., Hadfield, A. and Watt, S.M. (2000). Carcinoembryonic antigens are targeted by diverse strains of typable and non-typable *Haemophilus influenza*. *Molecular Microbiology* **36**, 784–795.

Vogel, U. and Frosch, M. (1999). Mechanisms of neisserial serum resistance. *Molecular Microbiology* **32**, 1133–1139.

Wang, J., Gray-Owen, S.D., Knorre, A., Meyer, T.F. and Dehio, C. (1998). Opa binding to cellular CD66 receptors mediates the transcellular traversal of *Neisseria gonorrhoeae* across polarized T84 epithelial cell monolayers. *Molecular Microbiology* **30**, 657–671.

Weiser, J.N., Goldberg, J.B., Pan, N., Wilson, L. and Virji, M. (1998). The phosphorylcholine epitope undergoes phase variation on a 43-kilodalton protein in *Pseudomonas aeruginosa* and on pili of *Neisseria meningitidis* and *Neisseria gonorrhoeae*. *Infection and Immunity* **66**, 4263–4267.

Wright, S.D., Levin, S.M., Jong, M.T., Chad, Z. and Kabbash, L.G. (1989). CR3 (CD11b/CD18) expresses one binding site for Arg-Gly-Asp-containing peptides and a second site for bacterial lipopolysaccharide. *Journal of Experimental Medicine* **169**, 175–183.

CHAPTER 3

Adhesive surface structures of oral streptococci

Roderick McNab, Pauline S. Handley and Howard F. Jenkinson

3.1 INTRODUCTION

3.1.1 The oral cavity

The human oral cavity is home to a large and diverse microbial population. The mouth provides a wide range of different habitats for bacterial colonization and growth, including hard, non-shedding surfaces (the teeth) as well as various epithelial surfaces such as the tongue dorsum and buccal mucosa. The composition of the bacterial flora at these different sites can vary considerably, reflecting the range of surfaces for attachment and environmental conditions that are available for growth. Nevertheless, streptococci, which comprise some 20% of the human normal oral flora, can be isolated from most oral sites (Nyvad and Kilian, 1987; Frandsen *et al.*, 1991). Streptococci, together with *Actinomyces* species, are the predominant organisms found in the early stages of biofilm (i.e. dental plaque) formation on tooth surfaces, and the interactions of these initial species with the salivary pellicle may define the strength of biofilm adhesion (Busscher *et al.*, 1995). Furthermore, the streptococci participate in a wide range of intra- and inter-generic coaggregation interactions that help to establish early biofilm communities (Kolenbrander and London, 1993; Kolenbrander, 2000). Consequently, streptococci play an important role in plaque development.

Oral streptococcal species have recently undergone considerable taxonomic upheaval (see Whiley and Beighton, 1998). This has resulted not only in the reclassification of species and the description of new species, but also the grammatical correction of Latin species epithets (thus *Streptococcus crista* has been renamed *S. cristatus*, and *Streptococcus parasanguis* is now *S. parasanguinis*). For this chapter we have chosen to retain the epithets in long-term usage. Currently, four species groups of oral streptococci have been described. These are the mitis, mutans, salivarius and anginosus groups. It is important to note that members of these groups are not necessarily restricted to the oral cavity, but may also be found at other body sites, and are often associated with disease.

3.1.2 Oral streptococcal adhesion

The growth and survival of streptococci within the oral cavity are dependent, at least in part, on bacterial adhesion to oral surfaces coated with salivary proteins and glycoproteins, and to other adherent bacteria (Whittaker *et al.*, 1996). Since adhesion is a prerequisite of oral colonization, it is not surprising that most oral streptococci express arrays of adhesins on their cell surfaces and consequently exhibit a wide range of adhesion properties (Jenkinson and Lamont, 1997). Thus the substrates to which streptococci can bind include salivary and serum components, host cells, tissue matrix components and other microbial cells. The expression of multiple adhesins would allow enhanced binding through co-operativity, and also allow a range of surfaces carrying the cognate receptors to be colonized. Conversely, the range of adhesins expressed will also account, by-and-large, for the tissue specific tropisms displayed by some oral streptococci. For example, *S. sanguis*, a colonizer of the tooth surface, is not found in the oral cavity of infants until the teeth erupt (Carlsson *et al.*, 1970; Caulfield *et al.*, 2000), and is rapidly lost from the mouth of edentulous persons not wearing dentures (Carlsson *et al.*, 1969). In contrast, *S. salivarius*, which colonizes the tongue and other mucosal surfaces, can be detected as early as three weeks after birth (Smith *et al.*, 1993).

Molecular genetic techniques have allowed great advances to be made in the understanding of the composition, structure and function(s) of cell surface-located adhesins, particularly amongst members of the mitis group (including *S. mitis*, *S. sanguis*, *S. parasanguis*, *S. oralis*, *S. gordonii* and *S. pneumoniae*) and mutans group (including *S. mutans* and *S. sobrinus*) streptococci (for a review, see Jenkinson and Lamont, 1997). Thus a number of genes encoding cell surface-associated proteins that function in adhesion have been cloned, sequenced and studied in detail (Table 3.1). In many cases, the role of these proteins in adhesion and colonization has been assessed using isogenic knockout mutant strains. So, what do we know about the adhesins of oral streptococci? In general, surface proteins of oral streptococci can be divided into two categories based on the means by which the protein is anchored at the cell surface. Thus a surface protein may be anchored by covalent modification at the C- or the N-terminus.

The principle mechanism of cell surface anchorage of adhesins

Table 3.1. *Adhesins of oral streptococci for which gene sequences are available*

Adhesin	Surface anchorage	Size (kDa)	Species	Substrate(s)	References
Antigen I/II family (Ag I/II, P1 (SpaP), PAc, Sr, SpaA, SspA, SspB, SoaA)	LPXTG wall anchor	160–175	*S. mutans, S. sobrinus, S. gordonii, S. sanguis, S. oralis*	Salivary components, collagen, *A. naeslundii, P. gingivalis, S. mutans, C. albicans*	Jenkinson and Demuth, 1997; Love *et al.*, 1997
CshA	LPXTG wall anchor	259	*S. gordonii*	Fibronectin, *A. naeslundii, S. oralis, C. albicans*	Holmes *et al.*, 1996; McNab *et al.*, 1996
Fap1	LPXTG wall anchor	261	*S. parasanguis*	Salivary component(s)	Wu and Fives-Taylor, 1999
Emb	Not yet known	>215	*S. defectivus*	Extracellular matrix	Manganelli and van de Rijn, 1999
GbpC	LPXTG wall anchor	64	*S. mutans*	Glucan	Sato *et al.*, 1997
WapA	LPXTG wall anchor	48	*S. mutans*	Glucan	Qian and Dao, 1993
GtfG (glucosyltransferase)	C-terminal 14 amino acid residues	174	*S. gordonii*	Human endothelial cells	Vacca-Smith *et al.*, 1994; Vickerman and Clewell, 1997

involves the covalent attachment of protein to peptidoglycan via a specialized C-terminal signal. Over 40 streptococcal cell surface proteins have now been identified that carry this cell-wall-sorting signal, including the majority of known adhesins of oral streptococci (Fischetti, 1996; Navarre and Schneewind, 1999). The cell-wall-sorting signal comprises the sequence motif LPXTG (Lys-Pro-X-Thr-Gly, where X can be any amino acid) followed by a hydrophobic sequence of approximately 20 amino acid residues and a short, charged cytoplasmic tail. The latter two components serve as a 'stop transfer' signal to protein translocation. An enzyme, sortase, then is proposed to cleave between the threonine and glycine residues within the LPXTG motif, and the free carboxyl group of threonine is subsequently amide-linked to a free amino group present within the peptidoglycan cross-bridge (Schneewind et al., 1995; Ton-That et al., 2000). Cell-wall-anchored proteins in this group have a characteristic molecular architecture and are frequently modular in design, comprising a mosaic of repetitive and non-repetitive amino acid sequences (Navarre and Schneewind, 1999). Oral streptococcal adhesins that possess a C-terminal wall anchor include CshA of *S. gordonii* and Fap1 of *S. parasanguis* that form large fibrillar structures (Table3. 1). Insertional inactivation of the gene encoding sortase in *S. gordonii* resulted in reduced adhesion of mutant cells to fibronectin and impaired colonization of the murine oral cavity (Bolken et al., 2001), properties conferred, at least in part, by wall-anchored CshA fibrils (McNab et al., 1994, 1996). Similar experiments in *Staphylococcus aureus* resulted in a failure of cells to process and display normally wall-anchored proteins, and this was associated with a corresponding defect in their ability to establish infection in experimental mice (Mazmanian et al., 2000).

Lipoproteins are anchored within the outer leaflet of the cytoplasmic membrane by lipid modification of an N-terminal cysteine residue (Braun and Wu, 1994). Virtually all of the streptococcal lipoproteins characterized to date are substrate-binding proteins serving ATP-binding cassette (ABC) transport systems (Sutcliffe and Russell, 1995). Nevertheless, some lipoproteins have been implicated in adhesion of oral streptococci to various substrata (Jenkinson, 1994).

Other surface anchor mechanisms do exist amongst streptococci. For example, a family of pneumococcal proteins, some of which function as adhesins, are held at the cell surface non-covalently via interaction of C-terminal amino acid residue repeat blocks with choline-containing teichoic acid or lipoteichoic acid (Yother and White, 1994; Rosenow et al., 1997).

3.2 ORAL STREPTOCOCCAL FIBRILS AND FIMBRIAE

Although most oral streptococci produce surface structures, it has not, in the majority of instances, been possible to establish unequivocally a link between these structures and bacterial adhesion. However, molecular evidence is now accumulating for a link between adhesion and the presence of a range of surface structures on oral streptococci, particularly members of the mitis group streptococci (Table 3.2). Two classes of surface structure, fibrils and fimbriae, have been described on oral streptococci on the basis of morphological descriptions of bacterial surfaces visualized by electron microscopy. Fimbriae are thin, flexible structures up to 3 μm in length, but with a defined width of 3–5 nm (Fig. 3.1a). Fimbriae are found on 50% of *Streptococcus salivarius* strains (Handley *et al.*, 1984) and on *S. parasanguis* (formerly classified as *S. sanguis*) (Elder *et al.*, 1982) but have not been reported consistently on other mitis-group streptococci (Handley *et al.*, 1985). A wealth of information is available on the structure, assembly and function of adhesive fimbriae of non-oral bacteria (for a review, see Soto and Hultgren, 1999). By contrast, fimbriae of oral streptococci remain relatively poorly defined in molecular terms. Nevertheless, adhesive fimbriae have been described in some detail for *S. parasanguis* (see below) and for certain other oral bacteria including Gram-negative periodontal pathogens *Porphyromonas gingivalis*, *Prevotella* spp. and *Actinobacillus actinomycetemcomitans*, and for the commensal Gram-positive bacterium *Actinomyces naeslundii* (for reviews, see Hamada *et al.*, 1998; Handley *et al.*, 1999).

A large proportion of oral streptococcal isolates display surface structures that are morphologically distinct from fimbriae (Handley, 1990). These structures, termed fibrils, appear more rigid and are shorter than fimbriae, usually extending up to 200 nm from the cell surface (Fig. 3.1b). The length, distribution and density of fibrils on the surface of oral streptococci may vary in a species- or strain-dependent manner, and frequently a strain may express two or more fibril types, each with characteristic length and properties. Fibrils are seen on strains of *S. salivarius* (Handley *et al.*, 1984), mitis group streptococci (Handley *et al.*, 1985), and on *S. mutans* (Hogg *et al.*, 1981). On some strains, fibrils are arranged in prominent polar or lateral tuft arrangements (Handley *et al.*, 1991; Jameson *et al.*, 1995) (Fig. 3.1c).

Evidence is accumulating, as will be described here, which suggests that the morphological differences between fibrils and fimbriae of oral bacteria, observed by electron microscopy, may reflect differences in composition and organization of the proteinaceous components of these structures.

Table 3.2. *Protein composition of oral streptococcal surface structures*

Protein	Structure type and dimensions	Mol mass (kDa)	Species and strain	Adhesion substrate(s)	References
CshA	Fibril, 61 nm long	259	*S. gordonii* DL1-Challis	Fibronectin, *A. naeslundii*, *S. oralis*, *C. albicans*	Holmes *et al.*, 1996; McNab *et al.*, 1996
Fap1	Fimbria, 3–4 nm wide, 216 nm long	261	*S. parasanguis* FW213	Salivary component(s)	Wu and Fives-Taylor, 1999
Emb	Fibril, ~90 nm long	>215	*S. defectivus* NVS-47	Extracellular matrix	Manganelli and van de Rijn, 1999
P1 (Ag I/II) GTF	Fimbria, 100–200 nm long	175 (P1) 157 (GTF)	*S. mutans* TH16	Salivary components	Perrone *et al.*, 1997
AgB (VBP)	Fibril, 91 nm long	320	*S. salivarius* HB	*V. parvula*	Weerkamp *et al.*, 1986a,b
AgC (HAF)	Fibril, 72 nm long	220–280	*S. salivarius* HB	BECs, erythrocytes, salivary component(s)	Weerkamp *et al.*, 1986a,b
LFP	Fibril, ~200 nm long	>300	*S. sanguis* 12	Salivary component(s), BECs	Morris *et al.*, 1987; Willcox *et al.*, 1989
TEPs	Tuft fibril, 290 nm long	175–250 (multiple)	*S. oralis* CN3410	Not known	Jameson *et al.*, 1995

Notes: BECs, buccal epithelial cells; GTF, glucosyltransferase; HAF, host-association factor; LFP, long fibril protein; TEPs, trypsin-extracted polypeptides; VBP, *Veillonella*-binding protein.

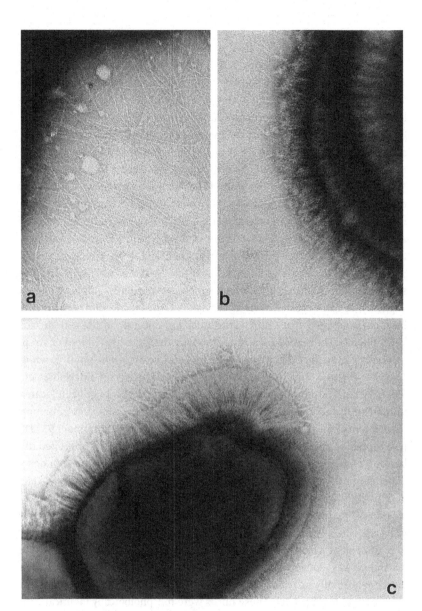

Figure 3.1. Cells were negatively stained with 1% (w/v) methylamine tungstate. (a) *S. salivarius* CHR (Lancefield K⁻ strain) carries flexible, peritrichous fimbriae 3.0–4.0 nm wide and 0.5–1.0 μm long. 176 610×. (b) *S. salivarius* HB (Lancefield K⁺ strain) carries a dense fringe of peritrichous fibrils. Sparse long fibrils (168 ± (5.0) nm) project through a dense mass of shorter fibrils (90 ± (4.0) nm). 117 450×. (a) and (b) are reprinted with permission from Handley *et al.*, 1984.) (c) *S. oralis* CN3410 carries a lateral tuft of fibrils. Long fibrils (289 ± (15) nm long) project through the shorter, more densely packed tuft fibrils (159 ± (5) nm long). 67 860×.

3.3 FIBRILS OF *STREPTOCOCCUS GORDONII*

Perhaps the best evidence linking large cell-wall-anchored proteins, surface structures and adhesion has come from studies on the CshA adhesive polypeptide of *S. gordonii*. CshA polypeptide (259 kDa), which is covalently attached to the streptococcal cell wall via its C-terminus, appears to be the structural and functional component of short fibrils on the surface of *S. gordonii* (McNab *et al.*, 1999). CshA polypeptide can be divided into four sections (Fig. 3.2): (i) a 41 residue N-terminal signal sequence directing export; (ii) a non-repetitive region (residues 42–878); (iii) an extensive amino acid repeat block region (residues 879–2417), containing 13 repeat blocks of 101 residues, and three incomplete repeat blocks, and (iv) a C-terminal cell wall anchor domain (residues 2418–2508) that is responsible for covalent attachment of CshA to cell wall peptidoglycan as described above. The mature polypeptide has numerous potential sites for N-linked glycosylation (the sequence Asn-X[Ser/Thr]), although there is currently no evidence to suggest that CshA is extensively glycosylated. Gene inactivation experiments have demonstrated that CshA contributes significantly to the cell surface hydrophobicity of *S. gordonii* and additionally mediates attachment of streptococcal cells to other oral microorganisms (*Actinomyces naeslundii*, *S. oralis*, *Candida albicans*) and to immobilized human fibronectin (McNab *et al.*, 1994, 1996). The adhesion-mediating sequences of CshA have been located to the N-terminal non-repetitive region on the basis of adhesion-inhibition experiments (McNab *et al.*, 1996). Antibodies raised to a recombinant 93 kDa polypeptide representing the N-terminal non-repetitive region inhibited the binding of *S. gordonii* to *A. naeslundii* and to human fibronectin in a dose-dependent manner. Electron microscopic analysis of negatively stained cells of *S. gordonii* DL1-Challis revealed thin, sparsely distributed fibrils that projected 60.7 ($\pm$ 14.5) nm from the cell surface (Fig. 3.3a). The fibrils bound CshA specific antibodies (Fig. 3.3b) and were absent from isogenic mutants of *S. gordonii* in which the *cshA* gene was inactivated. Furthermore, when the *cshA* gene was cloned and expressed in *Enterococcus faecalis* JH2–2, which does not normally carry fibrils, thin cell surface structures of 70.3 ($\pm$ 9.1) nm were expressed (Fig. 3.3c). These fibrils were morphologically identical with native *S. gordonii* fibrils, and bound CshA-specific antibodies (Fig. 3.3d) (McNab *et al.*, 1999). Recombinant *E. faecalis* concomitantly acquired surface and adhesion properties associated with CshA fibrils on *S. gordonii*. Thus the recombinant *E. faecalis* cells demonstrated significantly increased cell surface hydrophobicity and adhesion to human fibronectin, and acquired the ability to co-aggregate with *A. naeslundii* T14V.

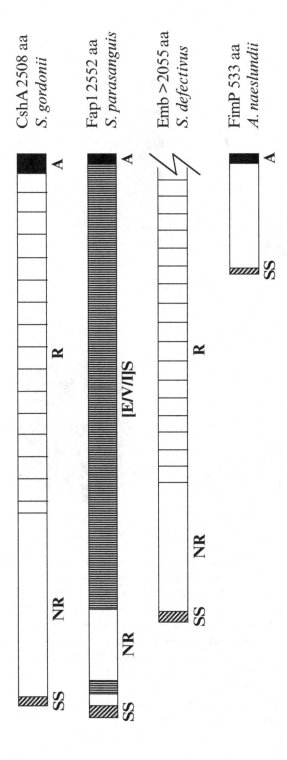

Figure 3.2. Schematic diagram of the features of fibril- or fimbrial-associated proteins of oral streptococci. Protein features described in the text have been labelled below each protein (aa, amino acid residues; SS, signal sequence; NR, non-repetitive region; R, amino acid repeat block region; A, LPXTG motif-containing anchor domain; [E/V/I]S, dipeptide repeat region; see text). *A. naeslundii* T14V type 1 fimbrial subunit, FimP, has been included for comparison.

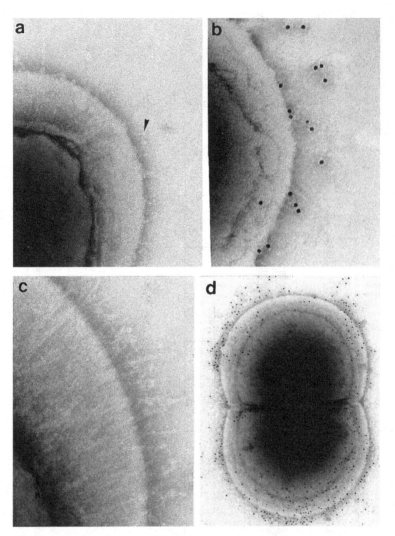

Figure 3.3. Cells were stained with 1% methylamine tungstate (MT) only (a and c) or with 1% MT together with CshA-specific antibodies and 10 nm gold-conjugated secondary antibody (b and d). (a) *S. gordonii* DL1-Challis with sparse, 60.7 ± (14.5) nm long, fibrils. Fibrils sometimes have globular ends (see arrowhead). 102060×. (b) *S. gordonii* DL1-Challis showing gold particles and antiserum labelling the CshA molecules, 61.3 ± (19.2) nm from the cell surface. Some fibrils are visible that have not been labelled. 149940×. (c) *E. faecalis* recombinant strain OB516 expresses *S. gordonii* CshA polypeptide and presents dense, 70.3 ± (9.1) nm long, peritrichous fibrils. 101640×. (d) *E. faecalis* OB516 cell probed with CshA-specific antiserum demonstrates gold particles 61.7 ± (12.2) nm from the cell surface. Fibrils are masked by the gold and antiserum. 102060×. (a–d) are reproduced with permission from McNab *et al.*, 1999.)

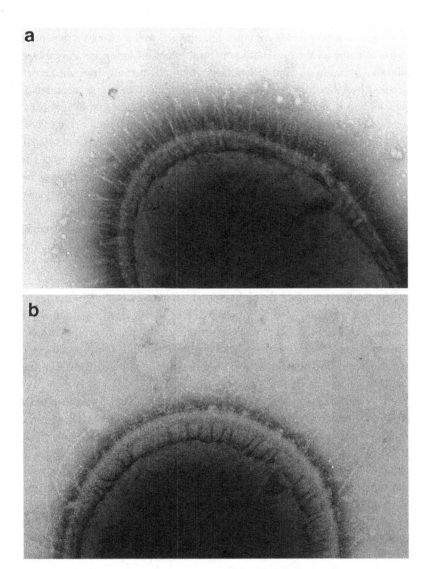

Figure 3.4. Cells were stained with 1% methylamine tungstate. (a) *S. sanguis* GW2 carries peritrichous fibrils (75.3 ± (22.3) nm). 156 600×. (Courtesy of Emma Harrison.) (b) *S. sanguis* FC1 carries a range of peritrichous short and long surface structures. 148 770×. (Courtesy of David Elliott.)

These results provide evidence that the cell-wall-anchored polypeptide CshA contains all the necessary structural information for the assembly of adhesive fibrils in an appropriate host, although currently it is not known whether each fibril is composed of one or multiple CshA molecules. For *S. gordonii* and *E. faecalis* expressing CshA, antibodies raised to the N-terminal region of CshA bound at a distance of 60–70 nm from the cell surface. This distance correlated with that measured for the ends or tips of fibrils visualized by negative staining (McNab *et al.*, 1999). Furthermore, some of the CshA fibrils on *E. faecalis* have globular ends, similar to those noted in micrographs of purified fibrillar proteins AgB and AgC of *S. salivarius* (Weerkamp *et al.*, 1986b; see below). These observations have led to the hypothesis that the N-terminal non-repetitive region, which carries the adhesive domain(s) of CshA, is held distal from the cell surface, possibly as a folded globular domain, via the extensive amino acid repeat block region. However, analysis of the sequence composition of the amino acid repeat region suggests that CshA is unlikely to adopt the alpha-helical coiled-coil structure of fibrous M and M-like proteins from *S. pyogenes* (Fischetti, 1989), nor is it likely to assume a collagen-like structure despite the high glycine and proline content of this region. Instead, it is speculated that the repeat region adopts an open, extended conformation, with the N-terminal non-repetitive region forming an adhesive tip domain (McNab *et al.*, 1999).

The correlation between CshA and fibrils may extend to other mitis group streptococci including strains of *S. oralis* and *S. sanguis*. Many of these strains possess short (48–75 nm) sparse fibrils (Fig. 3.4) (Handley *et al.*, 1985, 1999) and also express large surface proteins that are antigenically related to CshA of *S. gordonii* (McNab *et al.*, 1995). A number of these strains were investigated by immunoelectron microscopy using CshA-specific antiserum. In most cases, gold particles were distributed at a distance from the cell surface that corresponded to the tips of the short fibrils (D. Elliott, P.S. Handley and R. McNab, unpublished data) as was seen for *S. gordonii* DL1-Challis.

3.4 FIMBRIAE OF *STREPTOCOCCUS PARASANGUIS*

Streptococcus parasanguis strain FW213 expresses a dense array of peritrichous fimbriae (Fig. 3.5a), which is an unusual feature in oral streptococci. Spontaneous afimbriate mutants of FW213 no longer adhere to saliva-coated hydroxylapatite (SHA, a model of the tooth surface) (Fives-Taylor and Thompson, 1985) and antibodies raised to purified fimbriae block adhesion of *S. parasanguis* to SHA (Fachon-Kalweit *et al.*, 1985). It has

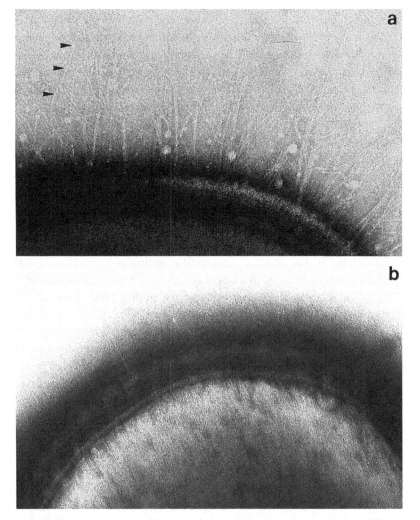

Figure 3.5. Cells were stained with 1% methylamine tungstate. (a) *S. parasanguis* FW213 with peritrichous, flexible fimbriae (see arrowheads), 3–4 nm wide and 216 ± (28) nm long. 151 380×. Reproduced with permission from Harty *et al.* (1990). (b) *S. parasanguis* *fap1*⁻ strain VT1393. No fimbriae are present; however, sparse fibrils are now visible. (Courtesy of Nicola Mordan.) 150 075×.

not proved possible to dissociate these fimbriae into constituent subunits (Elder and Fives-Taylor, 1986). Two putative adhesin genes, *fimA* and *fap1*, have now been identified in *S. parasanguis* and their products appear to be associated with the structure and function of adhesive fimbriae. The first gene to be cloned, *fimA*, encoded a 35 kDa protein that was shown by immunoelectron microscopy to be associated with the tips of fimbriae (Fenno *et al.*, 1995). FimA is a lipoprotein and the *fimA* gene is part of an operon that demonstrates significant sequence identity with operons encoding manganese-specific ABC transporter systems in *S. pneumoniae* (Dinthilac *et al.*, 1997) and *S. gordonii* (Kolenbrander *et al.*, 1998). By homology to these systems, FimA may be the metal ion-binding component of an ABC transporter. While *fimA* mutants are deficient in binding to fibrin (Burnette-Curley *et al.*, 1995), it is apparent that FimA is not required for the formation of fimbriae, since isogenic *fimA* mutants of *S. parasanguis* possess fimbriae (Fenno *et al.*, 1995).

Recent evidence indicates that the *S. parasanguis* fimbriae are composed of a novel protein, designated Fap1, that has an unusual, highly repetitive sequence (Wu *et al.*, 1998; Wu and Fives-Taylor, 1999). The large Fap1 protein (2552 residues) comprises four distinct regions (Fig. 3.2): (i) a 50 residue signal sequence; (ii) a non-repetitive region of approximately 400 residues that is interrupted by a run of 28 repeats of the dipeptide sequence [Glu/Val/Ile]Ser; (iii) an extensive region containing 1000 copies of the dipeptide repeat unit; (iv) a C-terminal wall anchor domain that contains the consensus LPXTG sequence motif for covalent attachment to cell-surface peptidoglycan. The N-terminal non-repetitive region has been reported to carry the adhesive epitope(s) for salivary receptor(s) (Wu and Fives-Taylor, 1999). Insertional inactivation of *fap1* resulted in loss of fimbriae (Fig. 3.5b) and reduced adhesion of cells to SHA (Wu and Fives-Taylor, 1999). Lack of fimbriae on Fap1⁻ mutant cells made it easier to visualize short, sparse, fibril-like structures on *S. parasanguis* that are currently of unknown composition and function (Fig. 3.5b). In contrast to the effect of *cshA* gene inactivation on *S. gordonii* cell surface properties, loss of fimbriae did not affect cell surface hydrophobicity of *S. parasanguis* (Wu and Fives-Taylor, 1999). A mutant, in which the wall anchor domain of Fap1 was deleted, secreted a truncated Fap1 protein into the extracellular medium. Mutant cells lacked fimbriae and had properties similar to those of cells of a null mutant, indicating that anchorage of Fap1 on the cell surface is required for fimbriae formation.

The functional significance of the dipeptide repeats is not understood. Similar dipeptide repeats (X-Ser) occur extensively in a number of *Staphylococcus aureus* wall-anchored proteins including the fibrinogen-binding clumping factors ClfA and ClfB (McDevitt *et al.*, 1994; Ní Eidhin *et al.*, 1998; see Chapter 1). The dipeptide repeats of ClfA are required for spanning the cell wall peptidoglycan and presenting the N-terminal fibrinogen-binding domain distal from the cell surface (Hartford *et al.*, 1997).

It is not clear how Fap1 polypeptides assemble to form fimbriae of *S. parasanguis*, although a mechanism similar to that involved in curli formation in *Escherichia coli* (Hammar *et al.*, 1996) has been proposed. The assembly of curli fibres occurs by the precipitation of secreted CsgA (curlin) monomers into thin aggregative fibres. This reaction requires cell-surface-attached CsgB polypeptide that acts as a nucleator for CsgA polymerization, although CsgB is also incorporated as a minor component of the fibres (Bian and Normark, 1997). Curli structures are very stable and cannot easily be dissociated into constituent monomers. In *S. parasanguis*, however, a precipitation model has to take into account that each Fap1 molecule has a cell wall anchor motif. Thus a mechanism must be evoked whereby sortase-cleaved Fap1 molecules are protected from cell wall anchorage and instead are secreted for polymerization into fimbriae.

An alternative method for Fap1 polymerization may be derived from analysis of fimbrial assembly in *A. naeslundii* that may involve covalent attachment of fimbrial subunits via the LPXTG motif (Yeung *et al.*, 1998). As for *S. parasanguis*, it has not proved possible to dissociate actinomyces fimbriae into constituent subunits. Nevertheless, molecular analysis has identified the genes encoding fimbrial subunits of type 1 (FimP) and type 2 (FimA) fimbriae of *A. naeslundii* T14V, and of type 2 (FimA) fimbriae of *A. naeslundii* WVU45 (Yeung, 1999). The fimbrial subunits (approximately 57 kDa) all possess, in addition to a signal sequence directing protein export, a typical wall anchor domain containing the consensus LPXTG motif. Antiserum that is specific to the C-terminal 20 amino acid residues of FimA of *A. naeslundii* T14V reacts with recombinant FimA purified from *E. coli*, as would be expected. However, this antiserum does not react with isolated fimbriae, suggesting that the FimA components have been processed at the C-terminus, presumably at the LPXTG motif. Gene inactivation experiments have implicated the open reading frame (ORF) lying immediately 3′ to the *fimA* and *fimP* fimbrial subunit genes in the assembly of fimbriae (Yeung and Ragsdale, 1997; Yeung *et al.*, 1998). Thus, although mutants in which the adjacent ORF is inactivated express fimbrial subunits that can be detected in cell wall preparations, these units do not assemble into fimbriae and consequently cells lack the associated adhesive properties. The gene immediately 3′ to *fimA*, *orf365*, encodes a protein with significant sequence similarity to sortase proteins (Pallen *et al.*, 2001). The data were used to propose the hypothesis that, once cleaved at the LPXTG motif in the wall anchor, subunits were covalently linked to another subunit either directly or else via a peptidoglycan fragment that has yet to be detected in purified fimbriae. This hypothesis is still highly speculative, but would account for the inability to resolve individual FimA subunits using detergents or chaotropic agents. What is becoming clear, however, is that surface anchoring via covalent attachment to cell wall peptidoglycan may be only one of the possible fates for a protein possessing an LPXTG wall anchor motif.

3.5 SURFACE STRUCTURES OF NUTRITIONALLY VARIANT STREPTOCOCCI

Nutritionally variant streptococci (NVS) are members of normal oral flora of humans but are also responsible for up to 10% of infective endocarditis cases. As the name suggests, NVS have unusual nutritional requirements, and are unable to grow in most normal laboratory media without the addition of pyroxidal hydrochloride or L-cysteine. Consequently, NVS may be an important cause of the blood culture-negative form of endocarditis, which is often associated with a poor clinical outcome. Three serotypes are recognized, with serotype I (*S. defectivus*) and serotype II (*S. adjacens*) together comprising 97% of isolates. These strains have recently been placed in a new genus called *Abiotrophia*, based on 16 S ribosomal DNA sequence analysis (Kawamura *et al.*, 1995).

Bacterial adhesion to damaged heart valves is a primary step in infective endocarditis, and substrates for attachment include the platelet–fibrin matrix as well as extracellular tissue matrix (ECM) components exposed on the damaged tissue (Herzberg, 1996). A large (>200 kDa) surface protein of *S. defectivus* was shown to mediate adhesion of cells to ECM (Tart and van de Rijn, 1993). Cloning and partial sequencing of the gene encoding Emb has identified an ORF that may be at least 14 kb (Manganelli and van de Rijn, 1999). The structure of Emb revealed by sequencing is shown in Fig. 3.2. The incomplete 2055 amino acid residue sequence is dominated by repeat blocks of 77 amino acid residues at the C-terminus of the protein. At least 50 repeats are predicted to be present in Emb, indicating that the protein may be greater than 500 kDa in size (Manganelli and van de Rijn, 1999). The N-terminus of the protein possesses a signal sequence directing export of the protein, and between this and the repeat region is a stretch of approximately 600 amino acid residues of non-repetitive sequence that carries the adhesive domain(s) of Emb. Thus antibodies specific for the non-repetitive region demonstrated 90% inhibition of *S. defectivus* binding to ECM. In contrast, antibodies specific for the repeat region blocked adhesion by only 10% (Manganelli and van de Rijn, 1999). Sequence is not available for the C-terminus of the protein, so it is not clear whether an LPXTG-containing wall anchor domain is present. Nevertheless, Emb demonstrates the typical modular design of

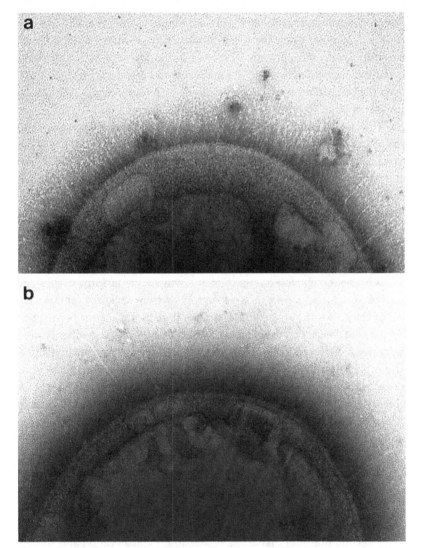

Figure 3.6. Cells were stained with 1% methylamine tungstate. (a) *S. defectivus* strain NVS-47 with dense peritrichous fibrils. 150075×. (b) *S. defectivus* mutant strain WF*emb*1 demonstrates significantly reduced fibril density. (Courtesy Nicola Mordan.) 150075×.

C-terminal wall-anchored proteins, and is surface associated and released from whole cells only following digestion of the cell wall with mutanolysin and lysozyme (Tart and van de Rijn, 1993). The repeats show homology to repeats of proteins in other pathogenic streptococci and are predicted to adopt an alpha-helical coiled-coil structure, at least for the first half of each repeat. This is in contrast to other fibrillar proteins such as M and M-like proteins of *S. pyogenes*, where the coiled-coil structure is conserved along almost the entire length of the protein. Nevertheless, the data suggest that the N-terminal non-repetitive region of the mature Emb protein may be held distal from the cell surface by a scaffold of amino acid repeat blocks (Manganelli and van de Rijn, 1999).

The fibrillar nature of Emb was confirmed by electron microscopic analysis of *S. defectivus* wild-type and isogenic Emb⁻ mutant cells (Fig. 3.6). Wild-type cells possess relatively densely distributed fibrils approximately 90 nm in length (Fig. 3.6a). These fibrils were absent from the surface of isogenic Emb⁻ cells, although sparse surface structures were revealed on some cells (Fig. 3.6b).

3.6 EXPORT OF FUNCTIONAL SURFACE STRUCTURES

Whether or not these large proteins polymerize to form multimeric surface structures, a pattern is emerging whereby extensive amino acid repeat blocks, which can constitute over two-thirds of the mature polypeptide but may not themselves function in adhesion, serve to present adhesive non-repetitive domains distal from the cell surface. A similar molecular architecture is presented by an ever-growing list of streptococcal and staphylococcal surface proteins (Navarre and Schneewind, 1999), although with the exception of M protein of *S. pyogenes* (Fischetti, 1989) and aggregation substance of *Enterococcus faecalis* (Olmsted *et al.*, 1993) detailed electron microscopic studies have not been performed. There are no obvious sequence similarities between the repeat block regions of CshA, Fap1 or Emb, and secondary structure predictions differ for each. How do the repeats fold to produce the characteristic fibril-like structures visualized by electron microscopy? This question remains to be addressed and provides a strong challenge to structural microbiologists.

A second question that arises relates to the translocation and surface location of these structures. The current model for cell wall anchoring via sortase and the LPXTG sequence motif invokes the covalent attachment of translocated proteins to lipid II (carrier)-linked peptidoglycan precursor molecules that are subsequently incorporated into the cell wall (Ton-That and Schneewind, 1999). An obvious consequence arising from this theory is that cell wall anchoring of proteins by sortase would be expected to be restricted to the site of new wall synthesis. In Gram-positive cocci, the site of new wall synthesis is the septal region (Cole and Hahn, 1962; Higgins & Shockman, 1970). Consequently, older wall is present towards the ends of dividing cells and, upon separation, daughter cell hemispheres comprise one-half new wall and one-half old wall, separated by the nascent cross-wall. Indeed, early immunofluorescence studies on the presentation of M protein at the surface of *S. pyogenes* cells indicated that M protein was incorporated at the site of new wall synthesis (Cole and Hahn, 1962). In contrast, the CshA fibrils of *S. gordonii* or recombinant *E. faecalis* normally exhibit an asymmetric distribution over the cell surface. Thus fibrils are more densely associated with the ends of cells (corresponding to older wall) and only sparsely distributed in the septal region, the site of new wall synthesis. This suggests that CshA fibrils are inserted into older, preformed wall. Alternatively, CshA polypeptides may be inserted at the site of new wall synthesis, but may only fold to present fibril structures, and thus to expose CshA for antibody binding, once the cell wall has 'aged' somewhat and is consequently distal to the septal region. It seems likely, nevertheless, that accessory protein(s) are required for presentation of CshA fibrils at older wall. However, the mechanism would not be CshA specific, since *E. faecalis*-expressed CshA structures demonstrate the same polar bias (McNab *et al.*, 1999). Indeed, *E. faecalis* produces a fibrillar protein, termed 'aggregation substance', in response to pheromone induction that is inserted preferentially into old wall (Wanner *et al.*, 1989; Olmsted *et al.*, 1993). Aggregation substance has a typical C-terminal wall anchor domain containing the LPXTG motif.

The genes that encode fimbrial subunits in Gram-negative bacteria are frequently organized into operons that include also genes for chaperones and export/assembly proteins. In contrast, it is clear that both *cshA* and *fap1* genes are monocistronic. Furthermore, flanking *cshA* in *S. gordonii* DL1-Challis are *aldB*, encoding acetolactate dehydrogenase, and *flpA*, encoding a protein that shows homology to a putative fibronectin-binding protein of *S. pyogenes* (McNab and Jenkinson, 1998). Gene inactivation of *flpA* did not affect cell surface presentation of CshA at late exponential phase of growth (McNab *et al.*, 1994). It seems likely, therefore, that any gene(s) encoding accessory protein(s) involved in the anchorage of proteins within mature wall are at genetic loci distinct from the genes encoding structural (sub)units. The isolation and analysis of mutants impaired in surface expression of proteins and/or structures may prove useful in elucidating the pathways involved in the surface presentation of fibrils and fimbriae. The mechanism of cell wall anchoring of the large structures present in lateral or polar tufts of mitis group streptococci is also not yet known but, again, a mechanism must be evoked for location-specific insertion and/or anchorage of the fibrils in these remarkable surface features.

3.7 OTHER PROTEIN STRUCTURE RELATIONSHIPS

CshA, Fap1 and Emb remain, at the time of writing, the best-described adhesive structures of oral streptococci in molecular terms. Nonetheless, there are other surface structures on most oral streptococci isolates, many of which are known to be involved in adhesion. These lack either detailed molecular analysis or else the association between protein and surface structure has not yet been confirmed. These examples illustrate well the complexity of cell wall structures of oral streptococci and the continued challenge of identifying their functions and compositions.

3.7.1 Surface structures of *Streptococcus mutans*

The predominant adhesin of *S. mutans* is the high molecular mass cell-wall-anchored protein variously known as antigen I/II, antigen B, SpaA, SpaP, or Pac (for a review, see Jenkinson and Demuth, 1997). This protein is present on all of the mutans group streptococci, but it is now clear that antigen I/II-like polypeptides are produced by virtually all species of streptococci that are indigenous to the human oral cavity. The antigen I/II polypeptides are multifunctional and exhibit diverse binding properties. Thus antigen I/II polypeptides bind to salivary agglutinin glycoprotein and mediate the attachment of streptococci to other oral microorganisms and to collagen (Jenkinson and Demuth, 1997; Love *et al.*, 1997). Antigen I/II polypeptides (165–170 kDa) are held at the cell surface via an LPXTG-containing wall anchor motif but have a more complex modular design than either CshA or Fap1 described above. Distinct domains within the linear sequence of antigen I/II polypeptides have been associated with different adhesive functions. Gene inactivation of *spaP*, encoding antigen I/II, in *S. mutans* strain NG8 resulted in the loss of a fibrillar fuzzy coat from the streptococcal cell surface (Lee *et al.*, 1989).

Fontana *et al.* (1995) have described the presence of peritrichous fimbrial-like structures 100–200 nm in length that protruded from a 'fuzzy coat' of constant width around cells of *S. mutans* serotype c strain TH16. A 'fimbrial-enriched preparation' was obtained by a shearing technique and yielded a number of proteins including one that reacted with antiserum to antigen I/II polypeptide, and a second protein that was recognized by antiserum to glucosyltransferase (Perrone *et al.*, 1997). It is not clear whether the fimbriae on the surface of *S. mutans* TH16 comprise the antigen I/II protein, either alone or complexed with other cell-wall-anchored polypeptides. Nevertheless, antigen I/II polypeptides SspA or SspB of *S. gordonii* DL1-Challis are not thought to be fibrillar, since mutant cells of *S. gordonii* no longer expressing the CshA fibrils were devoid of surface structures yet still anchored SspA and SspB at the cell surface (McNab *et al.*, 1999).

3.7.2 *Streptococcus salivarius* fibrils

Streptococcus salivarius isolates can be divided into Lancefield K$^+$ strains that possess a dense peritrichous fringe of fibrils, and Lancefield K$^-$ strains that lack fibrils but instead express longer, flexible peritrichous fimbriae (Handley *et al.*, 1984). The fibrillar K$^+$ strains generally exhibit a wider range of adhesive interactions than the fimbriate K$^-$ strains (Weerkamp and McBride, 1980; Handley *et al.*, 1987). Thus K$^+$ strains can participate in interbacterial coaggregation with *Veillonella parvula* and with *Fusobacterium nucleatum*. A third category of adhesin mediates host-related adhesion and aggregation including adhesion to buccal epithelial cells (BECs), saliva-induced aggregation, and haemagglutination. In contrast, fimbriate K$^-$ strains show reduced affinity for red blood cells and are not aggregated by saliva, although these strains do undergo intergeneric coaggregation (Weerkamp and McBride, 1980; Handley *et al.*, 1987).

Most *S. salivarius* K$^+$ strains have a dense fringe of short fibrils (76–111 nm in length, depending on the strain) through which project sparse longer fibrils (159–209 nm). The isolation and study of adhesion-deficient mutants of *S. salivarius* HB, a typical Lancefield K$^+$ strain, has revealed a complex picture of the cell surface in which the dense peritrichous fringe consists of three distinct classes of fibril that differ in length and composition (Weerkamp *et al.*, 1986a,b). Two cell surface protein antigens exhibiting adhesive functions were purified and characterized from *S. salivarius* HB. Antigen B (AgB), a 320 kDa glycoprotein, is the *Veillonella*-binding protein (VBP), while antigen C (AgC), a glycoprotein of 220–280 kDa, is responsible for adhesion to BECs, haemagglutination and salivary aggregation and is known as the host association factor (HAF). Both AgB, which constitutes the 91 nm long fibril subclass, and AgC, which constitutes the 72 nm long fibril subclass, are released from whole cells of *S. salivarius* HB by mutanolysin treatment. Purified proteins have a fibrillar morphology when visualized by low angle rotary shading (Weerkamp *et al.*, 1986b). The locations of the adhesive domain(s) on VBP and HAF fibrils are not known. However, given the dense packing of the fibrils within the short fringe, it seems likely that the adhesive domains reside at the tips of the fibril structures. It is also unclear how the shorter AgC (VBP) fibrils function in adhesion when coexpressed with the longer AgB (HAF) fibrils. The structure and function of the longer fibrils (178 nm in *S. salivarius* strain HB) is not known.

3.7.3 *Streptococcus sanguis* 12 fibrils

Cells of *S. sanguis* strain 12 carry three distinct types of surface structure (Morris *et al.*, 1987). Short fibrils (approximately 70 nm long) and long fibrils (approximately 200 nm long) are found on most cells and demonstrate a preference for the poles of cells. Occasional cells carry fimbriae that have a defined and constant width of 3–4 nm and are up to 1 μm in length. Freeze thaw followed by shear treatment of whole cells of *S. sanguis* 12 released long fibrils which, when analysed by sodium dodecylsulphate–polyacrylamide gel electrophoresis (SDS-PAGE), demonstrated a single protein band in excess of 300 kDa. This long fibril protein (LFP) was shown to comprise a glycoprotein that consisted of approximately equal amounts of protein and carbohydrate. LFP could be released from whole cells by mutanolysin treatment, which may indicate that the protein is anchored at the cell surface via a mechanism that involves peptidoglycan. Antibodies to a purified preparation of fibrils reacted with LFP and also blocked adhesion of *S. sanguis* to SHA by 85%. Furthermore, monoclonal antibodies to LFA specifically recognized the long fibrils when cells of *S. sanguis* 12 were analysed by immunoelectron microscopy. Finally, a naturally occurring mutant, 12na, lacked both long and short fibrils and showed a 50% reduction in adhesion to BECs (Willcox *et al.*, 1989). These cells also lacked LFP (Morris *et al.*, 1987), further linking LFP with cell surface fibrils and adhesive properties. However, these associations have not been confirmed by molecular methods.

3.7.4 Fibril tufts

A number of strains of the mitis group of streptococci have been shown to possess tufts of fibrils rather than a peritrichous fibril fringe. This group includes *S. mitis*, *S. oralis* and *S. crista* strains, and many have two or more lengths of fibril in the tuft (Handley *et al.*, 1999).

The lateral fibril tufts of *S. crista* strains are involved in specific co-aggregation reactions with *Corynebacterium matruchotii* and with *Fusobacterium nucleatum* (Lancy *et al.*, 1983) to form characteristic corn-cob formations where the streptococci adorn the central filamentous rod (Listgarten *et al.*, 1973). The fibrils probably contain protein component(s), since protease treatment removes fibril tufts completely (Hesketh *et al.*, 1987), although the *S. crista* fibril tufts react with antiserum to the glycerolphosphate backbone of lipoteichoic acid (Mouton *et al.*, 1980).

No further biochemical data are available for the adhesive tuft fibrils of *S. crista*. However, Jameson *et al.* (1995) investigated the nature of the fibril tufts of *S. oralis* CN3410. The cells produce lateral tufts of short (approximately 160 nm), densely packed fine fibrils through which protrude longer (290 nm) sparser fibrils. Protease treatment appeared to degrade the protein components of short fibrils, but released high molecular mass (>175 kDa) trypsin-resistant proteins that were components of the long fibrils (Jameson *et al.*, 1995). An adhesive function for the fibril tufts of *S. oralis* CN3410 has not yet been demonstrated.

3.8 CONCLUSIONS

The sequence of each new gene encoding a streptococcal surface structure reveals a very different polypeptide sequence. Nevertheless, each demonstrates a common molecular architecture that is exemplified by CshA of *S. gordonii*, where adhesion-mediating epitopes reside within the N-terminal non-repetitive region of the polypeptide that is held distal from the cell surface by an extensive amino acid repeat block region. A number of questions remain to be answered, including how the individual regions of each polypeptide fold to generate the functional domains of these surface appendages and whether additional gene products are required for their translocation and assembly, perhaps into higher-order structures, at the cell surface.

ACKNOWLEDGEMENTS

We thank Ivo van de Rijn for providing the *S. defectivus* strains and for permission to use the electron micrographs of wild-type and mutant strains, and Eunice Froeliger for provision of *S. parasanguis fap1⁻* strain. We also thank Nicola Mordan, David Elliott and Emma Harrison for their expert electron microscopic work.

REFERENCES

Bian, Z. and Normark, S. (1997). Nucleator function of CsgB for the assembly of adhesive surface organelles in *Escherichia coli*. *EMBO Journal* **16**, 5827–5836.

Bolken, T.C., Franke, C.A., Jones, K.F., Zeller, G.O., Jones, C.H., Dutton, E.K. and Hruby, D.E. (2001). Inactivation of the *srtA* gene in *Streptococcus gordonii* inhibits cell wall anchoring of surface proteins and decreases *in vitro* and *in vivo* adhesion. *Infection and Immunity* **69**, 75–80.

Braun, L. and Wu, H.C. (1994). Lipoproteins, structure, function, biosynthesis and model for protein export. In *Bacterial Cell Wall*, ed. by J.-M. Ghuysen and R. Hackenbeck, pp. 319–341. New York: Elsevier Science.

Burnette-Curley, D., Wells, V., Viscount, H., Munro, C.L., Fenno, J.C., Fives-Taylor, P. and Macrina, F.L. (1995). FimA, a major virulence factor associated with *Streptococcus parasanguis* endocarditis. *Infection and Immunity* **63**, 4669–4674.

Busscher, H.J., Bos, R and van der Mei, H.C. (1995). Initial microbial adhesion is a determinant for the strength of biofilm adhesion. *FEMS Microbiology Letters* **128**, 229–234.

Carlsson, J., Söderholm, G. and Almfeldt I. (1969). Prevalence of *Streptococcus sanguis* and *Streptococcus mutans* in the mouth of persons wearing full dentures. *Archives of Oral Biology* **14**, 243–249.

Carlsson, J., Grahnen, H., Jonsson, G. and Wickner, S. (1970). Establishment of *Streptococcus sanguis* in the mouths of infants. *Archives of Oral Biology* **15**, 1143–1148.

Caufield, P.W., Dasanayake, A.P., Li, Y., Pan, Y., Hsu, J. and Hardin, J.M. (2000). Natural history of *Streptococcus sanguinis* in the oral cavity of infants: evidence for a discrete window* of infectivity. *Infection and Immunity* **68**, 4018–4023.

Cole, R.M. and Hahn, J.J. (1962). Cell wall replication in *Streptococcus pyogenes*. *Science* **135**, 722–724.

Dinthilac, A., Alloing, G., Granadel, C. and Claverys, J.P. (1997). Competence and virulence of *Streptococcus pneumoniae*: Adc and PsaA mutants exhibit a requirement for Zn and Mn resulting from inactivation of putative ABC metal permeases. *Molecular Microbiology* **25**, 727–739.

Elder, B.L. and Fives-Taylor, P. (1986). Characterization of monoclonal antibodies specific for adhesion: isolation of an adhesin of *Streptococcus sanguis* FW213. *Infection and Immunity* **54**, 421–427.

Elder, B.L., Boraker, D. and Fives-Taylor, P. (1982). Whole bacterial cell enzyme linked immunosorbent assay for *Streptococcus sanguis* fimbrial antigens. *Journal of Clinical Microbiology* **16**, 141–144.

Fachon-Kalweit, S., Elder, B.L. and Fives-Taylor, P. (1985). Antibodies that bind to fimbriae block adhesion of *Streptococcus sanguis* to saliva-coated hydroxyapatite. *Infection and Immunity* **48**, 617–624.

Fenno, J.C., Shaikh, A., Spatafora, G. and Fives-Taylor, P. (1995). The *fimA* locus of *Streptococcus parasanguis* encodes an ATP-binding membrane transport system. *Molecular Microbiology* **15**, 849–863.

Fischetti, V.A. (1989). Streptococcal M protein: molecular design and biological behavior. *Clinical Microbiological Review* **2**, 285–314.

Fischetti, V.A. (1996). Gram-positive commensal bacteria deliver antigens to elicit mucosal and systemic immunity. *ASM News* **62**, 405–410.

Fives-Taylor, P.M. and Thompson, D.W. (1985). Surface properties of

Streptococcus sanguis FW213 mutants non-adherent to saliva-coated hydroxy-apatite. *Infection and Immunity* **47** 752–759.

Fontana, M., Gfell, L.E. and Gregory, R.L. (1995). Characterization of preparations enriched for *Streptococcus mutans* fimbriae: salivary immunoglobulin A antibodies in caries-free and caries-active subjects. *Clinical and Diagnostic Laboratory Immunology* **2**, 719–725.

Frandsen, E.V.G., Pedrazzoli, V. and Kilian, M. (1991). Ecology of viridans streptococci in the oral cavity and pharynx. *Oral Microbiology and Immunology* **6**, 129–133.

Hamada, S., Amano, A., Kimura, S., Nakagawa, I., Kawabata, S. and Morisaki, I. (1998). The importance of fimbriae in the virulence and ecology of some oral bacteria. *Oral Microbiology and Immunology* **13**, 129–138.

Hammar, M., Bian, Z. and Normark, S. (1996). Nucleator-dependent intercellular assembly of adhesive curli organelles in *Escherichia coli*. *Proceedings of the National Academy of Sciences, USA* **93**, 6562–6566.

Handley, P.S. (1990). Structure, composition and functions of surface structures on oral bacteria. *Biofouling* **2**, 239–264.

Handley, P.S., Carter, P. and Fielding, J. (1984). *Streptococcus salivarius* strains carry either fibrils or fimbriae on the cell surface. *Journal of Bacteriology* **157**, 64–72.

Handley, P.S., Carter, P., Wyatt, J.E. and Hesketh, L. (1985). Surface structures (peritrichous fibrils and tufts of fibrils) found on *Streptococcus sanguis* strains may be related to their ability to coaggregate with other oral genera. *Infection and Immunity* **47**, 217–227.

Handley, P.S., Harty, D.W.S., Wyatt, J.E., Brown, C.R., Doran, J.P. and Gibbs, A.C.C. (1987). A comparison of the adhesion, coaggregation and hydrophobicity properties of fibrillar and fimbriate strains of *Streptococcus salivarius*. *Journal of General Microbiology* **133**, 3207–3217.

Handley, P.S., Coykendale, A., Beighton, D., Hardie, J.M. and Whiley, R.A. (1991). *Streptococcus crista* sp. nov., a viridans streptococcus with tufted fibrils, isolated from the human oral cavity and throat. *International Journal of Systematic Bacteriology* **41**, 543–547.

Handley, P.S., McNab, R. and Jenkinson, H.F. (1999). Adhesive surface structures on oral bacteria. In *Dental Plaque Revisited: Oral Biofilms in Health and Disease*, ed. H.N. Newman and M. Wilson, pp. 145–170. Cardiff: BioLine.

Hartford, O., François, P., Vaudaux, P. and Foster, T.J. (1997). The dipeptide repeat region of the fibrinogen-binding protein (clumping factor) is required for functional expression of the fibrinogen-binding domain on the *Staphylococcus aureus* cell surface. *Molecular Microbiology* **25**, 1065–1076.

Harty, D.W.S., Willcox, M.D.P., Wyatt, J.E., Oyston, P.C.F. and Handley, P.S. (1990). The surface ultrastructure and adhesive properties of a fimbriate *Streptococcus sanguis* strain and six non-fimbriate mutants. *Biofouling* **2**, 75–86.

Herzberg, M.C. (1996). Platelet-streptococcal interactions in endocarditis. *Critical Reviews in Oral Biology and Medicine* **7**, 222–236.

Hesketh, L.M., Wyatt, J.E. and Handley, P.S. (1987). Effect of protease on cell surface structure, hydrophobicity and adhesion of tufted strains of *Streptococcus sanguis* biotypes I and II. *Microbios* **50**, 131–145.

Higgins, M.L. and Shockman, G.D. (1970). Model for cell wall growth of *Streptococcus faecalis. Journal of Bacteriology* **101**, 643–648.

Hogg, S.D., Handley, P.S. and Embery, G. (1981). Surface fibrils may be responsible for the salivary glycoprotein-mediated aggregation of the oral bacterium *Streptococcus sanguis. Archives of Oral Biology* **26**, 945–949.

Holmes, A.R., McNab, R. and Jenkinson, H.F. (1996). *Candida albicans* binding to the oral bacterium *Streptococcus gordonii* involves multiple adhesin-receptor interactions. *Infection and Immunity* **64**, 4680–4685.

Jameson, M.W., Jenkinson, H.F., Parnell, K. and Handley, P.S. (1995). Polypeptides associated with tufts of cell-surface fibrils in an oral *Streptococcus. Microbiology* **141**, 2729–2738.

Jenkinson, H.F. (1994). Adherence and accumulation of oral streptococci. *Trends in Microbiology* **2**, 209–212.

Jenkinson, H.F. and Demuth, D.R. (1997). Structure, function and immunogenicity of streptococcal antigen I/II polypeptides. *Molecular Microbiology* **23**, 183–190.

Jenkinson, H.F. and Lamont, R.J. (1997). Streptococcal adhesion and colonization. *Critical Reviews in Oral Biology and Medicine* **8**, 175–200.

Kawamura, Y., Hou, X.G., Sultana, F., Liu, S., Yamamoto, H. and Ezaki, T. (1995). Transfer of *Streptococcus adjacens* and *Streptococcus defectivus* to *Abiotrophia* gen. nov. as *Abiotrophia adjacens* comb. nov. and *Abiotrophia defectiva* comb. nov., respectively. *International Journal of Systematic Bacteriology* **45**, 798–803.

Kolenbrander, P.E. (2000). Oral microbial communities: biofilms, interactions, and genetic systems. *Annual Review of Microbiology* **54**, 413–437.

Kolenbrander, P.E. and London, J. (1993). Adhere today, here tomorrow: oral bacterial adherence. *Journal of Bacteriology* **175**, 3247–3252.

Kolenbrander, P.K., Andersen, R.A., Baker, R.A. and Jenkinson, H.F. (1998). The adhesion-associated *sca* operon in *Streptococcus gordonii* encodes an inducible high-affinity ABC transporter for Mn^{2+} uptake. *Journal of Bacteriology* **180**, 290–295.

Lancy, P., DiRienzo, J.M., Appelbaum, B., Rosan, B. and Holt, S.C. (1983). Corncob formation between *Fusobacterium nucleatum* and *Streptococcus sanguis*. *Infection and Immunity* **40**, 303–309.

Lee, S.F., Progulske-Fox, A., Erdos, G.W., Piacentini, D.A., Ayakawa, G.Y., Crowley, P.J. and Bleiweis, A.S. (1989). Construction and characterization of isogenic mutants of *Streptococcus mutans* deficient in major surface protein antigen P1 (I/II). *Infection and Immunity* **57**, 3306–3313.

Listgarten, M.A., Mayo, H. and Amsterdam, M. (1973). Ultrastructure of the attachment device between coccal and filamentous microorganisms in 'corncob' formations of dental plaque. *Archives of Oral Biology* **18**, 651–656.

Love, R.M., McMillan, M.D. and Jenkinson, H.F. (1997). Invasion of dentinal tubules by oral streptococci is associated with collagen recognition mediated by antigen I/II family of polypeptides. *Infection and Immunity* **65**, 5157–5164.

Manganelli, R. and van de Rijn, I. (1999). Characterization of *emb*, a gene encoding the major adhesin of *Streptococcus defectivus*. *Infection and Immunity* **67**, 50–56.

Mazmanian, S.K., Liu, G., Jensen, E.R., Lenoy, E. and Schneewind, O. (2000). *Staphylococcus aureus* sortase mutants defective in the display of surface proteins and in the pathogenesis of animal infections. *Proceedings of the National Academy of Sciences, USA* **97**, 5510–5515.

McDevitt, D., François, P., Vaudaux, P. and Foster, T.J. (1994). Molecular characterization of the clumping factor (fibrinogen receptor) of *Staphylococcus aureus*. *Molecular Microbiology* **11**, 237–248.

McNab, R. and Jenkinson, H.F. (1998). Altered adherence properties of a *Streptococcus gordonii hppA* (oligopeptide permease) mutant result from transcriptional effects on *cshA* adhesin gene expression. *Microbiology* **144**, 127–136.

McNab, R., Jenkinson, H.F., Loach, D.M. and Tannock, G.W. (1994). Cell-surface-associated polypeptides CshA and CshB of high molecular mass are colonization determinants in the oral bacterium *Streptococcus gordonii*. *Molecular Microbiology* **14**, 743–754.

McNab, R., Tannock, G.W. and Jenkinson, H.F. (1995). Characterization of CshA, a high molecular mass adhesin of *Streptococcus gordonii*. *Developments in Biological Standardization* **85**, 371–375.

McNab, R., Holmes, A.R., Clarke, J.C., Tannock, G.W. and Jenkinson, H.F. (1996). Cell surface polypeptide CshA mediates binding of *Streptococcus gordonii* to other oral bacteria and to immobilized fibronectin. *Infection and Immunity* **64**, 4204–4210.

McNab, R., Forbes, H., Handley, P.S., Loach, D.M., Tannock, G.W. and

Jenkinson, H.F. (1999). Cell wall-anchored CshA polypeptide (259 kilodaltons) in *Streptococcus gordonii* forms surface fibrils that confer hydrophobic and adhesive properties. *Journal of Bacteriology* **181**, 3087–3095.

Morris, E.J., Ganeshkumar, N., Song, M. and McBride, B.C. (1987). Identification and preliminary characterization of a *Streptococcus sanguis* fibrillar glycoprotein. *Journal of Bacteriology* **169**, 164–171.

Mouton, C., Reynolds, H.S. and Genko, R.J. (1980). Characterization of tufted streptococci isolated from the 'corn cob' configuration of human dental plaque. *Infection and Immunity* **27**, 235–245.

Navarre, W.W. and Schneewind, O. (1999). Surface proteins of Gram-positive bacteria and mechanisms of their targeting to the cell wall envelope. *Microbiology and Molecular Biology Reviews* **63**, 174–229.

Ní Eidhin, D., Perkins, S., François, P., Vaudaux, P., Höök, M. and Foster T.J. (1998). Clumping factor B (ClfB), a new surface-located fibrinogen-binding adhesin of *Staphylococcus aureus*. *Molecular Microbiology* **30**, 245–257.

Nyvad, B. and Kilian, M. (1987). Microbiology of the early colonization of human enamel and root surfaces *in vivo*. *Scandinavian Journal of Dental Research* **95**, 369–380.

Olmsted, S.B., Erlandsen, S.L., Dunny, G.M. and Wells, C.L. (1993). High-resolution visualization by field emission scanning electron microscopy of *Enterococcus faecalis* surface proteins encoded by the pheromone-inducible conjugative plasmid pCF10. *Journal of Bacteriology* **175**, 6229–6237.

Pallen, M.J., Lam, A.C., Antonio, M. and Dunbar, K. (2001). An embarrassment of sortases – a richness of substrates? *Trends in Microbiology* **9**, 97–101.

Perrone, M., Gfell, L.E., Fontana, M. and Gregory, R.L. (1997). Antigenic characterization of fimbria preparations from *Streptococcus mutans* isolates from caries-free and caries-susceptible subjects. *Clinical and Diagnostic Laboratory Immunology* **4**, 291–296.

Qian, H. and Dao, M.L. (1993). Inactivation of the *Streptococcus mutans* wall-associated protein A gene (*wapA*) results in a decrease in sucrose-dependent adherence and aggregation. *Infection and Immunity* **61**, 5021–5028.

Rosenow, C., Ryan, P., Weiser, J.N., Johnson, S., Fontan, P., Ortqvist, A. and Masure, H.R. (1997). Contribution of novel choline-binding proteins to adherence, colonization and immunogenicity of *Streptococcus pneumoniae*. *Molecular Microbiology* **25**, 819–829.

Sato, Y., Yamamoto, Y. and Kizaki, H. (1997). Cloning and sequence analysis of the *gbpC* gene encoding a novel glucan-binding protein of *Streptococcus mutans*. *Infection and Immunity* **65**, 668–675.

Schneewind, O., Fowler, A. and Faull, K.F. (1995). Structure of the cell wall anchor of surface proteins in *Staphylococcus aureus*. *Science* **268**, 103–106.

Smith, D.J., Anderson, J.M., King, W.F., van Houte, J. and Taubman, M.A. (1993). Oral streptococcal colonization of infants. *Oral Microbiology and Immunology* **8**, 1–4.

Soto, G.E. and Hultgren, S.J. (1999). Bacterial adhesins: common themes and variations in architecture and assembly. *Journal of Bacteriology* **181**, 1059–1071.

Sutcliffe, I.C. and Russell, R.R.B. (1995). Lipoproteins of gram-positive bacteria. *Journal of Bacteriology* **177**, 1123–1128.

Tart, R.C. and van de Rijn, I. (1993). Identification of the surface component of *Streptococcus defectivus* that mediates extracellular matrix adherence. *Infection and Immunity* **61**, 4994–5000.

Ton-That, H. and Schneewind, O. (1999). Anchor structure of staphylococcal surface proteins. IV. Inhibitors of the cell wall sorting reaction. *Journal of Biological Chemistry* **274**, 24316–24320.

Ton-That, H., Liu, G., Mazmanian, S.K., Faull, K.F. and Schneewind, O. (2000). Anchoring of surface proteins to the cell wall of *Staphylococcus aureus*. Sortase catalyzed *in vitro* transpeptidation reaction using LPXTG peptide and NH_2-Gly_3 substrates. *Journal of Biological Chemistry* **275**, 9876–9881.

Vacca-Smith, A.M., Jones, C.A., Levine, M.J. and Stinson, M.W. (1994). Glucosyltransferase mediates adhesion of *Streptococcus gordonii* to human endothelial cells *in vitro*. *Infection and Immunity* **62**, 2187–2194.

Vickerman, M.M. and Clewell, D.B. (1997). Deletions in the carboxy-terminal region of *Streptococcus gordonii* glucosyltransferase affect cell-associated enzyme activity and sucrose-associated accumulation of growing cells. *Applied Environmental Microbiology* **63**, 1667–1673.

Wanner, G., Formanek, H., Galli, D. and Wirth, R. (1989). Localization of aggregation substance of *Enterococcus faecalis* after induction by sex pheromones. An ultrastructural comparison using immunolabelling, transmission and high resolution scanning electron microscopic techniques. *Archives of Microbiology* **151**, 491–497.

Weerkamp, A.H. and McBride, B.C. (1980). Characterisation of the adherence properties of *Streptococcus salivarius*. *Infection and Immunity* **29**, 459–468.

Weerkamp, A.H., Handley, P.S., Baars, A. and Slot, J.W. (1986a). Negative staining and immunoelectron microscopy of adhesion-deficient mutants of *Streptococcus salivarius* reveal that adhesive protein antigens are separate classes of cell surface fibril. *Journal of Bacteriology* **165**, 746–755.

Weerkamp, A.H., van der Mei, H.C. and Liem, R.S.B. (1986b). Structural properties of fibrillar proteins isolated from the cell surface and cytoplasm of *Streptococcus salivarius* (K^+) cells and nonadhesive mutants. *Journal of Bacteriology* **165**, 756–762.

Whiley, R.A. and Beighton, D. (1998). Current classification of the oral strepto-cocci. *Oral Microbiology and Immunology* **13**, 195–216.

Whittaker, C.J., Klier, C.M. and Kolenbrander, P.E. (1996). Mechanisms of adhe-sion by oral bacteria. *Annual Review of Microbiology* **50**, 513–552.

Willcox, M.D.P., Wyatt, J.E. and Handley, P.S. (1989). A comparison of the adhe-sive properties and surface ultrastructure of the fibrillar *Streptococcus sanguis* 12 and an adhesion deficient non-fibrillar mutant 12 na. *Journal of Applied Bacteriology* **66**, 291–299.

Wu, H. and Fives-Taylor, P.M. (1999). Identification of dipeptide repeats and a cell wall sorting signal in the fimbriae-associated adhesin, Fap1, of *Streptococcus parasanguis*. *Molecular Microbiology* **34**, 1070–1081.

Wu, H., Mintz, K.P., Ladha, M. and Fives-Taylor, P.M. (1998). Isolation and char-acterization of Fap1, a fimbriae-associated adhesin of *Streptococcus parasan-guis* FW213. *Molecular Microbiology* **28**, 487–500.

Yeung, M.K. (1999). Molecular and genetic analyses of *Actinomyces* spp. *Critical Reviews of Oral Biology and Medicine* **10**, 120–138.

Yeung, M.K. and Ragsdale, P.A. (1997). Synthesis and function of *Actinomyces naeslundii* T14V type 1 fimbriae require the expression of additional fimbria-associated genes. *Infection and Immunity* **65**, 2629–2639.

Yeung, M.K., Donkersloot, J.A., Cisar, J.O. and Ragsdale, P.A. (1998). Identification of a gene involved in assembly of *Actinomyces naeslundii* T14V type 2 fimbriae. *Infection and Immunity* **66**, 1482–1491.

Yother, J. and White, J.M. (1994). Novel surface attachment mechanism of the *Streptococcus pneumoniae* protein PspA. *Journal of Bacteriology* **176**, 2976–2985.

CHAPTER 4

Regulation and function of phase variation in *Escherichia coli*

Ian Blomfield and Marjan van der Woude

4.1 INTRODUCTION

⓹89

Expression of the majority of proteins and other cellular constituents within bacteria such as *Escherichia coli* appears to be homogeneous within the bacterial population, changing only in response to specific alterations in the environment. Moreover, for most cellular properties, environmental change appears to elicit a uniform response from each cell. However, for a subset of genes, expression is restricted to a fraction of the population. In this more unusual mode of gene regulation, termed phase variation, individual cells switch between expressing (ON) and non-expressing (OFF) states reversibly at frequencies that can be as high as one switch per cell per generation to one switch per 10 000 cells per generation. Thus, at the level of an individual cell, it appears that, in molecular terms, one or more of the steps that are required for phase switching occurs at a *low* frequency. Furthermore, phase variation appears to be a random or stochastic process. According to this hypothesis, the molecular events leading to phase switching occur with equal probability for each cell in a given phase over the course of the cell cycle. As described in more detail below, phase variation is nevertheless *regulated* in most (if not all) phase variable systems in *E. coli* in response to specific environmental signals. Where regulation does occur, it seems likely that switching still occurs at random, even though the regulation results in switching at a different frequency. In the related phenomenon of antigenic variation, different cells express variants of the same protein or other macromolecule.

Phase variation has been observed in a wide range of bacteria, and has been reported to affect the expression of many cell surface structures such as adhesins, lipopolysaccharide (LPS) and capsules (for an excellent recent review, see Henderson *et al.*, 1999). However, phase variation can also affect properties such as DNA restriction and modification (Dybvig *et al.*, 1998).

The demonstration that a specific property undergoes phase variation is often provided using techniques (such as immunofluorescence microscopy) that detect cell surface structures preferentially. It is possible that phase variation is more widespread among cytoplasmic and periplasmic factors than has been reported, simply because the phenomenon is more difficult to detect in these cases.

Thus far in *E. coli*, a variety of different fimbriae have been found to be controlled by phase variation, as has the expression of the cell agglutination factor Ag43. Progress has been made in uncovering the molecular mechanisms that control phase variation both in *E. coli* and in other bacteria. Only in one instance in *E. coli* (type 1 fimbriation) is phase variation determined by site-specific recombination. In the other characterized phase variation systems in this microorganism, control is mediated via the action of the DNA methylating enzyme Dam in conjunction with a DNA-binding protein. However, the mechanism of control of phase variation of some fimbrial systems in *E. coli* remains to be determined. In other bacterial species, slipped strand mispairing and general recombination are commonly occurring regulatory mechanisms (for a recent review, see Henderson *et al.*, 1999), but as yet these have not been found to occur in *E. coli*.

It seems likely that phase variation is a strategy to enhance bacterial survival in what is not only an inhospitable world but also an unpredictable one (Dybvig, 1993; van der Woude *et al.*, 1996; Henderson *et al.*, 1999; Norris and Baumler, 1999). According to this hypothesis, the bacterial cell is unable to foresee whether or not expression of a specific property will be a net benefit. Benefit might include the ability to bind to the mucosal surface, whereas risk may entail encounter with the host defences (both cell mediated and humoral), or simply the wasted metabolic cost of producing a cell structure that is not needed. In the face of this uncertainty, a bacterial population producing both phase ON and phase OFF cells will be prepared for all eventualities. In this chapter, the detailed mechanisms that control phase variation in *E. coli* will be described. In addition, possible functions for phase variation will be discussed.

4.2 MECHANISMS OF CONTROL OF PHASE VARIATION IN *ESCHERICHIA COLI*

4.2.1 DNA methylation

A regulatory mechanism of phase variation has been identified in *E. coli* that requires DNA modification, specifically methylation. This mechanism differs from all other characterized mechanisms in that a change in DNA sequence is not required to acquire a switch in expression state. Regulation of phase variation of the pyelonephritis-associated pilus operon (*pap*) is the paradigm for DNA methylation-dependent phase variation. Phase variation of *pap* requires deoxyadenosine methylase (Dam) and the leucine-responsive regulatory protein (Lrp) (Blyn *et al.*, 1990; Braaten *et al.*, 1992; van der Woude *et al.*, 1992). Dam is the maintenance methylase in *E. coli*, being required for cellular processes such as mismatch repair and timing of the initiation of replication of the chromosome (Marinus, 1996). Dam mediates the transfer of a methyl group from S-adenosine-L-methionine (SAM) to the adenine of the target sequence 5′-GATC. No cognate restriction enzyme exists for Dam, nor does there appear to be a demethylating enzyme. Lrp is a global regulatory protein that is involved in transcriptional activation and repression of a wide variety of genes, ranging from those encoding enzymes involved in metabolic processes to structural genes, such as those for fimbriae (for a review, see Calvo and Matthews, 1994).

Phase variation of *pap* is the reversible switching between an expressing phase (ON) and a non-expressing phase (OFF). Transcription from the pBA promoter is essential for expression of Pap fimbriae (Fig. 4.1) (Blyn *et al.*, 1989). Lrp is both an activator and a repressor of pBA transcription, and its role in regulating *pap* transcription depends on which of two Lrp-binding regions in the *pap* regulatory region is occupied (Fig. 4.1) (Braaten *et al.*, 1992; van der Woude *et al.*, 1995; Weyand and Low, 2000). When Lrp binding occurs at the pBA promoter distal region, Lrp functions as an activator of transcription and this is required for the ON phase (Braaten *et al.*, 1992; van der Woude *et al.*, 1995). In contrast, Lrp binding at a promoter proximal site that overlaps the pBA promoter results in transcriptional repression, which results in the OFF phase (Weyand and Low, 2000).

The alternation between activation and repression of transcription and, therefore, phase variation of the *pap* operon, is due to translocation of Lrp between its two binding sites in the *pap* regulatory region. Thus the OFF to ON transition requires translocation of Lrp from the pBA promoter-proximal region to the promoter-distal binding region. The OFF to ON switch is facilitated by interaction of the operon-specific protein PapI with Lrp, which increases the affinity of Lrp for the distal region (Kaltenbach *et al.*, 1995; Nou *et al.*, 1995). Whether PapI remains associated with Lrp in the ON phase is not known. In addition, the methylation pattern of the *pap* GATC sequences is also essential for phase variation via affecting Lrp binding, as described below.

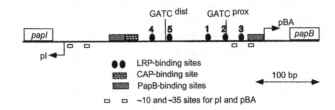

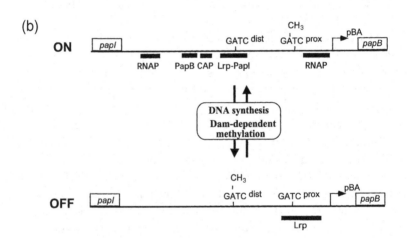

Figure 4.1. An overview of the genetic elements in the *pap* regulatory region. In (a), the location of the Dam target sites, GATC^{dist} and GATC^{prox}, the binding sites for Lrp, CAP and PapB, as well as the pI and pBA promoters are indicated. In (b), the sites occupied by regulatory proteins and RNA polymerase (RNAP) in the ON and OFF phases, respectively, are indicated by bars, with identification of the protein by the label (see text for details). The methylation states of GATC^{prox} and GATC^{dist} in the ON and OFF phases are shown as well. Note that the DNA sequence does not change between an ON and an OFF phase.

Both the promoter proximal and promoter distal binding regions of Lrp contain a Dam target sequence, designated 5'-GATC^{dist} and 5'-GATC^{prox}, respectively. *In vivo* and *in vitro* analyses have indicated that Lrp binding at either region is necessary and sufficient to block Dam-dependent methylation of the GATC sequence contained in the occupied Lrp-binding region (Blyn *et al.*, 1990; Braaten *et al.*, 1991, 1992, 1994; van der Woude *et al.*, 1998; Weyand and Low, 2000). The GATC sequence that is not in the occupied Lrp-binding region is accessible to Dam and is methylated (Braaten *et al.*, 1994; Weyand and Low, 2000). Thus Lrp-dependent blocking of methylation of the GATC sequences results in the characteristic converse methylation states found in cells in the ON and OFF phase (Fig. 4.1b) (Blyn *et al.*, 1990; Braaten *et al.*, 1994). These differential methylation states not only are characteristic of, but are essential for, *pap* phase variation. In an isolate with a GATC^{dist} point mutation that renders the site permanently non-methylated, the phenotype is locked ON (Braaten *et al.*, 1994). These data suggest that GATC^{dist} must be non-methylated for PapI-mediated Lrp translocation to the upstream binding site, which leads to Lrp-dependent activation of pBA and the ON phase. In agreement with this, the affinity of the Lrp binding for this region is decreased if the GATC^{dist} site is methylated, as indicated from *in vitro* analyses of protein-DNA interactions (Braaten *et al.*, 1994; Nou *et al.*, 1993, 1995). The switch in expression state is thought to occur immediately following DNA replication when DNA, including the *pap* GATC sequences, is temporarily hemimethylated (Nou *et al.*, 1995). To maintain the ON state after translocation has occurred to the GATC^{dist} region, methylation of the GATC^{prox} sequence is required. This is indicated by a locked OFF phenotype of an isolate containing a point mutation in the GATC^{prox} site, which renders it permanently non-methylated (Braaten *et al.*, 1994). Thus opposite methylation states of the two GATC sequences in the *pap* regulatory region are required for maintenance of the respective expression states in *pap* phase variation.

Phase variation of Pap is under the control of a complex autoregulatory loop mediated by the DNA-binding protein PapB, which is encoded by the first gene transcribed from the pBA promoter. PapB has a high as well as a low affinity binding site in the *pap* operon. The high affinity binding site lies upstream from pI, and occupation of this element activates pI. In contrast, the low affinity binding site overlaps with the pBA promoter, and PapB binding to this sequence inhibits initiation from pBA (Fig. 4.1) (Båga *et al.*, 1985). Thus, at low levels of PapB, the regulator binds upstream from pI to activate transcription from pI. The resulting rise in levels of PapI facilitates an increase in Lrp-dependent activation of transcription at pBA. The increase in PapB causes a positive feedback loop for pBA transcription until the DNA binding protein reaches higher levels, and the low affinity binding site upstream from pBA becomes occupied. Binding to this site leads to repression of transcription from pBA, and thus a decrease in PapB synthesis, initiating a regulatory loop that results in a decrease in the frequency of switching to the ON phase (Båga *et al.*, 1985; Forsman *et al.*, 1989).

Basic features of this *pap* regulatory mechanism are shared with a family of fimbrial operons that have been named the *pap*-like group of fimbrial operons. These operons have in common that 11 and 8 bp of DNA sequence containing the GATC^{dist} and GATC^{prox} sites, respectively, are conserved and each operon encodes a homologue to the PapI regulatory protein, as well as for a PapB homologue (van der Woude *et al.*, 1992). Indeed, it was shown that the amino acid residudes in PapB that are essential for oligomerization and DNA binding are conserved among the family of PapB-like proteins (Xia and Uhlin, 1999). On the basis of the DNA sequence homology of the regulatory region, the *sfa, daa, fae, clp* and *prf* fimbrial operons, as well as the *afa* operon encoding the afimbrial adhesin, are included in this family. Expression of the *sfa* and *daa* operons have been shown to be under the control of phase variation in a Dam- and Lrp-dependent manner (van der Woude and Low, 1994). However, not all operons that contain the conserved elements are controlled by phase variation. For example, even though the *fae* operon encoding K88 fimbriae includes two GATC sequences within a conserved region, and its expression is Dam and Lrp dependent, it is not control by phase variation. In addition, the *fae* homologue of PapI, FaeA, mediates repression instead of activation of transcription. Two IS*1* sequences in the regulatory region, and the presence of a third GATC sequence contained within the Lrp-binding region, seem to contribute to this altered Dam- and Lrp-dependent regulation (Huisman *et al.*, 1994; Huisman and de Graaf, 1995). Thus differences in regulatory elements outside the conserved regions may affect expression and the roles of Dam and Lrp in regulation. Other variations have also been described, such as an effect of leucine and alanine on regulation of the *clp* operon (Martin, 1996). Recently it was shown that a *pap*-like regulatory mechanism of phase variation also exists in *Salmonella typhimurium* but that additional factors are involved in transcriptional activation (Nicholson and Low, 2000). It seems likely that other variations on the *pap*-like regulatory theme will be identified.

A Dam-dependent phase variation mechanism also regulates expression of the outer membrane protein Ag43 in *E. coli*, but is clearly distinct from the

Pap phase variation mechanism. Ag43 expression is very common among *E. coli* isolates. Expression results in autoaggregation of cells and is required for mature biofilm formation under nutrient poor growth conditions (Diderichsen, 1980; Caffrey and Owen, 1989; Henderson *et al.*, 1997; Danese *et al.*, 2000). Phase variation of Ag43, which is encoded by the *agn43* gene, requires Dam and the global regulator OxyR (Henderson and Owen, 1999). OxyR is a DNA-binding protein that becomes oxidized in the presence of hydrogen peroxide. In the oxidized form, the regulator activates transcription of genes involved in the survival of oxidative stress induced by hydrogen peroxide. The reduced form of OxyR also can bind DNA and is known to repress transcription at some operons (for a review, see Zheng and Storz, 2000).

A 39 bp stretch in the *agn* regulatory region contains three GATC sequences and shows high homology to an OxyR binding site. This region is immediately downstream from a sequence that is required for phase variation and contains a σ^{70} promoter (Fig. 4.2) (M. van der Woude, unpublished data). OxyR is a repressor of *agn43* transcription, as is evident from the fact that in an *oxyR* mutant background phase variation is abrogated and all cells are in the expressing phase (ON) (Henderson and Owen, 1999; Haagmans and van der Woude, 2000). The precise mechanism of OxyR-dependent repression is not known. Several lines of evidence indicate that transcription, and thus the ON phase, is a result of abrogation of OxyR-dependent repression, and that this is mediated by methylation of the three GATC sequences in the regulatory region. First, the phenotype of cells with a high intracellular level of Dam resembles that of an *oxyR* mutant, in that Ag43 expression and transcription are locked in the ON phase. Second, the methylation state of the *agn43* GATC sequences on the chromosome varies with the expression state. In cells that are in the ON phase, the three GATC sequences are methylated, whereas in cells in the OFF phase they are non-methylated. The latter methylation protection requires a functional OxyR protein (Haagmans and van der Woude, 2000). Finally, OxyR binds to non-methylated *agn* DNA *in vitro*, but not to methylated *agn43* DNA (Haagmans and van der Woude, 2000). This is true for the oxidized form of OxyR as well as mutant OxyR(C119S), which has the properties of the reduced form (Kullik *et al.*, 1995; Haagmans and van der Woude, 2000). A model for *agn43* phase variation is shown in Fig. 4.2, indicating that Dam is required for *agn43* transcription by abrogating OxyR-mediated repression. However, Dam is also required for full transcriptional activation in an *oxyR* mutant background for an as yet unidentified reason (Haagmans and van der Woude, 2000). No other regulators have been identified for *agn43* transcription, and

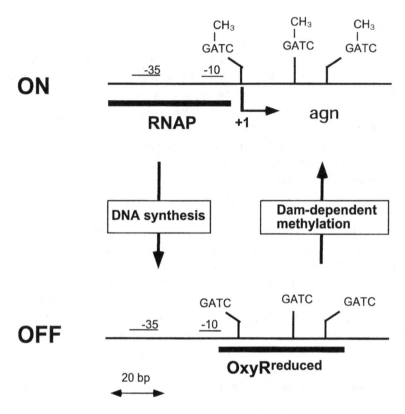

Figure 4.2. Model for Dam- and OxyR-dependent Ag43 phase variation. The location of the three genetic elements known to be required for *agn43* phase variation and transcription, specifically RNA polymerase (RNAP), three Dam target sites and the proposed OxyR-binding site, are shown. Indicated is the proposed occupation of these sites and the methylation state of the three GATC sequences in the ON and OFF phases, respectively. See text for more details. Note that the DNA sequence does not change between an ON and an OFF phase.

DNA upstream from the -35 sequence of the identified σ^{70} promoter is not required for transcription or phase variation. Whether oxidative stress plays a role in *agn43* regulation as it does for other members of the OxyR regulon is not clear. OxyR(C119S) complements an *oxyR* mutant for transcriptional repression *in vivo*, suggesting that oxidized OxyR is not required for phase variation, but preliminary data also indicate that sustained oxidative stress leads to a bias in the switch frequency of *agn43* towards the ON phase (Haagmans and van der Woude, 2000; M. van der Woude, unpublished data).

4.2.2 DNA inversion and the phase variation of type 1 fimbriation

In contrast to the situation for other fimbrial systems in *E. coli* for which the mechanism of phase variation has been elucidated, switching in expression of type 1 fimbriation (*fim*) is controlled by site-specific recombination (Abraham *et al.*, 1985). Transcription of the fimbrial structural genes initiates at a promoter that lies within a short (314 bp) invertible element of DNA, and fimbriae are expressed when the element is in one orientation (ON), but not the other (OFF) (Olsen and Klemm, 1994). Surprisingly, however, the invertible region (also known as the *fim* switch) in the ON orientation is necessary, but not sufficient for, fimbrial expression (McClain *et al.*, 1993). Thus mutants that contain the *fim* switch locked in the ON orientation continue to alternate between fimbrial and non-fimbrial phases. Although the mechanism of this second level of control has not been characterized in detail, replacement of the *fim* promoter with the *tac* promoter does not circumvent this effect. Thus the phase variation of *fim* expression is controlled at both transcriptional and post-transcriptional initiation steps.

Inversion of the *fim* element is catalysed by two site-specific recombinases, FimB and FimE, which appear to act independently to bring about the recombination reactions (Klemm, 1986; McClain *et al.*, 1991; Gally *et al.*, 1996). In addition, both FimB and FimE recombinations are stimulated by IHF (integration host factor) and by Lrp (Dorman and Higgins, 1987; Eisenstein *et al.*, 1987; Blomfield *et al.*, 1993). The *fim* recombinases, which are members of the lambda integrase family of site-specific recombinases, are encoded by genes that lie upstream from the *fim* switch in the ON orientation (Fig. 4.3a) (Klemm, 1986). Like other site-specific recombinases, FimB and FimE bind to half-binding sites that flank, and overlap with, the actual sites of strand cleavage and exchange (which lie somewhere within the inverted repeats 5'-TTGGGGCCA, termed IRL and IRR) (Gally *et al.*, 1996). Whereas FimB is able to catalyse inversion of the *fim* switch in either direction, FimE exhibits a high degree of specificity for inversion from the ON to the OFF orientation (Klemm, 1986). The specificity of FimE is determined by differences in the nucleotide sequences of the recombinase half-binding sites, although the mechanism by which this controls specificity remains undetermined (Kulasekara and Blomfield, 1999; Smith and Dorman, 1999). As described in more detail below, however, the specificity associated with *fimE* is also affected by differential expression of the recombinase depending on the orientation of the invertible element. An additional important feature of the *fim* invertible element is that the -10 region of the fimbrial promoter

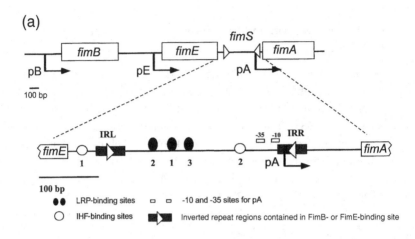

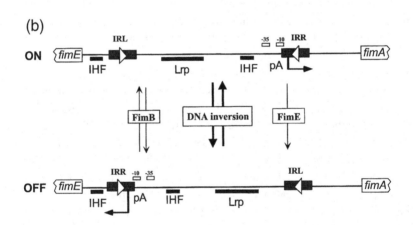

Figure 4.3. An overview of the genetic elements in the *fim* regulatory region and of the ON and OFF phases. In (a), an overview of the genetic organization of the *fimB*, *fimE* and *fimA* genes with their promoters is shown, as well as a detail of the *fimS* region. The location of the −10, −35 region of the RNA polymerase-binding site and the known binding sites for IHF and Lrp, as well as the inverted repeat regions right and left (IRR and IRL, respectively), are indicated. In (b), the genetic organization of the *fim* region in the ON and OFF phases is shown. Note that the DNA region flanked by IRR and IRL has been inverted. See text for details. The binding sites proposed to be occupied by regulatory proteins in the ON and OFF phases are indicated by a heavy bar.

overlaps with the internal half-binding site for the *fim* recombinases. Not surprisingly, both FimB and FimE inhibit transcription from the fimbrial promoter (Dove and Dorman, 1996; Smith and Dorman, 1999). Furthermore, it seems likely that recombination of the *fim* switch and transcription of the fimbrial subunit genes are mutually exclusive.

Although the steps of strand cleavage and exchange that are requisite for recombination to occur have not been determined for either FimB or FimE, both of these proteins contain a tetrad of conserved amino acid residues that are known to be important for catalysis by members of the lambda integrase family of site-specific recombinases (Arg-47 and Arg-41, His-141 and His-136, Arg-144 and Arg-139 and Tyr-176 and Tyr-171 for FimB and FimE, respectively) (Smith and Dorman, 1999; Burns et al., 2000). Mutations in each of these conserved residues in FimB and FimE do indeed result in a loss of recombinase activity. On the basis of similarity to the lambda integrase and other recombinases of this family, FimB and FimE are expected to catalyse recombination via the formation of a Holliday intermediate (for a review, see Nash, 1996). Members of the integrase family form transient 3'-DNA-tyrosine covalent complexes at one side of the region of DNA homology (Pargellis et al., 1988). Free 5'-hydroxyl ends on each strand then mount a nucleophilic attack on the complementary strand to displace the protein–DNA covalent linkage to complete the first round of strand exchanges (Pargellis et al., 1988). Following this, branch migration of the DNA in the overlap region (expected to be all or part of the region of nucleotide identity between IRL and IRR of *fim*), and then a second round of strand breakage and rejoining, completes the recombination reaction. An important feature of the *fim* switch is that, like many other site-specific recombinases and their substrates, mutations that destroy sequence homology in the overlap region at IRL and IRR of the *fim* switch block recombination (McClain et al., 1993; Nash, 1996; Kulasekara and Blomfield, 1999). From the practical point of view, this property has been particularly useful insofar as it has allowed the construction of mutants in which the *fim* switch is 'locked' in either the ON or OFF orientation in strains that are wild type for the *fim* recombinases.

For inversion of the *fim* switch to occur, the sites of strand exchange and cleavage must presumably become juxtaposed to form a synaptic complex at some stage in the recombination. IHF and Lrp, both of which can stimulate the *fim* inversion, bind to sites that lie within the invertible element (Fig. 4.3b) (Gally et al., 1996). Moreover, both Lrp and IHF introduce hairpin-like loops into DNA and, provided that these protein-induced bends are in phase, it seems likely that these DNA-binding proteins act in concert to facilitate synapses and hence recombination (Blomfield *et al.*, 1993) (Fig. 4.3b). As is discussed in more detail below, it also seems likely that factors that affect the tight structural organization of the *fim* switch play a role in controlling the recombination reaction in response to at least some environmental signals. In addition to IHF and Lrp, *fim* phase variation is also affected by the abundant nucleoid-associated protein H-NS (histone-like non-structural protein) (Spears *et al*, 1986). In contrast to the other DNA-binding proteins, however, H-NS appears to inhibit the action of both of the *fim* recombinases. How H-NS controls the *fim* inversion is somewhat unclear. Thus, although H-NS does bind to the *fim* switch region, mutations that abolish the DNA-binding activity of H-NS still inhibit *fim* recombination (Donato and Kawula, 1999). Moreover, H-NS also inhibits the expression of both *fimB* and *fimE* (Olsen and Klemm, 1994; Donato *et al.*, 1997). On the basis of these results, it seems likely that H-NS plays a complex role in controlling *fim* phase variation, exerting both direct and indirect effects on the system and perhaps acting at multiple steps in the recombination reactions themselves.

In addition to differences in specificity noted above, the activities associated with *fimB* and *fimE* also differ markedly; whereas *fimB* promotes recombination in both directions at quite low frequencies (around 10^{-3} per cell per generation), *fimE* is associated with inversion from the ON to OFF phases at very high frequencies indeed (up to 0.8 per cell per generation) (Blomfield *et al.*, 1991; Gally *et al.*, 1993). Thus, under typical laboratory growth conditions (rich, aerated media incubated at 37 °C), the prevailing activity of *fimE* ensures that only a small fraction of cells are in the fimbriate phase. It is important to note that production of different colony morphology types depending on the phase of type 1 fimbrial expression is, at least in *E. coli* K-12, a phenotype that is typically associated with mutations in *fimE* (Blomfield *et al.*, 1991). Wild-type bacteria switch from the fimbriate to afimbriate phase far too rapidly on rich agar medium to produce phase-variant colonies.

The existence of the two *fim* recombinases, differing as they do both in specificity and activity, leads to the idea that FimB and FimE are antagonists, and that the balance in levels of the two proteins controls the overall frequency of phase switching (Klemm, 1986). However, FimE recombinase activity is undetectable, and levels of *fimE* mRNA reduced many fold in phase OFF cells (Kulasekara and Blomfield, 1999; H.D. Kulasekara and I.C. Blomfield, unpublished data). The mechanism of this 'orientational control' remains to be determined, but could involve changes in *fimE* transcription initiation, or in mRNA degradation, depending on the orientation of the *fim*-invertible element. Irrespective, it seems that the *fim* recombinases are unlikely to compete for binding to the *fim* switch in the OFF orientation. Notwithstanding these observations, however, transcription from within *fimE* inhibits FimB recombination (O'Gara and Dorman, 2000). Thus factors that activate *fimE* transcription should both stimulate FimE-mediated phase variation from the ON to OFF phase, and block FimB-mediated inversion of the *fim* switch from the OFF orientation to the ON orientation in the first place.

4.3 THE REGULATION OF PHASE VARIATION

The regulation of gene expression is an adaptive response that allows the bacterium to control the synthesis of its constituents appropriately to maximize growth and/or survival in a wide range of situations. A fundamental feature of the phase-variable systems studied in most detail in *E. coli* is that they are all controlled in response to specific changes in the environment. Identifying the environmental signals that control phase-variable systems seems likely to offer insights into the biological function of the structures that are under this regulatory control mechanism. Moreover, these studies may also help us to determine the function of phase variation itself.

As noted earlier, the molecular events that determine phase variation must occur at very low frequencies. Understanding how phase variation is regulated, then, should lead to the identity of the key rate-limiting steps that are required for phase variation to occur. A detailed review of the regulation of phase variation in *E. coli* lies outside of the scope of this chapter, but the examples below should serve to illustrate the principles involved.

4.3.1 Regulation of *pap* in response to carbon source and temperature

Pap fimbriae are not expressed in the presence of glucose, but are expressed in media containing a poor carbon source such as glycerol. This regulation requires binding of the catabolite-activating protein (CAP) in complex with cyclic AMP (cAMP) to a site centred 215 bp upstream from the pBA promoter (Fig. 4.1) (Goransson and Uhlin, 1984; Båga *et al.*, 1985; Goransson *et al.*, 1989; Weyand *et al.*, 2001). In a detailed *in vivo* and *in vitro* analysis of transcription originating at pBA, Weyand and co-workers have shown that cAMP-CAP is required for activation of the pBA promoter, independently of a possible requirement of cAMP-CAP for PapI expression. Activation of Pap by CAP requires a functional AR1 region in the pBA proximal subunit of the CAP dimer (Weyand *et al.*, 2001). The AR1 region was shown previously to interact with RNA polymerase in class I CAP-dependent promoters (Zhou *et al.*, 1993). For Pap, the AR1 region is required for activation even in the absence of H-NS (Weyand *et al.*, 2001). Thus CAP plays a more direct role in *pap* regulation than in disrupting H-NS-mediated repression (Forsman *et al.*, 1992). In addition, cAMP-CAP-dependent activation of pBA requires a whole number of helical turns between the CAP-binding site and the pBA promoter, and is thus helical phase dependent. This cAMP-CAP-dependent activation also requires Lrp, but CAP and Lrp bind independently to the *pap* region. Together, these data show that CAP is directly involved in transcriptional activation of pBA, possibly through a direct interaction of CAP with RNA polymerase (Weyand *et al.*, 2001). Through the PapB-mediated autoregulatory loop described above, cAMP-CAP will also facilitate a switch to the ON phase.

Like the expression of many other virulence factors, Pap is affected by temperature. At 37 °C Pap phase variation occurs, whereas at temperatures below 26 °C Pap is not expressed owing to transcriptional regulation (Goransson and Uhlin, 1984; Blyn *et al.*, 1989). This temperature-dependent repression of transcription requires the histone-like protein H-NS (Goransson *et al.*, 1990; White-Ziegler *et al.*, 1998). White-Ziegler and co-workers showed that H-NS binds the *pap* regulatory region *in vitro* (White-Ziegler *et al.*, 1998). Transcription of *pap* is suppressed within one generation following a temperature downshift. Furthermore, over a period of 10 generations following a temperature shift, the DNA methylation state of the GATC sequences of *pap* changes gradually from the pattern characteristic of the ON phase to that of the OFF phase. These data imply that pBA is inactivated at low temperature by H-NS even in cells that retain the methylation pattern characteristic of ON phase cells. In addition, these data suggest that H-NS mediates repression by blocking Lrp binding at the GATC[dist] region and thereby prevents the Lrp translocation that is required for the switch to the ON phase (White-Ziegler *et al.*, 1998). An *hns* mutation also affects expression of *pap* under growth conditions with varying osmolarity, oxygen levels, and medium composition (White-Ziegler *et al.*, 2000). H-NS is not required for phase variation, and it remains to be determined whether these effects are direct or indirect (van der Woude *et al.*, 1995).

4.3.2 Regulation of the *fim* inversion in response to the branched-chain amino acids and alanine

The DNA inversion that controls the phase variation of type 1 fimbriation in *E.coli* is, unlike Pap phase variation, stimulated by the presence of the branched-chain amino acids (leucine, and to a lesser extent, isoleucine and valine) and alanine in growth media (Gally *et al.*, 1993). Remarkably, both FimB and FimE recombination are affected, so that inversion from OFF to ON *and* from ON to OFF is activated. Although the physiological significance of the regulation is unclear, the amino acids involved are a major constituent of type 1 fimbriae (40% of the major structural component, FimA (Klemm, 1984)). Since type 1 fimbriation must impose a significant drain on the cellular pools of alanine and the branched-chain amino acids, it is tempting to speculate that the regulation serves to control consumption of these nutrients. Perhaps consistent with this hypothesis is the fact that *fim* is part of the Lrp regulon (Blomfield *et al.*, 1993). As described in more detail below, control of *fim* by the branched-chain amino acids and leucine involves changes in the interaction of the global regulatory protein Lrp with the *fim* switch (Roesch and Blomfield, 1998).

Lrp binds with high affinity to the *fim*-invertible element, and mutations that diminish binding of the regulator to two adjacent sites within the *fim* switch (sites 1 and 2, Fig. 4.3a) *in vitro*, produce a parallel decrease in recombination *in vivo* (Gally *et al.*, 1994). In contrast, a mutation that blocks Lrp binding to a third site (site 3), proximal to site 1, stimulates recombination (Roesch and Blomfield, 1998). Loss of Lrp binding to site 3 results in a loss (for *fimB*) or a decrease (for *fimE*) in amino acid control of recombination (Roesch and Blomfield, 1998). Lrp binds co-operatively to all three sites within the *fim* switch (Lrp complex 1), but, in the presence of the amino acids, also forms an alternative nucleoprotein complex in which the regulator is bound to sites 1 and 2 only (Lrp complex 2).

Lrp complex 2 seems to be more proficient at supporting inversion of the *fim* switch than is complex 1, and the stimulation of inversion by alanine and the branched-chain amino acids may simply reflect this fact (Roesch and Blomfield, 1998). If the binding of Lrp and IHF within the *fim* switch serves to loop the DNA to bring the sites of recombination into a juxtaposition to favour productive synapses, then Lrp complex 2 may do this more effectively than complex 1 (Blomfield *et al.*, 1997). If this hypothesis is correct, it seems likely that synapse formation is a rate-limiting step in the *fim* recombinations, and its control an important regulatory event in *fim* phase variation.

4.3.3 REGULATORY CROSS-TALK BETWEEN FIMBRIAL OPERONS

The genomes of *E. coli* isolates often encode type 1 fimbriae, as well as one or more members of the *pap*-like family of fimbriae. For example, the *sfa* and *pap* operons often coexist in isolates of uropathogenic *E. coli* (Archambaud *et al.*, 1988). On the basis of our current understanding of the regulatory mechanisms it seems possible, and even likely, that phase variation of two members of the *pap*-like family of operons in a single cell are co-ordinated. For example, it is known that artificial changes in levels of the DNA methylase Dam repress expression of *pap* as well as of *daa* and *sfa* (van der Woude and Low, 1994). Thus, if levels of the methylase are controlled, then such regulation might produce co-ordinate regulation in fimbrial expression in *E. coli*.

Co-ordinating regulation may also occur through cross-talk that does not require a common global regulator. Cross-talk between the operons of the *pap*-like family may be mediated through PapB and PapI homologues, since functional complementation has been shown to occur (Goransson *et al.*, 1989; van der Woude and Low, 1994). Deletion of the *pap*-related *prf* operon decreased expression of the *sfa* operon in a pathogenic isolate, suggesting that this decrease occurs in natural isolates (Morschhäuser *et al.*, 1994). In addition, inversion of the *fim* element in the Nissle 1917 natural isolate from the OFF to ON was observed even in the absence of the *fimB* and *fimE* recombinases (Stentebjerg-Olesen *et al.*, 1999). This suggests that the phase variation of type 1 fimbriae can be mediated by other, non-*fim*-encoded site-specific recombinases.

It has been known for some time that the expression of *pap* and type 1 fimbriae is co-ordinately controlled (Nowicki *et al.*, 1984). More recently, it has been shown that the cross-regulation is mediated by the *pap* encoded DNA-binding protein PapB (Xia *et al.*, 2000). As discussed above, PapB is an activator of the pI promoter in *pap*, thereby facilitating the switch to the ON phase of the *pap* operon, as well as its own transcription. Xia *et al.* (2000) showed that PapB also modulates both *fimB* and *fimE* recombination to favour inversion to the OFF phase. The *fim* regulatory region contains PapB binding sites, and FimB-mediated recombination is decreased *in vitro* in the presence of PapB.

4.4 THE FUNCTION OF PHASE VARIATION OF ADHESINS

Control of a specific subset of genes by phase variation seems likely to allow survival or optimum growth in environments in which key elements in the bacterium's milieu change entirely at random. Thus, faced with the inability to judge whether or not expression of a specific phenotype will be an advantage, bacteria produce a mixed population in which at least some cells should thrive. If this hypothesis is indeed correct, then it seems that the expression of phase-variable phenotypes must be an *advantage* to the cell in some circumstances, but a distinct *disadvantage* to the cell in others. After all, if expression never conferred a disadvantage to the cell, then why not express the gene(s) concerned constitutively or solely in response to environmental cues? In the case of fimbrial adhesins, the advantages of attachment to the host mucosa are well documented, but in what situation might expression of the organelles be detrimental to the cell?

It has generally been assumed that phase variation is a mechanism of immune avoidance, allowing bacteria to evade the alternative pathway for complement activation and opsonin-enhanced phagocytosis. Since these antibacterial defence mechanisms kill bacteria, it is not difficult to understand how suppressing the expression of an antigen by phase variation might enhance survival of pathogenic microorganisms. In what is perhaps the clearest example of how phase variation is important in immune evasion, Norris and Baumler (1999) showed recently that phase variation of fimbrial antigens contributes to coexistence of multiple serotypes of *Salmonella*. Nevertheless, it is important to understand that the expression of many adhesins, and of Ag43, is also controlled by phase variation in commensal strains of *E. coli*. Such organisms may elicit the expression of secretory IgA via a non-T-cell-dependent pathway. However, although relatively little is known about the effects of the host's immune response on commensal *E. coli*, such a response appears unlikely to be bactericidal. Secretory IgA could block adhesin–receptor interactions by steric hindrance, so that production of the bacterial antigen would be futile. If so, then the major disadvantage of producing the surface structures concerned in the presence of neutralizing antibodies may be nothing more than wasted metabolic resources.

In some instances phase variation plays a role in cellular specialization as a key step in the interaction with the mammalian host. Although yet to be reported in *E. coli*, phase variation leads to the co-ordinate control of a set of cellular properties in both *Bordetella*, mediated by the BvgAS regulators. In this organism, the Bvg$^+$ phase is required for virulence whereas the Bvg$^-$ phase is needed for survival under nutrient deprivation (Stibitz *et al.*, 1989; Cotter and Miller, 1994; Akerley *et al.*, 1995). In other situations it seems likely that phase variation allows temporal control of gene expression when there is a requirement for an alternation in expression of a particular structure. McCormick *et al.* (1993) suggested that the expression of type 1 fimbriae in *E. coli* is an impediment to bacterial cell penetration of the mucosa, even thought the adhesin helps in attachment to the epithelial cell layer once it is reached. Furthermore, it also seems likely that phase variation from the

ON-to-OFF phase could serve to allow detachment from a primary site of colonization to facilitate spread to a new location, or even to enhance intracellular growth following invasion.

Bearing in mind the ideas discussed in this chapter, how can we interpret the fact that many phase-variable systems in *E. coli* are *regulated* in response to specific signals in the environment? For example, why should *pap* phase variation be stimulated in the OFF to ON direction, but not in the opposite direction, in the absence of glucose? Even if phase variation has evolved as a strategy to counteract variability in the environment, it still remains possible that microorganisms can, by monitoring key signals in the host milieu, acquire *limited* information about the probable benefits of switching the genes concerned on or off. According to this hypothesis, the enhanced phase variation of *pap* to the OFF phase in the presence of glucose occurs because the probable net benefit of producing the adhesin is lower in environments in which the sugar is present. Likewise, the increased frequencies of phase switching from the OFF to the ON phase that is seen for most fimbrial types in *E. coli* at 37 °C, as opposed to lower temperatures, seems to be an obvious adaptive response to growth in the mammalian host.

In one instance identified thus far (the control of the *fim* switch by the branched-chain amino acids and alanine), the environmental signal stimulates the frequency of phase switching in *both* directions. The existence of such a pattern of control seems to indicate that phase variation of type 1 fimbriation per se is important. Thus, as suggested above, the phase variation of type 1 fimbriation may be one step in a pathway in which both fimbriate and afimbriate cells participate to allow colonization of the mammalian host.

4.5. SUMMARY AND FUTURE DIRECTIONS

Phase variation in *E. coli* is commonly determined by the action of Dam in combination with DNA-binding proteins, at very specific Dam target sites in the regulatory region of the controlled operon. More unusually in this organism, this mode of gene regulation is determined by site-specific recombination. Thus far in *E. coli*, the identified examples of phase variation have been restricted to the control of cell surface adhesins (fimbriae and an autoagglutinin, Ag43). However, phase-variable systems are not easy to identify, particularly when they control factors other than cell surface structures, and when the frequency of switching is so rapid that bacteria do not form ON and OFF colony types on agar plates. Are there many unidentified phase-variable systems in *E. coli* and, if so, how will we recognize them? Phase variation, which produces heterogeneity in the bacterial population, may simply be a strategy to ensure that a fraction of the population is prepared for different eventualities at any given time. However, it seems likely that it also allows, at least in some instances, *temporal* control of gene expression when first one phase and then the other participate in different stages in host–parasite interactions. To what extent *does* phase variation play a role in temporal regulation in the different systems identified thus far? Moreover, if temporal control is a key function of phase variation, are additional factors co-ordinately controlled, and, if so, what are these? Phase variation seems to be a strategy to counteract unpredictability in the environment; yet the systems studied in *E. coli* are still affected by environmental factors such as temperature, carbon source and other nutrients. This being the case, it appears that bacteria alter the frequencies of phase variation in the light of the likely net benefit/disadvantage of producing the structures concerned. Phase variation does not seem to be regulated as a general rule in most bacterial species, however, and control of the process in *E. coli* could reflect the complex niche occupied by this microorganism. What exactly are the factors (both positive and negative) that *E. coli* (including commensal strains), as well as other bacteria, encounter within the mammalian host that drives phase variation? Phase variation appears to be an important property for bacteria, and the answers to these questions should enable us to develop better strategies to combat bacterial infection.

REFERENCES

Abraham, J.M., Freitag, C.S., Clements, J.R. and Eisenstein, B.I. (1985). An invertible element of DNA controls phase variation of type 1 fimbriae of *Escherichia coli*. *Proceedings of the National Academy of Sciences, USA* **82**, 5724–5727.

Akerley, B.J., Cotter, P.A. and Miller, J.F. (1995). Ectopic expression of the flagellar regulon alters development of the *Bordetella*–host interaction. *Cell* **80**, 611–620.

Archambaud, M., Courcoux, P. and Labigne-Roussel, A. (1988). Detection by molecular hybridization of *pap*, *afa*, and *sfa* adherence systems in *Escherichia coli* strains associated with urinary and enteral infections. *Annales de l'Institut Pasteur Microbiologie* **39**, 575–588.

Båga, M., Goransson, M., Normark, S. and Uhlin, B.E. (1985). Transcriptional activation of a *pap* pilus virulence operon from uropathogenic *Escherichia coli*. *EMBO Journal* **4**, 3887–3893.

Blomfield, I.C., McClain, M.S., Princ, J.A., Calie, P.J. and Eisenstein, B.I. (1991). Type 1 fimbriation and *fimE* mutants of *Escherichia coli* K-12. *Journal of Bacteriology* **173**, 5298–5307.

Blomfield, I.C., Calie, P.J., Eberhardt, K.J., McClain, M.S. and Eisenstein, B.I. (1993). Lrp stimulates phase variation of type 1 fimbriation in *Escherichia coli* K-12. *Journal of Bacteriology* **175**, 27–36.

Blomfield, I.C., Kulasekara, D.H. and Eisenstein, B.I. (1997). Integration host factor stimulates both FimB- and FimE-mediated site-specific DNA inversion that controls phase variation of type 1 fimbriae expression in *Escherichia coli. Molecular Microbiology* **23**, 705–713.

Blyn, L.B., Braaten, B.A., White-Ziegler, C.A., Rolfson, D.A. and Low, D.A. (1989). Phase-variation of pyelonephritis-associated pili in *Escherichia coli*: evidence for transcriptional regulation. *EMBO Journal* **8**, 613–620.

Blyn, L.B., Braaten, B.A. and Low, D.A. (1990). Regulation of *pap* pilin phase variation by a mechanism involving differential Dam methylation states. *EMBO Journal* **9**, 4045–4054.

Braaten, B.A., Blyn, L.B., Skinner, B.S. and Low, D.A. (1991). Evidence for a methylation-blocking factor *(mbf)* locus involved in *pap* pilus expression and phase variation in *Escherichia coli. Journal of Bacteriology* **173**, 1789–1800.

Braaten, B.A., Platko, J.V., van der Woude, M.W., Simons, B.H., de Graaf, F.K., Calvo, J.M. and Low, D.A. (1992). Leucine-responsive regulatory protein (Lrp) controls the expression of both the *pap* and *fan* pili operons in *Escherichia coli. Proceedings of the National Academy of Sciences, USA* **89** 4250–4254.

Braaten, B.A., Nou, X., Kaltenbach, L.S. and Low, D.A. (1994). Methylation patterns in *pap* regulatory DNA control pyelonephritis-associated pili phase variation in *E. coli. Cell* **76**, 577–588.

Burns, L.S., Smith, S.G.J. and Dorman, C.J. (2000). Interaction of the FimB integrase with the *fimS* invertible DNA element in *Escherichia coli in vivo* and *in vitro. Journal of Bacteriology* **182**, 2953–2959.

Caffrey, P. and Owen, P. (1989). Purification and N-terminal sequence of the α subunit of antigen 43, a unique protein complex associated with the outer membrane of *Escherichia coli. Journal of Bacteriology* **171**, 3634–3640.

Calvo, J.M. and Matthews, R.G. (1994). The leucine-responsive regulatory protein, a global regulator of metabolism in *Escherichia coli. Microbiological Reviews* **58**, 466–490.

Cotter, P.A. and Miller, J.F. (1994). BvgAS-mediated signal transduction: analysis of phase-locked regulatory mutants of *Bordetella bronchiseptica* in a rabbit model. *Infection and Immunity* **62**, 3381–3390.

Danese, P.N., Pratt, L.A., Dove, S. and Kolter, R. (2000). The outer-membrane protein, Ag43, mediates cell-to-cell interactions within *E. coli* biofilms. *Molecular Microbiology* **37**, 424–432.

Diderichsen, B. (1980). *flu*, a metastable gene controlling surface properties of *Escherichia coli. Journal of Bacteriology* **141**, 858–867.

Donato, G.M. and Kawula, T.H. (1999). Phenotypic analysis of random *hns* mutations differentiate DNA-binding activity from properties of *fimA* promoter inversion modulation and bacterial motility. *Journal of Bacteriology* **181**, 941–948.

Donato, G.M., Lelivelt, M.J. and Kawula, T.H. (1997). Promoter-specific repression of *fimB* expression by the *Escherichia coli* nuleoid-associated protein H-NS. *Journal of Bacteriology* **179**, 6618–6625.

Dorman, C.J. and Higgins, C.F. (1987). Fimbrial phase variation in *Escherichia coli*: dependence on integration host factor and homologies to other site-specific recombinases. *Journal of Bacteriology* **169**, 3840–3843.

Dove, S.L. and Dorman, C.J. (1996). Multicopy *fimB* gene expression in *Escherichia coli*: binding to the inverted repeats *in vivo*, effect of *fimA* transcription and DNA inversion. *Molecular Microbiology* **21**, 1161–1173.

Dybvig, K. (1993). DNA rearrangements and phenotypic switching in prokaryotes. *Molecular Microbiology* **10**, 465–471.

Dybvig, K., Ramakrishnan, S. and French, C.T. (1998). A family of phase-variable restriction enzymes with differing specificities generated by high-frequency gene rearrangements. *Proceedings of the National Academy of Sciences, USA* **95**, 13923–13928.

Eisenstein, B.I., Sweet, D.S., Vaughn, V. and Friedman, D.I. (1987). Integration host factor is required for the DNA inversion that controls phase variation in *Escherichia coli*. *Proceedings of the National Academy of Sciences, USA* **84**, 6506–6510.

Forsman, K., Gorannson, M. and Uhlin, B.E. (1989). Autoregulation and multiple DNA interactions by a transcriptional regulatory protein in *E. coli* pili biogenesis. *EMBO Journal*, **8**, 1271–1277.

Forsman, K., Sonden, B., Goransson, M. and Uhlin, B.E. (1992). Antirepression function in *Escherichia coli* for the cAMP–cAMP receptor protein transcriptional activator. *Proceedings of the National Academy of Sciences, USA* **89**, 9880–9884.

Gally, D.L., Bogan, J.A., Eisenstein, B.I. and Blomfield, I.C. (1993). Environmental regulation of the *fim* switch controlling type 1 fimbrial phase variation in *Escherichia coli* K-12: effects of temperature and media. *Journal of Bacteriology* **175**, 6186–6193.

Gally, D.L., Rucker, T.J. and Blomfield, I.C. (1994). The leucine-responsive regulatory protein binds to the *fim* switch to control phase variation of type 1 fimbrial expression in *Escherichia coli* K-12. *Journal of Bacteriology* **176**, 5665–5672.

Gally, D.L., Leathart, J. and Blomfield, I.C. (1996). Interaction of FimB and FimE with the *fim* switch that controls the phase variation of type 1 fimbriation in *Escherichia coli*. *Molecular Microbiology* **21**, 725–738.

Goransson, M. and Uhlin, B.E. (1984). Environmental temperature regulates transcription of a virulence pili operon in *E. coli. EMBO Journal* **3**, 2885–2888.

Goransson, M., Forsman, P., Nilsson, P. and Uhlin, B.E. (1989). Upstream activating sequences that are shared by two divergently transcribed operons mediate cAMP-CRP regulation of pilus-adhesin in *Escherichia coli. Molecular Microbiology* **3**, 1557–1565.

Goransson, M., Sonden, B., Nilsson, P., Dagberg, B., Forsman, K., Emanuelsson, K. and Uhlin, B.E. (1990). Transcriptional silencing and thermoregulation of gene expression in *Escherichia coli. Nature* **344**, 682–685.

Haagmans, W. and van der Woude, M. (2000). Phase variation of Ag43 in *E. coli:* Dam-dependent methylation abrogates OxyR binding and OxyR-mediated repression of transcription. *Molecular Microbiology* **35**, 877–887.

Henderson, I.R. and Owen, P. (1999). The major phase-variable outer membrane protein of *Escherichia coli* structurally resembles the immunoglobulin A1 protease class of exported protein and is regulated by a novel mechanism involving Dam and OxyR. *Journal of Bacteriology* **181**, 2132–2141.

Henderson, I.R., Meehan, M. and Owen, P. (1997). Antigen 43, a phase-variable bipartite outer membrane protein, determines colony morphology and auto-aggregation in *Escherichia coli* K-12. *FEMS Microbiology Letters* **149**, 115–200.

Henderson, I.R., Owen, P. and Nataro, J.P. (1999). Molecular switches – the ON and OFF of bacterial phase variation. *Molecular Microbiology* **33**, 919–932.

Huisman, T.T. and de Graaf, F.K. (1995). Negative control of *fae* (K88) expression by the 'global' regulator Lrp is modulated by the 'local' regulator FaeA and affected by DNA methylation. *Molecular Microbiology* **16**, 943–953.

Huisman, T.T., Bakker, D., Klaasen, P. and de Graaf, F.K. (1994). Leucine-responsive regulatory protein, IS1 insertions, and the negative regulator FaeA control the expression of the *fae* (K88). operon in *Escherichia coli. Molecular Microbiology* **11**, 525–536.

Kaltenbach, L.K., Braaten, B.A. and Low, D.A. (1995). Specific binding of PapI to Lrp-*pap* DNA complexes. *Journal of Bacteriology* **177**, 6449–6455.

Klemm, P. (1984). The *fimA* gene encoding the type-1 fimbrial subunit of *Escherichia coli. European Journal of Biochemistry* **143**, 395–399.

Klemm, P. (1986). Two regulatory *fim* genes, *fimB* and *fimE* control the phase variation of type 1 fimbriae in *Escherichia coli. EMBO Journal,* **5**, 1389–1393.

Kulasekara, H.D. and Blomfield, I.C. (1999). The molecular basis for the specificity of *fimE* in the phase variation of type 1 fimbriae of *Escherichia coli* K-12. *Molecular Microbiology* **31**, 1171–1181.

Kullik, I., Toledano, M.B., Tartaglis, L.A. and Storz, G. (1995). Mutational analysis of the redox-sensitive transcriptional regulator OxyR: regions important for oxidation and transcriptional activation. *Journal of Bacteriology* **177**, 1275–1284.

Marinus, M.G. (1996). Methylation of DNA. In Escherichia coli *and* Salmonella: *Cellular and Molecular Biology*, 2nd edn, ed. F.C. Neidhardt, R. Curtis III, J.L. Ingraham, E.C.C. Lin, K.B. Low, B. Magasanik, W.S. Reznikoff, M. Riley, M. Schaechter and H.E. Umbarger, pp. 782–791. Washington, DC: ASM Press.

Martin, C. (1996). The *clp* (CS31A) operon is negatively controlled by Lrp, Clp, and L-alanine at the transcriptional level. *Molecular Microbiology* **21**, 281–292.

McClain, M.S., Blomfield, I.C. and Eisenstein, B.I. (1991). Roles of *fimB* and *fimE* in site-specific inversion associated with phase variation of type 1 fimbriae in *Escherichia coli. Journal of Bacteriology* **173**, 5308–5314.

McClain, M.S., Blomfield, I.C., Eberhardt, K.J. and Eisenstein, B.I. (1993). Inversion-independent phase variation of type 1 fimbriae in *Escherichia coli. Journal of Bacteriology* **175**, 4335–4344.

McCormick, B.A., Klemm, P., Krogfelt, K.A., Burghoff, R.L., Pallesen, L., Laux, D.C. and Cohen, P.S. (1993). *Escherichia coli* F-18 phase locked 'on' for expression of type 1 fimbriae is a poor colonizer of the streptomycin-treated mouse large intestine. *Microbial Pathogenesis* **14**, 33–43.

Morschhäuser, J., Vetter, V., Emödy, L. and Hacker, J. (1994). Adhesin regulatory genes within large, unstable DNA regions of pathogenic *Escherichia coli*: cross-talk between different adhesin gene clusters. *Molecular Microbiology* **11**, 555–566.

Nash, H.A. (1996). Site-specific recombination: Integration, excision, resolution and inversion of defined DNA segments. In Escherichia coli *and* Salmonella: *Cellular and Molecular Biology*, 2nd edn, ed. F.C. Neidhardt, R. Curtis III, J.L. Ingraham, E.C.C. Lin, K.B. Low, B. Magasanik, W.S. Reznikoff, M. Riley, M. Schaechter and H.E. Umbarger, pp. 2363–2376. Washington, DC: ASM Press.

Nicholson, B. and Low, D. (2000). DNA methylation-dependent regulation of Pef expression in *Salmonella typhimurium. Molecular Microbiology* **35**, 728–742.

Norris, T.L. and Baumler, A.J. (1999). Phase variation of the *lpf* operon is a mechanism to evade cross-immunity between *Salmonella* serovars. *Proceedings of the National Academy of Sciences, USA* **96**, 13393–13398.

Nou, X., Skinner, B., Braaten, B., Blyn, L., Hirsch, D. and Low, D. (1993). Regulation of pyelonephritis-associated pili phase variation in *Escherichia coli*: binding of the PapI and Lrp regulatory proteins is controlled by DNA methylation. *Molecular Microbiology* **7**, 545–553.

Nou, X., Braaten, B., Kaltenbach, L. and Low, D. (1995). Differential binding of Lrp to two sets of *pap* DNA binding sites mediated by PapI regulates Pap phase variation in *Escherichia coli. EMBO Journal* **14**, 5785–5797.

Nowicki, B., Rhen, M., Väisänen-Rhen, V., Pere, A. and Korhonen, T.K. (1984). Immunofluorescence study of fimbrial phase variation in *Escherichia coli* KS71. *Infection and Immunity* **160**, 691–695.

O'Gara, J.P. and Dorman, C.J. (2000). Effects of local transcription and H-NS on inversion of the *fim* switch of *Escherichia coli*. *Molecular Microbiology* **36**, 457–466.

Olsen, P.B. and Klemm, P. (1994). Localization of promoters in the *fim* gene cluster and the effect of H-NS on the transcription of *fimB* and *fimE*. *FEMS Microbiology Letters* **116**, 95–100.

Pargellis, C.A., Nunes-Düby, S.E., Moitoso de Vargas, L. and Landy, A. (1988). Suicide recombination substrates yield covalent lambda integrase–DNA complexes and lead to identification of the active site tyrosine. *Journal of Biological Chemistry* **263**, 7678–7685.

Roesch, P.L. and Blomfield, I.C. (1998). Leucine alters the interaction of the leucine-responsive regulatory protein (Lrp) with the *fim* switch to stimulate site-specific recombination in *Escherichia coli*. *Molecular Microbiology* **27**, 751–761.

Smith, S.G.J. and Dorman, C.J. (1999). Functional analysis of the FimE integrase of *Escherichia coli* K-12: isolation of mutant derivatives with altered DNA inversion preferences. *Molecular Microbiology* **34**, 965–979.

Spears, P.A., Schauer, D.A. and Orndorff, P.E. (1986). Metastable regulation of type 1 piliation in *Escherichia coli* and isolation and characterization of a phenotypically stable mutant. *Journal of Bacteriology* **168**, 179–185.

Stentebjerg-Olesen, B., Chakraborty, T. and Klemm, P. (1999). Type 1 fimbriation and phase switching in a natural *Escherichia coli fimB* null strain, Nissle 1917. *Journal of Bacteriology* **181**, 7470–7478.

Stibitz, S., Aaronson, W., Monack, D. and Falkow, S. (1989). Phase-variation in *Bordetella pertussis* by frameshift mutation in a gene for a novel two-component system. *Nature* **338**, 266–269.

van der Woude, M.W. and Low, D.A. (1994). Leucine-responsive regulatory protein and deoxyadenosine methylase control the phase variation and expression of the *sfa* and *daa* pili operons in *Escherichia coli*. *Molecular Microbiology* **11**, 605–618.

van der Woude, M.W., Braaten, B.A. and Low, D.A. (1992). Evidence for a global regulatory control of pilus expression in *Escherichia coli* by Lrp and DNA methylation: model building based on analysis of *pap*. *Molecular Microbiology* **6**, 2429–2435.

van der Woude, M.W., Kaltenbach, L.S. and Low, D.A. (1995). Leucine-responsive regulatory protein plays dual roles as both an activator and a repressor of the *Escherichia coli pap* fimbrial operon. *Molecular Microbiology* **17**, 303–312.

van der Woude, M.W., Braaten, B. and Low, D. (1996). Epigenetic phase variation of the *pap* operon in *Escherichia coli*. *Trends in Microbiology* **4**, 5–9.

van der Woude, M.W., Hale, W.B. and Low, D.A. (1998). Formation of DNA methylation patterns: nonmethylated GATC sequences in *gut* and *pap* operons. *Journal of Bacteriology* **180**, 5913–5920.

Weyand, N. and Low, D. (2000). Lrp is sufficient for the establishment of the phase OFF *pap* DNA methylation pattern and repression of *pap* transcription *in vitro*. *Journal of Biological Chemistry* **275**, 3192–3200.

Weyand, N., Braaten, B., van der Woude, M., Tucker, J. and Low, D. (2001). The essential role of the promoter-proximal subunit of CAP in *pap* phase variation: Lrp- and helical phase-dependent activation of *papBA* by CAP from −215. *Molecular Microbiology* **39**, 1504–1522.

White-Ziegler, C.A., Angus Hill, M.L., Braaten, B.A., van der Woude, M.W. and Low, D.A. (1998). Thermoregulation of *E. coli pap* transcription: H-NS is a temperature-dependent DNA methylation blocking factor. *Molecular Microbiology* **28**, 1121–1138.

White-Ziegler, C.A., Villapakkam, A., Ronaszeki, K. and Young, S. (2000). H-NS controls *pap* and *daa* fimbrial transcription in *Escherichia coli* in response to multiple environmental cues. *Journal of Bacteriology* **182**, 6391–6400.

Xia, Y. and Uhlin, B.E. (1999). Mutational analysis of the PapB transcriptional regulator in *Escherichia coli*. Regions important for DNA binding and oligomerization. *Journal of Biological Chemistry* **274**, 19723–19730.

Xia, Y., Gally, D., Forsman-Semb, K. and Uhlin, B.E. (2000). Regulatory cross-talk between adhesin operons in *Escherichia coli*: inhibition of type 1 fimbriae expression by the PapB protein. *EMBO Journal* **19**, 1450–1457.

Zheng, M. and Storz, G. (2000). Redox sensing by prokaryotic transcription factors. *Biochemical Pharmacology* **59**, 1–6.

Zhou, Y., Zhang, X. and Ebright, R.H. (1993). Identification of the activating region of catabolite gene activator protein (CAP): isolation and characterization of mutants of CAP specifically defective in transcription activation. *Proceedings of the National Academy of Sciences, USA* **90**, 6081–6085.

CHAPTER 5

The regulation of capsule expression

Clare Taylor and Ian S. Roberts

5.1 BACTERIAL CAPSULES

The production of extracellular polysaccharide (EPS) molecules is a common feature of many bacteria (Whitfield and Valvano, 1993; Roberts, 1996). These molecules may be linked to the cell surface and organized into a discrete structure termed the capsule or, alternatively, may comprise an amorphous slime layer that is easily sloughed off from the cell surface. In essence, EPS provides a hydrated negatively charged gel that surrounds the bacterium and it is the physicochemical properties of this gel that account for the biological properties of bacterial capsules. A striking feature of bacterial capsular polysaccharides is their diversity, both in terms of component sugars and the glycosidic bond between repeating sugar residues. Even within a single bacterial species, there can be enormous structural diversity. For instance, in the case of *Streptococcus pneumoniae* there are in excess of 90 capsular serotypes. This diversity has important implications for the design of vaccine formulations that are based on capsular polysaccharides. Paradoxically, amongst this array of structural diversity, there are capsular polysaccharide molecules that are conserved across different bacterial species, such as the *Escherichia coli* K1 and *Neisseria meningitidis* serogroup B capsular polysaccharide (Jennings, 1990). Both the diversity of capsular polysaccharides and the conservation of certain polysaccharide structures across species barriers raise questions about the evolution of capsule gene clusters and the selective pressures that drive structural diversity.

For nearly 70 years, from the pioneering experiments of F. Griffith on the transformation of avirulent unencapsulated pneumococci to encapsulation and virulence (Griffith, 1928), it has been known that the expression of a capsule is an essential virulence factor. In invasive bacterial infections, interactions between the capsule and the host's immune system may be vital in deciding the outcome of an infection (Moxon and Kroll, 1990). In the absence of specific antibody, a capsule offers protection against the non-specific arm of the host's immune system by conferring increased resistance to complement-mediated killing and complement-mediated opsonophago-cytosis (Michalek et al., 1988; Moxon and Kroll, 1990). As a consequence of similarity to polysaccharide moieties present within the host (Vann et al., 1981; Finne, 1982), a small set of capsular polysaccharides are poorly immu-nogenic and elicit a poor antibody response in infected individuals. As such, these capsules also confer some measure of resistance to the host's adaptive humoral response and cannot be used as possible vaccine candidates.

In addition to mediating interactions with the host, it has been sug-gested that expression of a hydrated capsule around the cell surface may protect the bacteria from the harmful effects of desiccation and aid in the transmission of encapsulated pathogens from one host to the next (Ophir and Gutnick, 1994). This may be particularly important in highly host-adapted pathogens for which there are no alternative hosts and which are unable to survive in the environment.

5.2 BACTERIAL CAPSULES AND ADHESION

The polysaccharide capsule represents the outermost layer of the cell that mediates interactions between the bacterium and its immediate envi-ronment. In the case of abiotic surfaces, it has been demonstrated that EPS may promote the formation of biofilms and stimulate interspecies coaggre-gation, thereby enhancing the colonization of a variety of ecological niches. These include the colonization of industrial pipelines, food preparation machinery, waterpipes, indwelling catheters and prostheses (Costerton et al., 1987). In such instances, the extracellular polysaccharide may present a permeability barrier to decontaminating agents and antibiotics and hinder the effective eradication of the bacteria (Costerton et al., 1999).

5.2.1 *Staphylococcus epidermidis*

In the case of *Staph. epidermidis*, the expression of a capsular polysaccha-ride has been shown to be vital for initial attachment to abiotic and medical surfaces (Müller et al., 1993a,b; Shiro et al., 1994). The polysaccharide termed PS/A is a polymer of β-1,6-linked *N*-acetylglucosamine residues sub-stituted at the amino group with succinate residues (McKenney et al., 1998). The identification of a second lower molecular weight polysaccharide intra-cellular adhesin (PIA) believed to be involved in intercellular adhesion (Heilmann et al., 1996; Ziebuhr et al., 1997) has lead to some confusion with respect to the role of PS/A in biofilm formation. However, it is now clear that expression of PS/A is correlated with initial adherence of *Staph. epidermidis* to biomaterials, while the development of cellular aggregates is associated with the establishment of a biofilm (McKenney *et al.*, 1998). The expression of PS/A also conveys resistance to phagocytosis and virulence in animal models of infection (Shiro *et al.*, 1995). In the case of other human pathogens, the role of capsular polysaccharides in the initial adhesion to abiotic surfaces is less well proven. In the case of *Pseudomonas aeruginosa*, *Vibrio cholerae* El Tor and *Escherichia coli*, the expression of a capsular polysaccharide has been implicated post initial adhesion in stabilizing the three-dimensional stratified biofilm architecture rather than being essential for bacterial attachment (Davies *et al.*, 1993; Watnick and Kolter, 1999; Danese *et al.*, 2000).

The regulation of polysaccharide capsule expression during the different stages of a successful infection cycle represents a paradox to invasive pathogens of humans. On the one hand, during the initial stages of colonization when bacteria–host cell interactions are vital for adhesion, the expression of a polysaccharide capsule may be disadvantageous by hindering the receptor–ligand interactions vital for binding to host tissue. Indeed, it has been demonstrated for a number of human pathogens that capsule-minus mutants bind to epithelial cell lines to a greater degree than does the encapsulated progenitor strain (Runnels and Moon, 1984; Virji *et al.*, 1992; Stephens *et al.*, 1993; St Geme and Cutter, 1996). However, following transgression of the epithelial surface and the development of a subsequent bacteraemia during dissemination from the initial focus of infection, then maximum capsule expression will be essential to confer resistance to non-specific host defences (Moxon and Kroll, 1990). As such, an ability to down- or up-regulate the level of cell surface capsular polysaccharide will be vital in ensuring the successful outcome of different stages of an infection.

5.2.2 *Streptococcus pyogenes*

Adhesion to host tissues is likely to be a multifactorial process involving a number of bacterial cell surface structures that may act co-operatively in the binding of bacteria to host cells. Capsules have been implicated in the adhesion of a number of human pathogens to host tissue and successful colonization of infected animals. *Streptococcus pyogenes* or group A *Streptococcus* (GAS) is responsible for a large number of infections including streptococcal pharyngitis, skin infections, acute rheumatic fever, streptococcal toxic shock syndrome and invasive syndromes such as necrotizing fasciitis (Hoge *et al.*, 1993; Kaul *et al.*, 1997). Colonization of the pharynx is not only the first stage in the development of pharyngitis, but also serves as a possible reservoir for GAS, from which it can cause invasive infections and be disseminated to other non-carriers. As a consequence, colonization of pharyngeal epithelial cells is a vital stage in the life cycle of GAS. It has been demonstrated that the hyaluronic acid capsule of GAS can bind to the CD44 molecule on the surface of keratinocytes and act as a receptor for colonization of the pharynx by GAS (Cywes *et al.*, 2000). The binding *in vitro* to keratinocytes and pharyngeal colonization of mice could be blocked by exogenous hyaluronic acid and by the administration of anti-CD44 monoclonal antibody (Cywes *et al.*, 2000). In addition, transgenic mice lacking CD44 were not colonized by GAS and GAS failed to adhere to keratinocytes lacking CD44 (Cywes *et al.*, 2000). These data provide strong evidence for a role for the hyaluronic acid capsule in the adhesion of GAS to host tissue and the colonization of the orthopharynx. A number of putative adhesins have been described for GAS (Jenkinson and Lamont, 1997) and it is likely that adhesion will be multifactorial involving cell surface structures in addition to the hyaluronic acid capsule. Indeed, acapsular mutants of GAS do bind to the pharynx via CD44-independent mechanisms, although, in contrast to encapsulated strains, the acapsular mutants were rapidly cleared, probably as a consequence of increased sensitivity to phagocytosis. It is likely that the hyaluronic acid capsule is involved in the initial interactions with the CD44 receptor and, subsequently, other receptor–ligand interactions involving a number of cell surface adhesins may then stabilize the interactions (Cywes *et al.*, 2000). Following tissue damage, CD44 expression is increased and it is possible that this provides additional binding sites for hyaluronic acid capsule-mediated attachment of GAS during invasion of damaged host tissue. Hyaluronic acid capsule expression is regulated by the CsrR/CsrS two-component system (Levin and Wessels, 1998), with CsrR acting to repress transcription of the *has* (capsule biosynthesis) operon by binding to *has* operon promoter (Bernish and van de Rijn 1999). The role of the CsrR/CsrS two-component system in mediating hyaluronic acid capsule expression following adhesion to, and invasion of, pharyngeal host tissue is not, as yet, clear.

5.2.3 *Klebsiella pneumoniae*

In *K. pneumoniae*, it has been shown that expression of a polysaccharide capsule is essential for the colonization of the large intestine of mice and that an isogenic capsule-minus mutant was rapidly out-competed by the encapsulated wild-type strain (Favre-Bonte et al., 1999b). Analysis of the interaction between K. pneumoniae and epithelial cell lines in vitro demonstrated, as predicted, that a capsule-minus mutant adhered to a greater extent than the encapsulated wild-type strain to epithelial cells (Favre-Bonte et al., 1999a). However, surprisingly, when adhesion to a mucus-producing cell line was compared, then the encapsulated strain adhered to a greater extent than did the capsule-minus mutant (Favre-Bonte et al., 1999a). One interpretation of these data is that the capsule is required for the initial steps of colonization by interacting with the mucus layer and this interaction is vital for successful colonization in vivo. Subsequently, the interaction between the bacteria and the underlying epithelial cell is inhibited, but not abolished, by the presence of a polysaccharide capsule. This suggests that, following initial interaction with the mucus layer, capsule expression may be down-regulated to facilitate interactions between bacterial cell surface adhesins and the underlying epithelial cell. The observation that adhesin expression is inversely correlated to that of capsule expression (Favre-Bonte et al., 1995, 1999a) supports the notion that there is some form of co-ordinated regulation of the expression of these cell surface structures to maximize adhesion. In K. pneumoniae, capsule expression is regulated by the Rcs system, with additional input in the highly mucoid strain from the ancillary activator RmpA (Wacharotayankun et al., 1993). The central feature of this regulatory circuit is a two-component regulatory system in which RcsC is the sensor and RcsB is the response regulator whose activity is modulated by phosphorylation and dephosphorylation by the RcsC protein in response to environmental stimuli (Fig. 5.1) (Gottesman, 1995). The activity of the phosphorylated RcsB protein is potentiated by the unstable RcsA protein, the availability of which is regulated at the level of transcription and also by proteolytic degradation by the Lon protease (Fig. 5.1) (Gottesman, 1995). Superimposed on this basic framework, there are a number of other inputs which impinge on the regulation of the capsule gene cps expression (Fig. 5.1). It is clear that such a myriad of interacting regulatory pathways provides an opportunity to fine-tune the expression of capsular polysaccharide in response to environmental signals. In the case of colanic acid (slime) expression in E. coli, which is the best studied Rcs-regulated system (Gottesman, 1995), temperature has been identified as the key environmental factor, with reduced colanic acid expression at 37 °C. However, at 37 °C K. pneumoniae expresses copious amounts of capsular polysaccharide indicating that, in this case, the RcsBC regulatory circuit is not responding to growth at 37 °C to down-regulate capsule expression. One possibility is that following the initial capsule-mediated interaction with host mucus, the bacteria respond to the new environment (mucus) via the

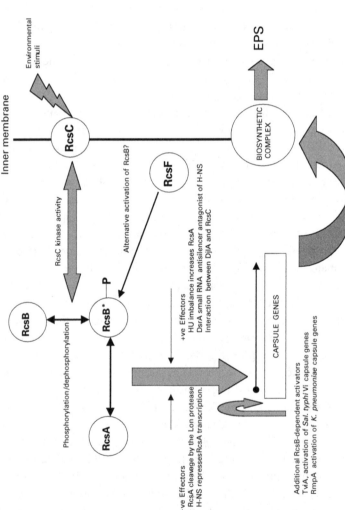

Figure 5.1. A schematic representation of the Rcs regulatory pathway. The conserved core of the pathway was established by studies on colanic acid expression in *Escherichia coli* (Gottesman, 1995). On top of this core framework, additional regulatory inputs from other microorganisms are shown. −ve, negative; +ve, positive; RcsB*, phosphorylated activated RcsB; EPS, extracellular polysaccharide; HU, histone-like protein; H-NS, histone-like non-structural protein.

Rcs system and reduce capsule expression. The concomitant increase in adhesin expression would have the net effect of enhancing bacteria–epithelial cell interactions essential for long-term colonization. There may be a role for the Rcs system in co-ordinating the regulation of these different cell surface structures in *K. pneumoniae* in response to environmental changes, but it is still conjecture and needs to be established experimentally.

5.2.4 *Salmonella typhi*

In the case of expression of the *Sal. typhi* Vi capsular polysaccharide, invasion proteins and flagellin are all differentially regulated by the RscBC two-component system (Arricau *et al.*, 1998). In this case, under conditions of low osmolarity that enhance Vi antigen expression, the transcription of the *iagA, invF and sipB* genes encoding invasion-associated proteins is negatively regulated by RcsB, probably acting in conjunction with TviA protein, which activates Vi capsule gene expression (Virlogeux *et al.*, 1996). In addition, under these conditions the increased expression of a Vi capsule inhibits the secretion of the Sip proteins and flagellin (Arricau *et al.*, 1998). However, increasing osmolarity in the growth medium results in enhanced transcription of the *iagA, invF* and *sipB* genes, with reduced transcription of the Vi capsule biosynthesis genes (Arricau *et al.*, 1998). One interpretation of this co-ordinated regulation of capsule, motility and invasion genes is that it permits the adaptation of *Sal. typhi* to the different environments encountered during the infection cycle. When outside the host in a low osmolarity environment (wastewater for instance), the Vi capsule is switched on to offer protection from environmental insults, while functions required for host invasion are switched off. Upon ingestion into the host, *Sal. typhi* in the intestinal lumen will be exposed to increased osmolarity, causing a decease in Vi capsule expression and a concomitant increase in expression of the genes essential for adhesion and invasion of host cells. At this stage, it is unclear what role, if any, the Vi antigen may play in mediating the interaction between *Sal. typhi* and the mucus surrounding host epithelial cells. Subsequently, following exit from the epithelial cells, blood-borne carriage of *Sal. typhi* will occur. The osmolarity of blood is in the order of 150 mM (Miller and Mekalanos, 1988), under which conditions Vi capsule expression would be increased while expression of invasion-associated proteins would decrease. The net effect of this will be to protect the systemic *Sal. typhi* from the host's specific and non-specific defences. As a consequence, the RcsBC system provides a method for the co-ordinated expression of cell surface structures involved in adhesion and survival in *Sal. typhi*.

5.2.5 *Neisseria meningitidis*

In terms of capsule expression in *N. meningitidis*, a similar paradox exists. Specifically, how to regulate the level of capsule expression apposite for initial mucosal surface colonization and then subsequent systemic infection. Colonization is mediated by pili and two outer membrane proteins, Opa and Opc, and adhesion is enhanced by a loss or reduction in capsule expression (Virji, 2000). Capsule expression in *N. meningitidis* is modulated in at least three ways. First, phase variation through a poly(dC) repeat that introduces a $+1/-1$ frameshift in the *siaD* gene, causing premature translation (Hammerschmidt *et al.*, 1996b). Second, Rho-dependent intracistronic transcription termination in the *siaD* gene (Lavitola *et al.*, 1999). Third, by the site-specific insertion of IS*1301* in the *siaA* gene encoding CMP-NeuNAc synthetase (Hammerschmidt *et al.*, 1996a). The consequence of the insertion of IS*1301* in the *siaA* gene is to abolish capsule expression and sialylation of the LOS molecules, both of which will promote interaction with epithelial cells. This stochastic approach to gene regulation, in which host selection drives the emergence of mutated/reverted pathogens expressing the appropriate phenotype, is an approach typical of many highly adapted human pathogens. These pathogens appear to rely less on complex regulatory control mechanisms for sensing and responding to environmental stimuli, which probably reflects the relatively small number of environments to which these pathogens are exposed and in which they survive.

5.3 *ESCHERICHIA COLI* CAPSULES AND THEIR ROLE IN ADHESION

Escherichia coli produces more than 80 chemically and serologically distinct capsules, called K antigens (Jann and Jann, 1992). These capsules have been separated into four groups, 1, 2, 3 and 4 (Table 5.1), on the basis of capsule gene organization, regulation of expression and the biosynthetic mechanism (Whitfield and Roberts, 1999). The expression of particular K antigens is associated with specific infections, with the majority of extra-intestinal isolates of *E. coli* expressing group 2 capsules (Roberts, 1996). Group 2 K antigens have been most intensively studied, both in terms of their role in disease and the biochemistry and genetics of expression. This chapter will focus on *E. coli* group 2 capsules and the reader is referred to recent reviews that describe in detail the other *E. coli* capsule groups (Roberts, 2000; Whitfield *et al.*, 2000).

Group 2 capsules are very heterogeneous in composition and, in terms

Table 5.1 *Classification of E. coli capsules*

Characteristic	Group			
	1	2	3	4
Former K antigen group	IA	II	I/II or III	IB (O antigen capsules)
Coexpressed with O serogroups	Limited range (08, 09, 020, 0101)	Many	Many	Often 08, 09 but sometimes none
Coexpressed with colanic acid	No	Yes	Yes	Yes
Thermostability	Yes	No	No	Yes
Terminal lipid moiety	Lipid A-core in K_{LPS}; unknown for capsular K antigen	α-Glycerophosphate	α-Glycerophosphate	Lipid A-core in K_{LPS}; unknown for capsular K antigen
Direction of chain growth	Reducing terminus	Non-reducing terminus	Non-reducing terminus?	Reducing terminus
Polymerization system	Wzy dependent	Processive	Processive?	Wzy-dependent
Transplasma membrane export	Wzx	ABC-2 exporter	ABC-2 exporter	Wzx
Elevated levels of CMP-Kdo synthetase at 37°C	no	yes	no	no
Genetic locus	cps near *his* and *rfb*	kps near *serA*	kps near *serA*	rfb near *his*
Thermoregulated – (i.e. not expressed below 20°C)	No	Yes	No	No
Model system	Serotype K30	Serotypes K1, K5	Serotypes K10, K54	Serotypes K40, 0111
Similar to	*Klebsiella, Erwinia*	*Neisseria, Haemophilus*	*Neisseria, Haemophilus*	Many genera

Note: K_{LPS}, low molecular weight K antigen oligosaccharide.

of structure and cell surface assembly, they closely resemble the capsular polysaccharides of *N. meningitidis* and *Haemophilus influenzae*. Group 2 capsular polysaccharides are linked via their reducing terminus to α-glycerophosphatidic acid, which is believed to play a role in the formation and stabilization of the capsule structure, possibly by anchoring the polysaccharide to the outer membrane via hydrophobic interactions (Jann and Jann, 1990). The most studied *E. coli* group 2 K antigens are K1 and K5, both of which are frequently found among isolates causing extra-intestinal infections. *Escherichia coli* K1 isolates are commonly associated with neonatal meningitis, while *E. coli* K5 strains are implicated in sepsis and urinary tract infections (Gransden *et al.*, 1990). Both polymers closely resemble host polysaccharide moieties. The K1 polysaccharide is a polymer of α-2,8-linked *N*-acetyl neuraminic acid which is a mimic of sialic acid structures present on host glyco-conjugates (Finne, 1982), while the K5 polymer of -4)-βGlcA-(1,4)-αGlcNAc-(1– is identical with *N*-acetyl-heparosan a non-sulphated precursor of heparin (Vann *et al.*, 1981). The similarity between these capsular polysaccharides and host molecules results in a poor antibody response in individuals with *E. coli* K1 or K5 infections (Jennings, 1990).

Studies using gnotobiotic rats have demonstrated that expression of a K5 capsule enhances persistence of *E. coli* in the large intestine of the rat (Herias *et al.*, 1997). The likely mechanism by which the expression of a K5 capsule confers a selective advantage in colonization is, as yet, unclear. No increase in adherence to mucus was observed when isogenic K5$^+$ and K5$^-$ strains were compared (Herias *et al.*, 1997), suggesting that the effect of the K5 capsule is not in enhancing the interaction with the mucosal epithelium. It will be interesting to see whether the expression of other group 2 capsular polysaccharides confers an advantage to *E. coli* in the colonization of the host large intestine.

5.4 THE GENETIC ORGANIZATION AND REGULATION OF *ESCHERICHIA COLI* GROUP 2 CAPSULE GENE CLUSTERS

The cloning and analysis of a large number of *E. coli* group 2 capsule gene clusters established that group 2 capsule gene clusters have a conserved modular genetic organization consisting of three regions 1, 2 and 3 (Fig. 5.2) (Silver *et al.*, 1984; Roberts *et al.*, 1986, 1988; Boulnois *et al.*, 1987; Roberts, 1996). This modular organization, first demonstrated with *E. coli* group 2 capsule gene clusters, now appears to be applicable to capsule gene clusters from other bacteria (Roberts, 1996). Regions 1 and 3 are conserved in all of the group 2 capsule gene clusters analysed and encode proteins involved in

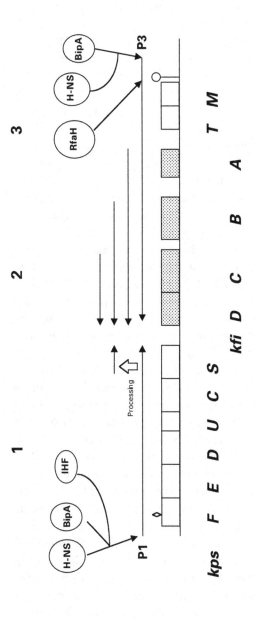

Figure 5.2. Diagrammatic representation of the *Escherichia coli* K5 capsule gene cluster. The numbers at the top refer to the three functional regions present in *E. coli* group 2 capsule genes clusters, with the region 2 genes shaded. P1 and P3 represent the region 1 and region 3 promoters and the arrows denote the major transcripts. The diamond within *kpsF* identifies the intragenic, Rho-dependent transcriptional terminator, while the stem–loop structure ◦ denotes the JUMPstart sequence.

the transport of group 2 polysaccharides from their site of synthesis on the inner face of the cytoplasmic membrane onto the cell surface. Region 2 is serotype specific and encodes enzymes for the polymerization of the polysaccharide molecule and, where necessary, for the biosynthesis of the specific monosaccharide components that make up the polysaccharide. The size of the specific region 2 is variable and, in part, reflects the complexity of the polysaccharide to be synthesized (Boulnois *et al.*, 1992). The region 2 DNA of the K5 and K1 capsule gene clusters have a high (66%) A + T content as compared to that of regions 1 (50%) and 3 (57%) (Roberts, 1996). This is typical of genes that encode enzymes for polysaccharide biosynthesis (Roberts, 1995) and would suggest that group 2 capsule diversity has been achieved, in part, through the acquisition of different region 2 sequences. Amplification by polymerase chain reaction (PCR) of sequences between regions 1 and 2, and between regions 2 and 3 from a number of group 2 capsule gene clusters, failed to find any evidence for insertion sequences or site-specific recombination events playing a role in this process (Roberts, 1996). Rather, the acquisition of new region 2 sequences may occur through homologous recombination between the flanking regions 1 and 3 of an incoming and resident capsule gene cluster. The observation that 3'-ends of the *kpsS* and *kpsT* genes that flank either side of region 2 (Fig. 5.2) show the greatest divergence amongst the conserved region 1 and 3 *kps* genes (Roberts, 1996) supports this hypothesis for acquiring and losing region 2 sequences. The mechanism by which region 2 diversity, and therefore the diversity of *E. coli* group 2 capsular polysaccharides, has been achieved is still unknown.

Region 1 contains six genes, *kpsFEDUCS*, organized in a single transcriptional unit (Fig. 5.2). The functions performed by these proteins in the transport of group 2 capsular polysaccharides are shown in Table 5.2. A single *E. coli* σ^{70} promoter is located 225 bp 5' of *kpsF* (Simpson *et al.*, 1996). Analysis of the promoter identified three integration host factor (IHF) binding site consensus sequences. One of these is located 80 bp 5' to the initiation codon of *kpsF*, while the other two are 60 bp and 110 bp 5' to the transcription start point. Gel retardation experiments using PCR fragments spanning the region 1 promoter have confirmed that IHF binds to the promoter (Rowe *et al.*, 2000). The observation that mutations in the *himA* and *himD* genes lead to a 20% reduction in expression of the KpsE protein (Simpson *et al.*, 1996) confirm that IHF plays a role in regulating the expression of region 1 at 37 °C. Transcription from the region 1 promoter generates an 8.0 kb polycistronic transcript that is subsequently processed to give a stable 1.3 kb *kpsS*-specific transcript. (Simpson *et al.*, 1996). The processing of this transcript would appear to be independent of either RNaseIII or

Table 5.2. *The functions of Kps proteins in group 2 capsular polysaccharide expression*

Protein	Function	Reference
KpsF	Regulator of capsule expression?	Cieslewicz and Vimr, 1997
KpsE	Inner membrane protein involved in the transport of group 2 polysaccharides across the periplasmic space	Arrecubieta *et al.*, 2001
KpsD	Periplasmic protein that cycles between the inner and outer membrane in the export of group 2 polysaccharides across the periplasmic space	Arrecubieta *et al.*, 2001
KpsU	CMP-Kdo synthetase	Pazzani *et al.*, 1993
KpsC, KpsS	Synthesis and attachment of phosphatidyl-Kdo to the reducing end of the nascent polysaccharide chain prior to export across the inner membrane	Rigg *et al.*, 1998
KpsT	ATPase component of the inner membrane ABC transporter for the export of group 2 polysaccharides across the inner membrane	Bliss and Silver, 1996
KpsM	Integral inner membrane component of the ABC transporter for the export of group 2 polysaccharides across the inner membrane.	Bliss and Silver, 1996

RNaseE (Simpson *et al.*, 1996). The processing of mRNA has been implicated in the differential expression of bacterial genes (Bilge *et al.*, 1993; Klug, 1993), and it is possible that the generation of a separate *kpsS*-specific transcript may enable the differential expression of KpsS from the other region 1 proteins. An intragenic Rho-dependent transcriptional terminator is located with the *kpsF* gene. Such intragenic terminators have been implicated in regulating transcription in response to physiological stress

(Richardson, 1991). In the case of region 1, under conditions of physiological stress in which the mRNA message is not being efficiently translated, transcription would cease at the intragenic terminator within *kpsF*, thereby switching off expression of region 1. The observation that mutations in region 1 genes that abolish polysaccharide export out of the cell reduce membrane transferase activity (Bronner *et al.*, 1993) means that the overall effect of terminating transcription within *kpsF* would be to reduce capsule expression under physiological stressful conditions.

Region 3 contains two genes, *kpsT* and *kpsM* organized in a single transcriptional unit (Fig. 5.2) (Bliss and Silver, 1996; Roberts, 1996). The promoter has been mapped to 741 bp 5′ to the initiation codon of the *kpsM* gene and the promoter has a typical *E. coli* σ^{70}-10 consensus sequence but no −35 region (Stevens *et al.*, 1997). No consensus binding sequences for other σ factors were detectable and no IHF binding sites were present in the region 3 promoter (Stevens *et al.*, 1997). However a *cis*-acting regulatory sequence termed *ops*, which is essential for the action of RfaH, was identified 33 bp 5′ to the initiation codon of the *kpsM* gene (Stevens *et al.*, 1997). The *ops* sequence of GGCGGTAG is contained within a larger 39 bp regulatory element called JUMPstart (Just Upstream from Many Polysaccharide-associated gene starts) (Hobbs and Reeves, 1994). RfaH regulates a number of gene clusters in *E. coli*, including the *cps*, *hly*, *rfa*, *rfb* and *tra* operons (Bailey *et al.*, 1997; Whitfield and Roberts, 1999). RfaH is a homologue of the essential transcription elongation factor, NusG, that is required for Rho-dependent transcription termination and bacteriophage λN-mediated antitermination. RfaH is believed to act as a transcriptional elongation factor that permits transcription to proceed over long distances. As such, mutations in *rfaH* result in increased transcription polarity throughout RfaH-regulated operons without affecting the initiation from the operon promoters (Bailey *et al.*, 1997). It is believed that *ops* sequence in the nascent mRNA molecule recruits RfaH, and possibly other proteins, to the transcription complex to promote transcriptional elongation. Recently, it has been proposed that the larger JUMPstart sequence may permit the formation of stem–loop structures in the 5′ mRNA that mediate the interactions between the mRNA molecule and RfaH (Marolda and Valvano, 1998). The observations that either a mutation in the *rfaH* gene or deletion of the JUMPstart sequence abolished K5 and K1 capsule production confirmed a role for RfaH in regulating group 2 capsule expression in *E. coli* (Stevens *et al.*, 1997). Analysis of the phenotype of an *rfaH* mutant demonstrated that expression of region 2 genes was dramatically reduced and, by quantitative reverse transcriptase (RT)/PCR, it was possible to show that this effect was due to a reduction in readthrough transcription across a Rho-dependent terminator in the *kpsT-kfiA* junction (Fig. 5.2) (Stevens *et al.*, 1997). This is in keeping with RfaH regulating group 2 capsule expression by permitting transcription originating from the region 3 promoter to proceed through into region 2. The co-regulation of a number of cell surface factors by RfaH is curious and it will be interesting to see how the expression of *rfaH* is regulated and how this relates to the expression of particular cell surface structures under specific environmental conditions.

The genetic organization of region 2 is serotype specific. In the case of the K5 capsule gene cluster, region 2 comprises four genes *kfiABCD* (Petit *et al.*, 1995), while there are six genes in region 2 of the K1 capsule gene cluster (Bliss and Silver, 1996). In both cases, transcription of region 2 is in the same direction as that of region 3, which is important in permitting the regulation of region 2 expression by RfaH (Stevens *et al.*, 1997; Whitfield and Roberts, 1999). In the K5 capsule gene cluster, promoters have been mapped, 5′ to *kfiA*, *kfiB* and *kfiC* genes. Transcription from the *kfiA* promoter generates a polycistronic transcript of 8.0 kb, while transcription from the *kfiB* or *kfiC* promoter results in transcripts of 6.5 and 3.0 kb, respectively (Petit *et al.*, 1995). This transcriptional organization is surprising, since it generates transcripts with two large untranslated intergenic regions, a gap of 340 bp between the *kfiA* and *kfiB* genes and a gap of 1293 bp between *kfiB* and *kfiC* genes – both of which appear to be untranslated (Petit *et al.*, 1995). The role, if any, of these regions in the mRNA molecule in regulating expression of the region 2 genes is currently unknown. The three region 2 promoters are not temperature regulated, with equivalent transcription at both 37 °C and 18 °C (Roberts, 1996). However, the region 2 promoters are weak and generate low levels of expression of the region 2 genes, which, in the absence of RfaH-mediated readthrough transcription from the region 3 promoter, is insufficient for synthesis of detectable K5 polysaccharide (Stevens *et al.*, 1997). This complex pattern of transcription raises the question of the role of these promoters in the expression of the K5 capsule. One possibility is that these promoters play a role in fine-tuning the expression of the *kfi* genes, or in allowing the bacteria to respond rapidly to temperature changes by maintaining a pool of *kfi*-specific mRNA. Equally, it is possible that these promoters play no role in regulating *kfi* gene expression, rather they may be remnants following the evolution of the K5 capsule gene cluster and the acquisition of the K5-specific region 2. This process may have occurred either in a single event from another bacterial species in which these promoters were functionally important, or in a piecemeal fashion, with each incoming region 2 gene(s) bringing with it its own promoter. Ultimately, provided the transcription of the acquired K5 region 2 was in the same direction as that of region 3, then whatever promoters were also inherited would be irrelevant. If the region 2 promoters play no functional role, it would suggest that acquisition by *E. coli* of the K5 region 2 was a relatively recent event.

Expression of group 2 capsules is temperature regulated, with capsule expression at 37°C but not at 18°C. Transcription from the region, 1 and 3 promoters is temperature regulated, with no transcription detectable at 18 0C (Cieslewicz and Vimr, 1996; Simpson *et al.*, 1996). Temperature regulation is in part controlled by the global regulatory protein H-NS (histone-like non-structural protein), since *hns* mutants show detectable transcription from the regions 1 and 3 promoters at 18 0C, albeit lower than that seen at 37°C (S. Rowe and I.S. Roberts, unpublished data). This is analogous to the H-NS-mediated thermoregulation of the *virB* promoter in *Shigella flexneri* (Dorman and Porter, 1998). In this system, activation of the *virB* promoter has an absolute requirement for the AraC-like protein VirF (Dorman and Porter, 1998). Whether an AraC-like transcriptional activator is involved in mediating transcription from the regions 1 and 3 promoters is as yet unclear.

Recently, it has been shown that mutations in *bipA* result in an increase in transcription of the capsule genes at 20°C and a reduction in transcription of the capsule genes at 37°C (Rowe *et al.*, 2000). Although this phenotype is analogous to that of a *hns* mutant, the effects of a *bipA* mutation cannot be accounted for by loss of H-NS function that is unaltered in a *bipA* mutant (Rowe *et al.*, 2000). In enteropathogenic *E. coli* (EPEC), *bipA* mutants do not trigger cytoskeletal rearrangements in host cells typical of EPEC, are hypersensitive to the host defence protein BPI, (bactericidal/permeability increasing protein) and demonstrate increased mobility and flagella expression (Farris *et al.*, 1998). BipA is a GTPase with homology to elongation factor G (EF-G) and the TetO resistance protein, both of which interact with the ribosome (Farris *et al.*, 1998). This has led to the suggestion that BipA may represent a new class of regulators that function at the level of the ribosome by regulating translation elongation (Farris *et al.*, 1998). If this is the case, then the effects on transcription from the regions 1 and 3 promoters at 37°C and 20°C are fascinating and are currently under investigation.

At 37°C the situation is further complicated by the interaction of IHF with the region 1 promoter. IHF tends to act as a facilitator, potentiating the activity of other regulatory proteins (Freundlich *et al.*, 1992) and, as such, it is likely that IHF interacts with as yet unidentified transcriptional activators that control transcription from the regions 1 and 3 promoters at 37°C. The lack of any IHF consensus binding sequences in the region 3 promoter (Stevens *et al.*, 1997) confirms that there is no absolute requirement for IHF.

Therefore, in summary, the control of expression of group 2 capsule gene clusters in *E. coli* is complex, involving several overlapping regulatory circuits (Fig. 5.2). The temperature regulation is achieved by temperature-dependent transcription from the regions 1 and 3 promoters (Fig. 5.2). This is mediated, at least in part, by H-NS and BipA. Superimposed on this system at 37°C is IHF acting at the region 1 promoter, the intragenic terminator within the *kpsF* gene and the processing of the region 1 mRNA to generate a stable *kpsS*-specific transcript. In addition, RfaH acts to allow transcription from the region 3 promoter to extend into region 2 and thereby results in sufficient expression of region 2 genes for capsular polysaccharide biosynthesis. The net effect of this is that expression of group 2 capsule gene clusters is regulated by two convergent promoters (Fig. 5.2). However, there are still many unanswered questions concerning the regulation of group 2 capsule gene clusters. In particular, what are the functions of the large untranslated regions of 225 bp and 741 bp in the 5' end of the respective regions 1 and 3 mRNA molecules? How are changes in temperature sensed and transduced to induce changes in gene expression? How is the expression of group 2 capsule gene clusters mediated in response to attachment to, and interaction with, host cells?

5.5. CONCLUSIONS

In summary, it is clear that the expression of a polysaccharide capsule plays a key role in the colonization of host animals. In certain cases, the capsule plays a significant role in mediating the initial bacteria–host interactions that are essential for any subsequent receptor–ligand interactions and intimate contact with the host epithelial cell. As such, encapsulation may participate directly in the initial attachment process. In addition, encapsulation is vital in resisting the subsequent host responses to attachment and thereby is essential for the maintenance of the adhered bacteria. Where a significant biofilm is established, then the expression of extracellular polysaccharides will be instrumental in the development of the correct three dimensional spatial architecture of the biofilm. Understanding the regulation of capsule expression under these different environmental conditions encountered during an infection is a fascinating challenge.

ACKNOWLEDGEMENTS

The work in the laboratory of I.S.R. is supported by the BBSRC, the Wellcome Trust and the EU framework V.

REFERENCES

Arrecubieta, C., Hammarton, T.C., Barrett, B., Chareonsudjai, S., Hodson, N., Rainey, D. and Roberts. I.S. (2001). The transport of group 2 capsular polysaccharides across the periplasmic space in *Escherichia coli*: roles for the KpsE and KpsD proteins. *Journal of Biological Chemistry* **276**, 4245–4250.

Arricau, N., Hermant, D., Waxin, H., Ecobichon, C., Duffey, P.S. and Popoff, M. Y. (1998). The RcsB–RcsC regulatory system of *Salmonella typhi* differentially modulates the expression of invasion proteins, flagellin and Vi antigen in response to osmolarity. *Molecular Microbiology* **29**, 835–850.

Bailey, M.J., Hughes, C. and Koronakis, V. (1997). RfaH and the *ops* element, components of a novel system controlling bacterial transcription elongation. *Molecular Microbiology* **26**, 845–851.

Bernish, B. and van de Rijn, I. (1999). Characterization of a two-component system in *Streptococcus pyogenes* which is involved in regulation of hyaluronic acid production. *Journal of Biological Chemistry* **274**, 4786–493.

Bilge, S.S., Apostol, J.M., Fullner, K.J. and Moseley, S.L. (1993). Transcription organisation of the F1854 fimbrial adhesin determinant of *Escherichia coli*. *Molecular Microbiology* **7**, 993–1006.

Bliss, J.M. and Silver, R.P. (1996). Coating the surface: a model for expression of capsular polysialic acid in *Escherichia coli* K1. *Molecular Microbiology* **21**, 221–231.

Boulnois, G.J., Roberts, I.S., Hodge, R., Hardy, K., Jann, K. and Timmis, K.N. (1987). Analysis of the K1 capsule biosynthesis genes of *Escherichia coli*: definition of three functional regions for capsule production. *Molecular and General Genetics* **208**, 242–246.

Boulnois, G.J., Drake, R., Pearce, R., and Roberts, I.S. (1992). Genome diversity at the *serA*-linked capsule locus in *Escherichia coli*. *FEMS Microbiology Letters* **100**, 121–124.

Bronner, D., Sieberth, V., Pazzani, C., Roberts, I.S., Boulnois, G.J., Jann, B. and Jann, K. (1993). Expression of the capsular K5 polysaccharide of *Escherichia coli*: biochemical and electron microscopic analyses of mutants with defects in region 1 of the K5 gene cluster. *Journal of Bacteriology* **175**, 5984–5992.

Cieslewicz, M. and Vimr, E. (1996). Thermoregulation of *kpsF*, the first region 1 gene in the *kps* locus for polysialic acid biosynthesis in *Escherichia coli* K1. *Journal of Bacteriology* **178**, 3212–3220.

Cieslewicz, M. and Vimr, E. (1997). Reduced polysialic acid capsule expression in *Escherichia coli* K1 mutants with chromosomal defects in *kpsF*. *Molecular Microbiology* **26**, 237–249.

Costerton, J.W., Chjeng, K.-J., Geesey, G.G., Ladd, T.I., Nickel, J.C., Dasgupta, M. and Marrie, T. (1987). Bacterial biofilms in nature and disease. *Annual Review of Microbiology* **41**, 435–464.

Costerton J.W., Stewart, P.S. and Greenberg E.P. (1999). Bacterial biofilms: a common cause of persistent infections. *Science* **284**, 1318–1322.

Cywes, C., Stamenkovic, I. and Wessels, M.R. (2000). CD44 as a receptor for colonization of the pharynx by group A *Streptococcus*. *Journal of Clinical Investigation* **106**, 995–1002.

Danese, P.N., Pratt, L.A., Dove, S.L. and Kolter, R. (2000). The outer membrane protein, antigen 43, mediates cell-to-cell interactions within *Escherichia coli* biofilms. *Molecular Microbiology* **37**, 424–432.

Davies, D.G., Chakrabarty, A.M. and Geesey, G.G. (1993). Exopolysaccharide production in biofilms: substratum activation of alginate gene expression by *Pseudomonas aeruginosa*. *Applied and Environmental Microbiology* **59**, 1181–1186.

Dorman, C.J. and Porter, M.E. (1998).The *Shigella* virulence gene regulatory cascade: a paradigm of bacterial gene control mechanisms. *Molecular Microbiology* **29**, 677–684.

Farris, M., Grant, A., Richardson, T.B. and O'Connor, C.D. (1998). BipA: a tyrosine-phosphorylated GTPase that mediates interactions between enteropathogenic *Escherichia coli* (EPEC) and epithelial cells. *Molecular Microbiology* **28**, 265–279.

Favre-Bonte, S., Darfeuille, A. and Forestier, C. (1995). Aggregative adherance of *Klebsiella pneumoniae* to human intestine-407 cell line. *Infection and Immunity* **63**, 1318–1338.

Favre-Bonte, S., Joly, B. and Forestier, C. (1999a). Consequences of reduction of *Klebsiella pneumoniae* capsule expression on interactions of this bacterium with epithelial cells. *Infection and Immunity* **67**, 554–561.

Favre-Bonte, S., Licht, T.R., Forestier, C. and Krogfelt, K.A. (1999b). *Klebsiella pneumoniae* capsule expression is necessary for colonisation of the large intestines of streptomycin treated mice. *Infection and Immunity* **67**, 6152–6156.

Finne, J. (1982). Occurrence of unique polysialosyl carbohydrate units in glycoproteins in developing brains. *Journal of Biological Chemistry* **257**, 11966–11970.

Freundlich, M., Ramani, N., Mathew, E., Sirko, A. and Tsui, P. (1992). The role of integration host factor in gene expression in *Escherichia coli*. *Molecular Microbiology* **6**, 2557–2563.

Gottesman, S. (1995). Regulation of capsule synthesis: modifications of the two-component paradigm by an accessory unstable regulator. In *Two Component*

Signal Transduction, ed. J.A. Hoch and T.J. Silhavy, pp. 253–262. Washington, DC: ASM Press.

Gransden, W.R., Eykyn, S.J., Phillips, I. and Rowe, B. (1990). Bacteremia due to *Escherichia coli*: a study of 861 episodes. *Reviews in Infectious Disease*. **12**, 1008–1018.

Griffith, F. (1928). The significance of pneumococcal types. *Journal of Hygiene* **27**, 113–159.

Hammerschmidt, S., Hilse, R., van Putten, J.P., Gerardy-Schahn, R., Unkmeir, A. and Frosch, M. (1996a). Modulation of cell surface sialic acid expression in *Neisseria meningitidis* via a transposable genetic element. *EMBO Journal* **15**, 192–198.

Hammerschmidt, S., Muller, A., Sillmann, H., Muhlenoff, M., Borrow, R., Fox, A., van Putten, J.P., Zollinger, W.D., Gerardy-Schahn, R. and Frosch, M. (1996b). Capsule phase variation in *Neisseria meningitidis* serogroup B by slipped strand mispairing in the polysialyl transferase gene (*siaD*): correlation with bacterial invasion and the outbreak of meningococcal disease. *Molecular Microbiology* **20**, 1211–1220.

Heilmann, C., Schweitzer, O., Gerke, C., Vanittanakom, N., Mack, D. and Gotz, F. (1996). Molecular basis of intercellular adhesion in the biofilm-forming *Staphylococcus epidermidis*. *Molecular Microbiology* **20**, 1083–1091.

Herias, M.V., Midtvedt, T., Hanson, L.A. and Wold, A.E. (1997). *Escherichia coli* K5 capsule expression enhances colonisation of the large intestine in gnotobiotic rat. *Infection and Immunity* **65**, 531–536.

Hobbs, M. and Reeves, P.R. (1994). The JUMPstart sequence: a 39 bp element common to several polysaccharide gene clusters. *Molecular Microbiology* **12**, 855–856.

Hoge, C.W., Schwartz, B., Talkington, D.F., Breiman, R.F., MacNeill, E. M. and Englender S.J. (1993). The changing epidemiology of invasive group A streptococcal infections and the emergence of streptococcal toxic shock-like syndrome. A retrospective population-based study. *Journal of the American Medical Association* **269**, 384–389.

Jann, B. and Jann, K. (1990). Structure and biosynthesis of the capsular antigens of *Escherichia coli*. In *Bacterial capsules, Current Topics in Microbiology and Immunology* **150**, 19–42.

Jann, K. and Jann, B. (1992). Capsules of *Escherichia coli*, expression and biological significance. *Canadian Journal of Microbiology* **38**, 705–710.

Jenkinson, H.F. and Lamont, R.J. (1997). Streptococcal adhesion and colonisation. *Critical Reviews in Oral Biology* **8**, 175–200.

Jennings, H.J. (1990). Capsular polysaccharides as vaccine candidates. *Current Topics in Microbiology and Immunology* **150**, 97–128.

Kaul, R., McGeer, A., Low, D.E., Green, K. and Schwartz, B. (1997). Population-based surveillance for group A streptococcal necrotizing fasciitis: clinical features, prognostic indicators, and microbiologic analysis of seventy-seven cases. Ontario Group A Streptococcal Study. *American Journal of Medicine* **103**, 18–24.

Klug, G. (1993). The role of mRNA degradation in regulated expression of bacterial photosynthesis genes. *Molecular Microbiology* **9**, 1–7.

Lavitola, A., Bucci, C., Salvatore, P., Maresca, G., Bruni, C.B. and Alifano, P. (1999). Intracistronic transcription termination in polysialyltransferase gene (*siaD*) affects phase variation in *Neisseria meningitidis*. *Molecular Microbiology* **33**, 119–127.

Levin, J.C. and Wessels, M.R. (1998). Identification of csrR/csrS, a genetic locus that regulates hyaluronic acid capsule synthesis in group A *Streptococcus*. *Molecular Microbiology* **30**, 209–219.

Marolda, C.L. and Valvano, M.A. (1998). Promoter region of the *Escherichia coli* O7-specific lipopolysaccharide gene cluster: structural and functional characterization of an upstream untranslated mRNA sequence. *Journal of Bacteriology* **180**, 3070–3079.

McKenney, D., Hubner, J., Muller, E., Wang, Y., Goldman, D.A. and Pier, G.B. (1998). The *ica* locus of *Staphylococcus epidermidis* encodes production of the capsular polysaccharide/adhesin. *Infection and Immunity* **66**, 4711–4720.

Michalek, M., Mold, C. and Brenner, E. (1988). Inhibition of the alternative pathway of human complement by structural analogues of sialic acid. *Journal of Immunology* **140**, 1588–1599.

Miller, V.L. and Mekalanos, J.J. (1988). A novel suicide vector and its use in construction of insertion mutations: osmoregulation of outer-membrane proteins and virulence determinants in *Vibrio cholerae* requires *toxR*. *Journal of Bacteriology* **170**, 2575–2583.

Moxon, E.R. and Kroll, J.S. (1990). The role of bacterial polysaccharide capsules as virulence factors. *Current Topics in Microbiology and Immunology* **150**, 65–86.

Müller, E., Huebner, J., Gutierrez, N., Takeda, S., Goldman, D.A. and Pier, G.B. (1993a). Isolation and characterisation of transposon mutants of *Staphylococcus epidermidis* deficient in capsular polysaccharide and slime. *Infection and Immunity* **61**, 551–558.

Müller, E., Takeda, S., Shiro, H., Goldman, D.A. and Pier, G.B. (1993b). Occurrence of capsular polysaccharide adhesin among clinical isolates of coagulase-negative staphylococci. *Journal of Infectious Disease* **168**, 1211–1218.

Ophir, T. and Gutnick, D. (1994). A role for exopolysaccharides in the protection of micro-organisms from dessication. *Applied and Environmental Microbiology* **60**, 740–745.

Pazzani, C., Rosenow, C., Boulnois, G.J., Bronner, D., Jann, K. and Roberts, I.S. (1993). Molecular analysis of region 1 of the *Escherichia coli* K5 antigen gene cluster: a region encoding proteins involved in cell surface expression of capsular polysaccharide. *Journal of Bacteriology* **175**, 5978–5983.

Petit, C., Rigg, G.P., Pazzani, C., Smith, A., Sieberth, V., Stevens, M., Boulnois, G., Jann, K. and Roberts, I.S. (1995). Region 2 of the *Escherichia coli* K5 capsule gene cluster encoding proteins for the biosynthesis of the K5 polysaccharide. *Molecular Microbiology* **17**, 611–620.

Richardson, J.P. (1991). Preventing the synthesis of unused transcripts by Rho factor. *Cell* **64**, 1047–1049.

Rigg, G.P., Barrett, B. and Roberts, I.S. (1998). The localization of KpsC, S and T and KfiA, C and D proteins involved in the biosynthesis of the *Escherichia coli* K5 capsular polysaccharide: evidence for a membrane-bound complex. *Microbiology* **144**, 2905–2914.

Roberts, I.S. (1995). Bacterial polysaccharides in sickness and in health. *Microbiology* **141**, 2023–2031.

Roberts, I.S. (1996). The biochemistry and genetics of capsular polysaccharide production in bacteria. *Annual Reviews in Microbiology* **50**, 285–315.

Roberts, I.S. (2000). The expression of polysaccharide capsules in *Escherichia coli*: a molecular genetic perspective. In *Glycomicrobiology*, ed. R. Doyle, pp 441–464. New York: Kluwer Academic/Plenum Publishers.

Roberts, I.S., Mountford, R., High, N., Bitter-Suerman, D., Jann, K., Timmis, K.N. and Boulnois, G.J. (1986). Molecular cloning and analysis of the genes for production of the K5, K7, K12, and K92 capsular polysaccharides of *Escherichia coli*. *Journal of Bacteriology* **168**, 1228–1233.

Roberts, I.S., Mountford, R., Hodge, R., Jann, K. and Boulnois, G.J. (1988). Common organization of gene clusters for the production of different capsular polysaccharides (K antigens) in *Escherichia coli*. *Journal of Bacteriology* **170**, 1305–1310.

Rowe, S., Hodson, N., Griffiths, G. and Roberts, I.S. (2000). Regulation of the *Escherichia coli* K5 capsule gene cluster; evidence for the role of H-NS, BipA and IHF in the regulation of group II capsule gene clusters in pathogenic *E. coli*. *Journal of Bacteriology* **182**, 2741–2745.

Runnels, P.L. and Moon, H.W. (1984). Capsule reduces adherence of enterotoxigenic *Escherichia coli* to isolated intestinal epithelial cells of pigs. *Infection and Immunity* **65**, 737–740.

Shiro, H., Muller, E., Gutierrez, N., Boisot, S., Grout, M., Tosterton, T.D., Goldman, D.A., and Pier, G.B. (1994). Transposon mutants of *Staphylococcus epidermidis* deficient in elaboration of capsular polysaccharide/adhesin and slime are avirulent in a rabbit model of endocarditis. *Journal of Infectious Diseases* **169**, 1042–1049.

Shiro, H., Meluleni, G., Groll, A., Müller, E., Tosterton, T.D., Goldman, D.A. and Pier, G.B. (1995). The pathogenic role of *Staphylococcus epidermidis* capsular polysaccharide/adhesin in a low-innoculum rabbit model of prosthetic valve endocarditis. *Circulation* **92**, 2715–2722.

Silver, R.P., Vann, W.F., and Aaronson, W. (1984). Genetic and molecular analyses of *Escherichia coli* K1 antigen genes. *Journal of Bacteriology* **157**, 568–575.

Simpson, D.A., Hammarton, T.C. and Roberts, I.S. (1996). Transcriptional organization and regulation of expression of region 1 of the *Escherichia coli* K5 capsule gene cluster. *Journal of Bacteriology* **178**, 6466–6474.

St Geme, J. W. III and Cutter, D. (1996). Influence of pili, fibrils and capsule on *in vitro* adherence by *Haemophilus influenzae* type b. *Molecular Microbiology* **21**, 21–31.

Stephens, D.S., Spellman, P.A. and Swartley, J.S. (1993). Effect of the (α2–8) linked polysialic acid capsule on adherence of *Neisseria meningitidis* to human mucosal cells. *Journal of Infectious Diseases* **167**, 475–479.

Stevens, M.P., Clarke, B.R. and Roberts, I.S. (1997). Regulation of the *Escherichia coli* K5 capsule gene cluster by transcription antitermination. *Molecular Microbiology* **24**, 1001–1012.

Vann, W.F., Schmidt, M., Jann, B., and Jann, K. (1981). The structure of the capsular polysaccharide (K5 antigen) of urinary tract infective *Escherichia coli* O10:k5:H4. A polymer similar to desulfo-heparin. *European Journal of Biochemistry* **116**, 359–364.

Virji, M. (2000). Glycans in meningococcal pathogenesis. In *Glycomicrobiology*, ed. R. Doyle, pp. 31–65. New York: Kluwer Academic/Plenum Publishers.

Virji, M., Makepeace, K., Peak, I.R.A., Fergusson, D.J.P., Jennings, M.P. and Moxon, E.R. (1992). Expression of the Opc protein correlates with invasion of epithelial and endothelial cells by *Neisseria meningitidis*. *Molecular Microbiology* **6**, 2785–2795.

Virlogeux, I., Waxin, H., Ecobichon, C., Lee, J.O. and Popoff, M.Y. (1996). Characterization of the *rcsA* and *rcsB* genes from *Salmonella typhi*: *rcsB* through *tviA* is involved in regulation of Vi antigen synthesis. *Journal of Bacteriology* **178**, 1691–1698.

Wacharotayankun, R., Arakawa, Y., Ohta, M., Tanaka, K., Akashi, T., Mori, M. and Kato, N. (1993). Enhancement of extracapsular polysaccharide synthesis in *Klebsiella pneumoniae* by RmpA2, which shows homology to NtrC and FixJ. *Infection and Immunity* **61**, 3164–3174.

Watnick, P.I., and Kolter, R. (1999). Steps in the development of a *Vibrio cholerae* El Tor biofilm. *Molecular Microbiology* **34**, 586–595.

Whitfield, C. and Roberts, I.S. (1999). Structure, assembly and regulation of expression of capsules in *Escherichia coli*. *Molecular Microbiology* **31**, 1307–1319.

Whitfield, C. and Valvano, M.A. (1993). Biosynthesis and expression of cell surface polysaccharides in Gram-negative bacteria. *Advances in Microbiology and Physiology* **35**, 135–246.

Whitfield, C., Drummelsmith, J., Rahn, A. and Wugeditsch, T. (2000). Biosynthesis and regulation of expression of group 1 capsules in *Escherichia coli* and related extracellular polysaccharides in other bacteria. In *Glycomicrobiology*, ed. R. Doyle, pp. 275–297. New York: Kluwer Academic/Plenum Publishers.

Ziebuhr, W., Heilmann, C., Gotz, F., Meyer, P., Wilms, K., Straube, E. and Hacker, J. (1997). Detection of the intercellular adhesion gene cluster (*ica*) and phase variation in *Staphylococcus epidermidis* blood culture strains and mucosal isolates. *Infection and Immunity* **65**, 890–896.

CHAPTER 6

Role of pili in *Haemophilus influenzae* adherence, colonization and disease

Janet R. Gilsdorf

6.1 INTRODUCTION

(139)

Haemophilus influenzae is a Gram-negative bacillus that causes significant human disease. This organism is classified on the basis of the presence and serological specificity (serotypes a through f) of its polysaccharide capsule. Systemic infections such as bacteraemia, meningitis, septic arthritis and pneumonia in young children are caused primarily by *H. influenzae* possessing the type b capsule (Hib), while respiratory infections such as pneumonia in patients with chronic pulmonary disease, otitis media, sinusitis, and bronchitis are caused primarily by non-encapsulated strains (the so-called non-typeable *H. influenzae*).

The organism lives exclusively on human mucosal surfaces, most commonly in the nasopharynx and occasionally on the conjunctivae or in the female genital tract, where it is carried asymptomatically. Respiratory tract colonization rates in children vary between 30% and 100% (Gratten *et al.*, 1989; Trottier *et al.*, 1989; Harabuchi *et al.*, 1994; Faden *et al.*, 1995; Fontanals *et al.*, 2000; St Sauver *et al.*, 2000), with point prevalence rates of about 25% (Bou *et al.*, 2000) to 77–84% (St Sauver *et al.*, 2000), and are higher among children as compared with adults (Turk, 1984).

The pathogenesis of *H. influenzae* infections is a complex, and incompletely understood, process. Transmission to new hosts is presumed to be via infected respiratory droplets, and the bacterial factors important in successful transmission are poorly defined (Lipsitch and Moxon, 1997). Upon entry into a new host, the organism adheres to respiratory tract structures by means of a number of adhesins that prevent removal of the bacteria by respiratory tract cleansing mechanisms such as mucin trapping and ciliary action. Among the bacterial factors that have been shown to facilitate adherence of the organism to respiratory tract tissues are fimbrial structures, such as haemagglutinating pili (sometimes called fimbriae) (Gilsdorf *et al.*, 1997),

P5 fimbriae (Bakaletz *et al.*, 1988; Sirakova *et al.*, 1994), and Hia (St Geme and Cutter, 1995; St Geme *et al.*, 1996a), as well as non-fimbrial structures, such as the high molecular adhesins HMW1 and HMW2 (St Geme *et al.*, 1993; Barenkamp and St Geme, 1996), Hap (St Geme *et al.*, 1994) a 46 kDa surface protein (Busse *et al.*, 1997), and OapA (Weiser *et al.*, 1995; Weiser, 2000). Of these adhesins, pili were the first described and have been most extensively studied.

6.2 PREVALENCE OF PILIATION OF *HAEMOPHILUS INFLUENZAE* STRAINS

The prevalance of pili on *H. influenzae* clinical isolates varies, depending on the source of the organisms studied and the technique used for determining pili. Upon subculture, most clinical isolates do not express pili. Within the population of bacteria from clinical isolates, however, piliated variants may be identified, selected for, and enriched by their agglutination of human erythrocytes (Connor and Loeb, 1983; Gilsdorf *et al.*, 1992), which is strongly correlated with the presence of pili (Guerina *et al.*, 1982; LiPuma and Gilsdorf, 1988). Of 25 Hib strains isolated from cerebral spinal fluid (Stull *et al.*, 1984), 22 could be enriched for haemagglutination. Gilsdorf *et al.* found that 12 of 38 randomly selected non-typeable isolates expressed haemagglutinating pili (Gilsdorf *et al.*, 1992), four *de novo* and eight more after red blood cell selection. On electron microscopy, Scott and Old (1981) detected fimbrial structures on three of six respiratory and four of five conjunctival isolates. Krasan *et al.* (2000) studied nasopharyngeal and middle ear isolates from 17 children with non-typeable acute otitis media due to *H. influenzae* and, by electron microscopy and Southern hybridization analysis, found that two of the 17 nasopharyngeal isolates and none of the 17 middle ear isolates expressed pili, in spite of the presence of the entire pilus gene cluster in the ear isolates from the two patients with piliated nasopharyngeal isolates. On the other hand, St Geme *et al.* (1991) identified pilus genes in all eight *H. influenzae* biogroup aegyptius (a subset of *H. influenzae* associated with conjunctivitis) strains studied; pili were expressed (as shown on electron microscopy and antibody reactivity) on four of these strains. Using Southern blotting with probes derived from Hib strains, Geluk *et al.* (1998) demonstrated pilus genes in 18% of non-typeable strains tested, ranging from 8% of strains colonizing healthy individuals, 13% of patients with otitis media, to 13–50% of patients with chronic bronchitis. These techniques may underestimate the presence of pilus genes because of strain-to-strain sequence variation, particularly if high stringency conditions were used in the assays.

Apicella *et al.* (1984) evaluated 21 *H. influenzae* strains from the sputa of adult patients; two of the strains possessed fimbrial structures on electron microscopy. Interestingly, 40–50% of the bacterial cells from the fresh isolates possessed these structures, and, after five serial passages on laboratory media, <1% of the cells did, demonstrating that pilus expression may be lost during passage on laboratory media. The number of pili per bacterium varies from none on non-piliated isolates to at least 100 on some piliated isolates (Scott and Old, 1981; Stull *et al.*, 1984; Gilsdorf *et al.*, 1992).

The data showing rapid loss of pili with laboratory passage and the ability of the bacteria to regain pilus expression with red blood enrichment confirms that, like the pili of other Gram-negative organisms, *H. influenzae* pili are phase variable. Environmental factors that may regulate the phase variation of pilus expression in this organism have not been identified.

6.3 *HAEMOPHILUS INFLUENZAE* PILUS STRUCTURE

Brinton *et al.* (1989) described four families of *H. influenzae* pili on the basis of their appearance on electron microscopy and characterized them on the basis of width and length. Only pili of the long, thick phenotype (LKP) were associated with haemagglutination. Electron microscopic measurements have shown haemagglutinating pili to be 4.7–18.0 nm in diameter and 200–1500 nm in length (Scott and Old, 1981; Stull *et al.*, 1984; Guerina *et al.*, 1985) (Fig. 6.1). When examined by most investigators by electron microscopy, the pili appear in a uniform, peritrichous distribution around the bacterial surface (Scott and Old, 1981; Pichichero *et al.*, 1982; Stull *et al.*, 1984; Brinton *et al.*, 1989; Gilsdorf *et al.*, 1992). Apicella *et al.* (1984), however, described polar distribution of pili.

Comparison of piliated and non-piliated variants of the same *H. influenzae* strains by sodium dodecyl sulphate-polyacrylamide gel electrophoresis (SDS-PAGE) analysis of their outer membrane proteins revealed an approximately 24 kDa band unique to piliated strains (Pichichero *et al.*, 1982; Stull *et al.*, 1984; Gilsdorf *et al.*, 1989a, 1992), which represents pilin, the major structural subunit of pili. Pili have been purified by mechanically shearing them from the bacterial cells, followed by repeated cycles of pilus precipitation at low pH and pilus resolubilization at high pH, on the basis of the ability of soluble pili to reassemble into insoluble pili under pilus-specific ionic strength and pH conditions (Stull *et al.*, 1984; Guerina *et al.*, 1985; Brinton *et al.*, 1989; Armes and Forney, 1990). Amino acid analysis of purified pili of *H. influenzae* strain Eagan revealed the pilin subunits to possess a molecular mass of 21 152 Da with 196 amino acid residues (Armes and

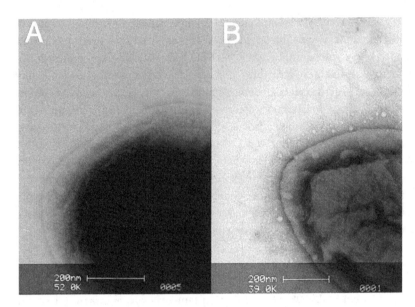

Figure 6.1. Transmission electron micrographs of *H. influenzae* strains stained with phosphotungstic acid. (A) Non-piliated strain AAr176, 41 600×. (B) Piliated strain AAr176, 31 200×.

Forney, 1990). This sequence determination was corroborated by the amino acid sequence derived by gene analysis (Forney *et al.*, 1991). These data are consistent with the structural models of other bacterial pili, in which pilus subunits (pilins) are assembled into long, multimeric pilus structures (pili) on the bacterial surface.

On negative stain electron microscopy of pili of *H. influenzae* strain Eagan, Stull *et al.* (1984) observed that pili possess a hollow core. On high magnification, quick-freeze, deep-etch electron microscopy, St Geme *et al.* (1996b) showed that pili are two-stranded, helical structures with regularly spaced horizontal striations with cross-over repeats. In addition, they possess a short, thin fibrillum at the tip, reminiscent of the fibrillar structure at the tip of *Escherichia coli* pap pili.

Early analysis of *H. influenzae* pili suggested that these proteins are complex structures, composed of three proteins (McCrea *et al.*, 1994, 1997; Watson *et al.*, 1994; St Geme *et al.*, 1996b). Pilins (also known as HifA) are the major structural subunit. HifD proteins are located at the pilus tip in short, thin, fibrillar structures, and HifE, also located at the pilus tip, possesses the epithelial cell adhesive domain (McCrea *et al.*, 1997).

Haemophilus influenzae pilins are members of so-called class I major fimbrial subunits (Low *et al.*, 1992), and, as such, possess the structural features common to these proteins, including two similarly placed cysteine residues in the N-terminal region (McCrea *et al.*, 1994). Recent analysis of the predicted two-dimensional structure of pilins of a variety of Gram-negative strains (Girardeau *et al.*, 2000) demonstrated that *H. influenzae* pilins are structurally related to the other class I major fimbrial subunits, sharing similarities in their secondary structure, and comprise one (type IE) of seven subfamilies of these fimbrial subunits.

Biogenesis of *H. influenzae* pili, like that of other members of the class I family such as type 1 and *pap* pili of *E. coli*, is dependent on a periplasmic chaperone (called HifB), which is a member of the *E. coli pap* pilus chaperone PapD family of immunoglobulin-like chaperones (Holmgren *et al.*, 1992; Hultgren *et al.*, 1993; St Geme *et al.*, 1996b). Although the crystal structures of the proteins involved in *H. influenzae* pilus biogenesis have not been solved to date, presumably the structure of HifB will resemble that of PapD (Sauer *et al.*, 1999) and will function similarly to PapD in pilus biogenesis, with the C-termini of the structural subunits binding to the conserved crevice of the PapD cleft via β-zipper interactions. In this configuration, PapD caps the interactive surfaces of the subunits, preventing their inappropriate aggregation and proteolytic degradation (Kuehn *et al.*, 1991, 1993). Crystal structural analysis of PapD complexed with PapK (a tip adapter of *E. coli* Pap pili) demonstrated both donor strand complementation, in which the G1β strand of PapD completes the immunoglobulin-like folding of the subunit by providing the integral G7β strand, as well as donor strand exchange, in which subunit strands complete the folding of neighbouring subunits (Sauer *et al.*, 1999; Barnhart *et al.*, 2000). Modelling of the *H. influenzae* HifA subunit structure based on the *E. coli* PapD-PapK structure showed high levels of sequence conservation in regions predicted to interact with HifB, the *H. influenzae* chaperone (Girardeau *et al.*, 2000; Krasan *et al.*, 2000).

6.4 FUNCTION OF *HAEMOPHILUS INFLUENZAE* PILI IN ADHERENCE AND COLONIZATION

The ability of *H. influenzae* to agglutinate human erythrocytes was first demonstrated in 1950 (Davis *et al.*, 1950), and red cell agglutination was subsequently shown to correlate strongly with adherence to human buccal epithelial cells and with the presence of pili detected by electron microscopy (Scott and Old, 1981; Guerina *et al.*, 1982). In 1986, van Alphen *et al.* identified the Anton antigen (now called AnWj antigen) as the red cell receptor for piliated *H. influenzae*. AnWj is borne on CD44, a member of the cartilage link protein family that is expressed on almost all tissues (Telen, 1995).

AnWj is present on the red cells of most individuals, except those possessing the dominant $In(Lu)$ gene, which inhibits its expression (Gilsdorf et al., 1989b). Expression of AnWj by red cells is developmentally regulated, as it is not expressed by cord erythrocytes (van Alphen et al., 1986), is present in only a minority of babies during the first 15 days of age, and is present in the majority of babies by age 40 days. AnWj, or a similar molecule that also binds *Haemophilus influenzae* pili, is also expressed on the buccal epithelial cells of human individuals whose erythrocytes express AnWj (van Alphen et al., 1987). *H. influenzae* possessing pili of varying antigenic specificities recognize the same red cell receptor, AnWj, suggesting that the adhesin is conserved among antigenically distinct pili (Gilsdorf et al., 1989b).

Although the red cell and buccal cell pilus receptors appear to be related to the AnWj and CD44 antigens, the precise pilus-binding sites on human cells remain to be defined. Several investigators (van Alphen et al., 1991; Gilsdorf et al., 1996) have demonstrated that sialylated glycosphingolipids, including the gangliosides GM1, GM2, GM3 and GD1a, inhibit pilus-mediated binding of the organism to human respiratory cells. In addition, Fakih et al. (1997) showed by thin-layer chromatography that piliated *H. influenzae* did not bind to desialylated HEp-2 gangliosides. Thus *H. influenzae* pilus receptors on human epithelial cells appear to possess sialylated ganglioside moities. Electron microscopy of human respiratory tissues in organ culture that were infected with piliated and non-piliated *H. influenzae* revealed that piliated strains preferentially attached to non-ciliated cells (Loeb et al., 1988; Farley et al., 1990) and to damaged cells (Read et al., 1991).

Early studies of adherence of *H. influenzae* documented the ability of pili to mediate binding of the organism to human respiratory cells, using human buccal epithelial cells as the model (Guerina et al., 1982; Gilsdorf and Ferrieri, 1984). Subsequent adherence studies have demonstrated cell type specificity of pilus-mediated binding: pili facilitate binding of the organism to human buccal epithelial cells and human bronchial epithelial cells but not to HeLa cells (of human cervical carcinoma origin), A549 cells (of human alveolar epithelial carcinoma origin), human nasal epithelial cells, human tracheal fibroblasts, KB cells (of human oral epithelial carcinoma origin), HEp-2 cells (of human laryngeal carcinoma origin), Chang epithelial cells (of human conjunctival origin), HaCaT cells (human keratinocytes), and human foreskin fibroblasts (Sable et al., 1985; Gilsdorf et al., 1996; St Geme and Cutter, 1996). Furthermore, pili do not mediate adherence to animal cells, including Buffalo green monkey cells, Madin–Darby canine kidney cells, mink epithelial cells, rat buccal or nasal epithelial cells (Kaplan et al., 1983;

Sable *et al.*, 1985; Gilsdorf *et al.*, 1996), which is synchronous with the observation that *H. influenzae* lives only in the respiratory tracts of humans.

In addition to its role in adherence to human respiratory cells, pili have also been shown to facilitate adherence of the organism to human respiratory mucins (Kubiet *et al.*, 2000), which may be an additional mechanism by which the organism adheres to respiratory tract tissues.

Although the contribution of pili to the adherence of *H. influenzae* to respiratory epithelial cells has been well established, their contribution to the establishment or maintenance of respiratory tract colonization is more difficult to document, particularly in the absence of adherence, a prerequisite of colonization, to respiratory cells of animal origin. Nevertheless, pili have been shown to facilitate the initiation of nasopharyngeal colonization of rats (Anderson *et al.*, 1985) and rhesus monkeys (Weber *et al.*, 1991).

6.5 FUNCTION OF PILI IN DISEASE DUE TO *HAEMOPHILUS INFLUENZAE*

Haemophilus influenzae appears to be able to invade respiratory tissues by both intracellular and intercellular routes, but pili seem to play little role in these processes (Farley *et al.*, 1990; Gilsdorf *et al.*, 1996; Ketterer *et al.*, 1999). Virkola *et al.* (2000), however, showed that pili from a type b strain of the organism bound to the heparin-binding extracellular matrix proteins fibronectin and heparin-binding growth-associated molecule. Indeed, following intranasal, intraperitoneal or intravascular inoculation of the organism, rats receiving non-piliated cells were more likely to develop bacteraemia and meningitis, and to die than those receiving piliated cells (Kaplan *et al.*, 1983; Gutierrez *et al.*, 1990; Miyazaki *et al.*, 1999). Furthermore, the organisms recovered from the blood of animals inoculated with piliated strains were non-piliated. This observation suggests that piliated cells are cleared from the bloodstream more efficiently than are non-piliated cells. Two possible mechanisms have been proposed to explain this increased clearance – increased antibody killing of piliated cells and increased opsonophagocytosis of piliated cells. Indeed, non-piliated cells have been shown to be more resistant to human serum and to bind complement less efficiently than do piliated cells (Miyazaki *et al.*, 1999). In addition, antibodies directed against non-pilus surface antigens bind to piliated *H. influenzae* more readily than to non-piliated cells, and these antibodies show greater complement-mediated bactericidal activity against piliated bacteria than against non-piliated organisms (Gilsdorf *et al.*, 1993). Also, piliated *H. influenzae* stimulates enhanced opsonization-dependent phagocytosis by neutrophils as compared with non-piliated cells (Tosi *et al.*, 1985). Thus, although pili appear to play an important role in the adherence of the organism to the human respiratory tract and possibly to facilitate respiratory tract colonization, they appear to play little or no role in invasion of the respiratory tract and to predispose the organisms to increased serum and neutrophilic killing.

Few studies have investigated the role of pili in the development of acute otitis media due to *H. influenzae*. Recently, however, Melhus *et al.* (1998) showed that pili appeared to play no role in the frequency or intensity of acute otitis media in a rat model using direct instillation of the organism into the middle ear space.

While the major role of pili in respiratory cell adherence is, most likely, to prevent the organism from being washed away by respiratory secretions, pilus-mediated adherence to respiratory cells may stimulate other biological processes related to immunity or disease. Clemans *et al.* (2000) demonstrated that respiratory epithelial cells respond to stimulation with the organism by secreting interleukins 6 and 8. Although lipopolysaccharide (LOS) accounted for only a portion of the stimulation, bacteria–epithelial cell interactions with the adhesins pili, Hia, HMW 1 or 2, or Hap could not explain the non-LOS stimulated response.

6.6 GENETICS OF *HAEMOPHILUS INFLUENZAE* PILI

Haemophilus influenzae pili are generated from the products of five genes, *hifA* to *E*, located contiguously in the pilus gene cluster between homologues of the *E. coli* housekeeping genes *purE* and *pepN* (van Ham *et al.*, 1989, 1994; Watson *et al.*, 1994; McCrea *et al.*, 1997; Gilsdorf, 1998) (Fig. 6.2). Most piliated strains possess one copy of the gene cluster; the exception is the *H. influenzae* biogroup aegyptius (Hae) strains, which are associated with conjunctivitis and include the Brazilian purpuric fever strains. Hae strains possess two complete pilus gene clusters (Read *et al.*, 1996), although variation in individual genes between these copies occurs (Read *et al.*, 1998). One copy is inserted in the same genomic location as the gene clusters of other *H. influenzae* strains – between the *E. coli purE* and *pepN* homologues – while the other copy is inserted between the *E. coli pmbA* and *hpt* homologues. The genomic DNA of *H. influenzae* strain Rd has been completely sequenced and does not contain the pilus gene cluster (Fleischmann *et al.*, 1995).

hifA, located upstream from *purE*, is approximately 650 bp in length and encodes the major pilus subunit pilin protein HifA. Each pilin possesses a leader sequence of between 18 and 20 amino acid residues, followed by a

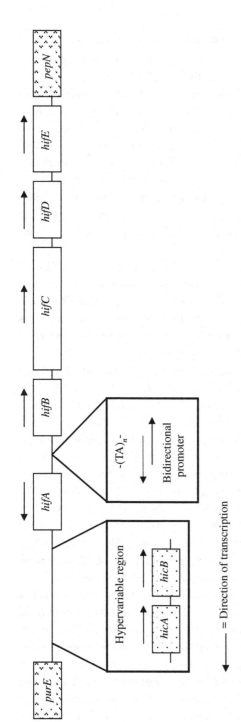

Figure 6.2. The *hif* pilus gene cluster of *H. influenzae*.

mature protein of between 191 and 196 amino acid residues (Clemans *et al.*, 1998). *Haemophilus influenzae* pilins show sequence homology to pilins of *E. coli* F17, type 1c, and Pap pili as well as pili of *Klebsiella pneumoniae, Bordetella pertussis* and *Serratia marcescens* (Gilsdorf *et al.*, 1990a). Among different strains of the organism, *hifA* shows significant amino acid sequence heterogeneity, ranging from 59% to 100% identity by pairwise comparison (Clemans *et al.*, 1998). When *hifA*s of 26 strains were subjected to restriction fragment length polymorphism (RFLP) analysis using the restriction enzymes *AluI* and *RsaI*, 10 unique patterns were identified, with moderate correlation between serotype and *hifA* genotype. On the basis of the adherence of *E. coli* transformed with *hifA* gene sequences, some investigators suggest that an epithelial cell-binding region of *H. influenzae* pili resides on HifA (van Ham *et al.*, 1995).

Upstream from *hifA*, but transcribed in the opposite direction, is *hifB*, which is 711 bp in length and encodes a 214 amino acid residue protein with strong nucleotide sequence homology to the immunoglobulin-like *E. coli* chaperone protein, PapD (Hultgren *et al.*, 1993). Like other bacterial pilus chaperones, HifB binds to pilus subunits, including pilin (HifA) and HifD (St Geme *et al.*, 1996b), capping these proteins to prevent premature assembly and stabilizing them against degradation. HifB possesses the Arg-8 and Lys-112 residues critical to the binding pocket domain that lends stability to the chaperone–subunit complex (Kuehn *et al.*, 1993). *Haemophilus influenzae* with mutations in *hifB* express neither pilin (HifA) in periplasmic and whole cell lysates nor pili on their surfaces and lack haemagglutinating activity (McCrea *et al.*, 1994; van Ham *et al.*, 1994; Watson *et al.*, 1994; St Geme *et al.*, 1996b). HifB proteins show a high level of amino acid sequence homology (95–98%) among different strains of the organism (Gilsdorf *et al.*, 1997).

Downstream from *hifB*, and transcribed in the same direction, is a 2514 bp open reading frame, *hifC*, that encodes a 837 amino acid residue protein with sequence homology to members of the pilus assembly platform (usher) protein family, including FhaA of *Bordetella pertussis*, MrkC of *Klebsiella pneumoniae*, and FimD and PapC of *E. coli* (Watson *et al.*, 1994). The usher proteins escort pilus subunits from chaperone–subunit complexes into pili as they become assembled, directing the correct order of the subunits (Dodson *et al.*, 1993). Mutations in *hifC* result in bacteria that produce pilins (HifAs), but not pili. In addition, these mutants do not agglutinate human red cells, suggesting that the function of the pilus adhesin has been altered (Watson *et al.*, 1994), possibly by failure of the HifE adhesin to be properly incorporated into the assembling pili. HifC proteins show a high level of amino acid sequence homology (97–98%) among strains of the organism (Gilsdorf *et al.*, 1997).

Immediately (16 bp) downstream from *hifC* is a 645 bp open reading frame, *hifD*, that encodes a 216 amino acid residue lipoprotein-like protein (van Ham *et al.*, 1994). Its gene product, HifD, is incorporated into pili, and is located at the pilus tip (St Geme *et al.*, 1996b; McCrea *et al.*, 1997). Mutants possessing in-frame deletions in *hifD* express pilins (HifAs), but are unable to assemble pilins into pili or to haemagglutinate (McCrea *et al.*, 1994). HifD has significant amino acid sequence homology to HifA, but not to pilus proteins of other organisms (McCrea *et al.*, 1994). The function of HifD has not been defined, but may serve adapter/initiator functions, similar to those of the PapK and PapF pilus tip proteins of *E. coli* (Hultgren *et al.*, 1993). HifD proteins show a modest level of amino acid sequence homology (74–92%) among different strains of the organism (McCrea *et al.*, 1998).

Downstream from *hifD* (27 bp) is a 1305 bp open reading frame, *hifE*, that encodes a 435 amino acid residue protein located at the pilus tip (McCrea *et al.*, 1994; 1997). Mutations in *hifE* result in mutants that express pilin, show reduced pilus expression, and exhibit no haemagglutination nor buccal epithelial cell adherence (McCrea *et al.*, 1994; van Ham *et al.*, 1994). These observations, plus the data showing that antibodies raised against HifE completely block binding of piliated *H. influenzae* to human red cells while antisera against HifA only partially blocks red cell binding (McCrea *et al.*, 1997), provide compelling evidence that the epithelial cell-binding domain resides in HifE. HifE proteins show a modest level of amino acid sequence homology (35–99%) among strains of the organism (McCrea *et al.*, 1998; Read *et al.*, 1998).

Hif A, D and E, the subunits of mature, surface-assembled pili, show relatively high amino acid sequence conservation at their C-termini, and possess the chaperone-binding motif common to class I pili (Hacker, 1990; Low *et al.*, 1992; McCrea *et al.*, 1994), including the penultimate tyrosine residues that fit into the cleft between the two chaperone β-barrel domains and the alternating hydrophobic residues in the C-terminal domain that, in interactions with complementary structures on the chaperone, form the β-zipper motif (McCrea *et al.*, 1994). Mutation of the penultimate tyrosine residue of HifA abrogates its binding to HifB (Krasan *et al.*, 2000). In addition, Hif A, D and E all possess the glycine located 14 residues from their C-termini that appears to be important in subunit–subunit interactions (Clemans *et al.*, 1998; McCrea *et al.*, 1998). Mutation of this glycine residue of HifA abrogates HifA oligomerization, but has no impact on HifA-HifB binding (Krasan *et al.*, 2000).

Three recent studies have detailed by DNA sequence analysis the genomic regions surrounding the pilus gene cluster from a variety of *H. influenzae*

(Geluk *et al.*, 1998; Mhlanga-Mutangadura *et al.*, 1998; Read *et al.*, 2000). All of these analyses have documented the hypervariable, complex genetic organization of these regions, identifying insertions and/or deletions and multiple direct and inverted repeat sequences. In addition, several strains displayed partial deletions of HifE (seen in all type f strains tested) and HifD. The association of these genetic rearrangements with repeat sequences suggest that these sequences may be recombination hot spots (Geluk *et al.*, 1998; Read *et al.*, 2000). Analysis of C + G content of the pilus gene cluster shows it to be 39%, similar to that of the *H. influenzae* genome (Fleischmann *et al.*, 1995), suggesting that it is not a 'pathogenicity island', nor has it been incorporated recently, in evolutionary terms, into *H. influenzae* from another bacterial species. Approximately 90% of strains of the organism possess open reading frames, named *hicA* and *hicB*, located between *purE* and *hifA* (Mhlanga-Mutangadura *et al.*, 1998; Read *et al.*, 2000); whether the gene products of these coding regions are expressed is unclear, as are their function.

Like some other bacterial pili, *H. influenzae* pili exhibit phase variation, with a piliated to non-piliated switch frequency of 3×10^{-4} per bacterium per generation and a non-piliated to piliated switch frequency of 7×10^{-4} per bacterium per generation (Farley *et al.*, 1990). The genetic mechanism governing pilus phase variation in this organism (van Ham *et al.*, 1993) appears to be slipped-strand mispairing occurring at a sequence of TA repeats in the bidirectional pilus promoter located between *hifA* and *hifB*. *Haemophilus influenzae* possessing 10, 11, or 12 TA repeats express pili, while those expressing nine repeats do not. Geluk *et al.* (1988) showed that red cell enrichment of non-piliated *H. influenzae* possessing nine TA repeats yielded piliated variants that possessed 10 TA repeats. In addition, one strain possessing four TA repeats was non-piliated. These data suggest that the overlapping -10 and -35 sequences in this region are optimally oriented for efficient transcription when the promoter region contains 10 to 12 TA repeats and fails to transcribe with nine or fewer TA repeats.

6.7 STRAIN TO STRAIN VARIATION OF *HAEMOPHILUS INFLUENZAE* PILI

Haemophilus influenzae pili, like other surface structures of the organism, differ greatly from strain to strain, a characteristic that may facilitate avoidance of local immunity and allow its survival in its environmental niche, which is restricted to the antibody-laden, human nasopharyngeal mucosa (Gilsdorf, 1998; Weiser, 2000). Rather than responding to changing environmental cues through two-component or other complex regulatory systems, the organism appears to rely on the principles of population genetics and evolutionary selection to survive. The large variety of genetic rearrangements described for the pilus gene cluster suggests a stochastic process in which gene rearrangements occur frequently, potentially arming each population of organisms with a number of variant types that possess survival advantage under varied circumstances.

Non-piliated organisms may lack the entire pilus gene cluster, may possess deletions in the pilus cluster genes, or may lack the optimal number of TA repeats in the *hifA-hifB* promoter region to effectively allow transcription of the pilus genes (i.e. in the phase-variable 'off' position) (Geluk *et al.*, 1998; Mhlanga-Mutangadura *et al.*, 1998; Read *et al.*, 2000). Type f strains, which appear to be fairly clonal on the basis of their HifA sequences (Clemans *et al.*, 1998), possess a conserved deletion encompassing most of *hifE* (Read *et al.*, 2000). On the other hand, Brazilian purpuric fever strains, a subset of *H. influenzae* biogroup aegyptius, show complete duplication of the pilus gene cluster, which explains why these strains are commonly piliated – phase variation of both clusters to the 'off' motif is required for the non-piliated phenotype. The circumstances for selection and persistence of these genetic mutations are unclear. Data showing that piliated isolates are selected against during mucosal colonization and bacteraemia (Kaplan *et al.*, 1983; Weber *et al.*, 1991; Miyazaki *et al.*, 1999) suggest that pili may play a role in transmission from host to host or in the early stages of microbial adherence to, and colonization of, mucosal sites.

In addition to heterogeneity of pilus genes from strain to strain, the organism demonstrates mosaicism among genes in the pilus cluster. Comparison of RFLP patterns of *hifA* and *hifD/E* from a number of strains revealed discordance – i.e. among strains whose *hifA* patterns were alike, their *hifE* patterns were variable and vice versa (McCrea *et al.*, 1998). This apparent 'mixing and matching' of pilus genes may be explained by horizontal gene transfer in which DNA is taken up by the organism by natural transformation (a well-known characteristic of *H. influenzae*) and incorporated into the bacterial genome by homologous recombination. Data showing similar rearrangements in the *purE-hifA* region among strains from disparate genetic backgrounds, such as typeable and non-typeable strains, as well as very different rearrangements in this region among type b strains that have identical *hifE* genes, is further evidence of possible horizontal gene transfer (Read *et al.*, 2000). Mosaicism within pilus genes, facilitated by horizontal transfer, may explain regional similarities of *hifE* gene sequence among strains with different leader sequences (McCrea *et al.*, 1998).

Besides demonstrating genetic heterogeneity, pili also demonstrate antigenic heterogeneity. Antibodies raised to Hib strain Eagan denatured pilins (HifA) – the 24 kDa protein cut from SDS-PAGE gels – bound to pilins of all type b and non-typeable strains tested (Gilsdorf et al., 1992). On the other hand, antibodies raised to Hib strain Eagan native pili (which are directed at pilin/ HifA) (Forney et al., 1992; McCrea et al., 1997) bound to 50% of 22 piliated Hib strains, while antibodies raised to Hib strain M43 native pili bound to 82% of 22 piliated Hib strains; neither of these antisera bound to 12 piliated non-typeable strains (Gilsdorf et al., 1992). Furthermore, these antibodies raised to native pili did not bind to denatured pilins of any strains, including the homologous strains, on Western blot assays, demonstrating that their antigenic specificity was directed at conformational epitopes not present on the denatured antigens (Gilsdorf et al., 1990b).

Antisera raised to native HifD and HifE of Hib strain Eagan bound to all piliated type b strains and to none of the non-typeable strains tested, suggesting that HifE epitopes are conserved on type b strains and heterogeneous on non-typeable strains (McCrea et al., 1998). This finding corroborates the finding that the RFLP patterns of hifEs from type b strains were identical, while those from non-typeable strains were variable (McCrea et al., 1998).

6.8 *HAEMOPHILUS INFLUENZAE* PILUS VACCINES

Because pili are located on the bacterial surface, are immunogenic, and are the target of bactericidal antibodies (LiPuma and Gilsdorf, 1988), they are attractive candidates for vaccines that inhibit bacterial adherence to the respiratory tract and/or stimulate bactericidal antibodies, thus preventing infection and disease. Indeed, antibodies against *H. influenzae* pili inhibit the adherence of strains possessing immunologically homologous pili (LiPuma and Gilsdorf, 1988; Forney et al., 1992). In animal studies using purified pili as the immunogen, 40% of vaccinated chinchillas, compared with 93% of control animals, developed otitis media after intrabullar challenge with the homologous strain (Karasic et al., 1989). Similarly, after nasopharyngeal challenge, 17% of vaccinated chinchillas, compared with 75% of control animals, developed otitis media. This vaccine has not progressed in development; on the basis of antibody binding data, it would not be expected to protect against challenge with heterologous strains.

6.9 THE FUTURE

Haemophilus influenzae is a highly adaptable organism, well suited for survival in its unique environment. The genetic mechanisms that foster this adaptability are being defined. Pili, which exhibit many of the basic mechanisms that underly the diversity of the organism's pathogenic factors, constitute an excellent model for further understanding important evolutionary aspects of bacterial survival.

Over the past two decades, considerable progress has been made in understanding the structure and function of *H. influenzae* pili. Yet, important questions remain. What is the epithelial cell receptor to which pili bind? What is the epithelial cell-binding domain on HifE? Do pili, or their peptides, have a place as vaccine antigens? What host epithelial cell responses do pili stimulate, if any? What are the forces that drive the dramatic heterogeneity of pili? What are the bacterial mechanisms that facilitate this heterogeneity? Future studies will undoubtedly lead to answers to these important questions.

(153)

ACKNOWLEDGEMENTS

The author is supported, in part, by Public Health Service grant AI25630 from the National Institute of Allergy and Infectious Diseases and would like to thank Dr Carl Marrs, Dr Kirk McCrea, and Dr Graham Krasan for insightful comments.

REFERENCES

Anderson, P.W., Pichichero, M.E. and Connor, E.M. (1985). Enhanced nasopharyngeal colonization of rats by piliated *Haemophilus influenzae* type b. *Infection and Immunity* 48, 565–568.

Apicella, M.A., Shero, M., Dudas, K.C., Stack, R.R., Klohs, W., LaScolea, L.J., Murphy, T.F. and Mylotte, J.M. (1984). Fimbriation of *Haemophilus* species isolated from the respiratory tract of adults. *Journal of Infectious Diseases* 150, 40–43.

Armes, L.G. and Forney, L.J. (1990). The complete primary structure of pilin from *Haemophilus influenzae* type b strain Eagan. *Journal of Protein Chemistry* 9, 45–52.

Bakaletz, L.O., Tallan, B.M., Hoepf, T., DeMaria, T.F., Birck, H.G. and Lim, D.J. (1988). Frequence of fimbriation of nontypable *Haemophilus influenzae* and its ability to adhere to chinchilla and human respiratory epithelium. *Infection and Immunity* 56, 331–335.

Barenkamp, S.J. and St Geme, J.W. III (1996). Identification of a second family of high-molecular-weight adhesion proteins expressed by non-typable *Haemophilus influenzae*. *Molecular Microbiology* 19, 1215–1223.

Barnhart, M.M., Pinkner, J.S., Soto, G.E., Sauer, F.G., Langermann, S., Waksman, G., Frieden, C. and Hultgren S.J. (2000). PapD-like chaperones provide the missing information for folding of pilin proteins. *Proceedings of the National Academy of Sciences, USA* **97**, 7709–7714.

Bou, R., Dominguez, A., Fontanals, D., Sanfeliu, I., Pons, I., Renau, J., Pineda, V., Lobera, E., Latorre, C., Majo, M. and Salleras, L. (2000). Prevalence of *Haemophilus influenzae* pharyngeal carriers in the school population of Catalonia. *European Journal of Epidemiology* **16**, 521–526.

Brinton, C. C. Jr, Carter, M.J., Derber, D.B., Kar, S., Kramarik, J.A., To, A.C., To, S.C. and Wood, S.W. (1989). Design and development of pilus vaccines for *Haemophilus influenzae* diseases. *Pediatric Infectious Disease Journal* **8**, S54–S61.

Busse, J., Hartmann, E. and Lingwood, C.A. (1997). Receptor affinity purification of a lipid-binding adhesion from *Haemophilus influenzae*. *Journal of Infectious Diseases* **175**, 77–83.

Clemans, D.L., Marrs, C.F., Patel, M., Duncan, M. and Gilsdorf, J. R. (1998). Comparative analysis of *Haemophilus influenzae hifA* (pilin) genes. *Infection and Immunity* **66**, 656–663.

Clemans, D.L., Bauer, R.J., Hanson, J.A., Hobbs, M.V., St Geme, J.W. III, Marrs, C.F. and Gilsdorf, J.R. (2000). Induction of proinflammatory cytokines from human respiratory epithelial cells after stimulation by nontypeable *Haemophilus influenzae*. *Infection and Immunity* **68**, 4430–4440.

Connor, E.M. and Loeb, M.R. (1983). A hemadsorption method for detection of colonies of *Haemophilus influenzae* type b expressing fimbriae. *Journal of Infectious Diseases* **148**, 855–860.

Davis, D.J., Pittman, M. and Griffitts, J.J. (1950). Haemagglutination by the Koch–Weeks bacillus (*Hemophilus aegyptius*). *Journal of Bacteriology* **59**, 427–431.

Dodson, K.W., Jacob-Dubuisson, F., Striker, R.T. and Hultgren, S.J. (1993). Outer-membrane PapC molecular usher discriminately recognizes periplasmic chaperone–pilus subunit complexes. *Proceedings of the National Academy of Sciences, USA* **90**, 3670–3674.

Faden, H., Duffy, L., Williams, A., Krystofik, D.A. and Wolf, J. (1995). Epidemiology of nasopharyngeal colonization with nontypeable *Haemophilus influenzae* in the first 2 years of life. *Journal of Infectious Diseases* **172**, 132–135.

Fakih, M.G., Murphy, T. F., Pattoli, M.A. and Berenson, C.S. (1997). Specific binding of *Haemophilus influenzae* to minor gangliosides of human respiratory epithelial cells. *Infection and Immunity* **65**, 1965–1700.

Farley, M.M., Stephens, D.S., Kaplan, S.L. and Mason, E.O. Jr (1990). Pilus-and

J. R. GILSDORF

non-pilus-mediated interactions of *Haemophilus influenzae* type b with human erythrocytes and human nasopharyngeal mucosa. *Journal of Infectious Diseases* **161**, 274–280.

Fleischmann, R.D., Adams, M.D., White, O., Clayton, R.A., Kirkness, E.F., Kerlavage, A.R., Bult, C.J., Tomb, J.F., Dougherty, B.A. and Merrick, J.M. (1995). Whole-genome random sequencing and assembly of *Haemophilus influenzae*. *Science* **269**, 496–512.

Fontanals, D., Bou, R., Pons, I., Sanfeliu, I., Dominguez, A., Pineda, V., Renau, J., Munoz, C., Latorre, C. and Sanches, F. (2000). Prevalence of *Haemophilus influenzae* carriers in the Catalan preschool population. *European Journal of Clinical Microbiology and Infectious Diseases* **19**, 301–304.

Forney, L.J., Marrs, C.F., Bektesh, S.L. and Gilsdorf, J.R. (1991). Comparison and analysis of the nucleotide sequences of pilin genes from *Haemophilus influenzae* type b strains Eagan and M43. *Infection and Immunity* **59**, 1991–1996.

Forney, L.J., Gilsdorf, J.R. and Wong, D.C. (1992). Effect of pili-specific antibodies on the adherence of *Haemophilus influenzae* type b to human buccal cells. *Journal of Infectious Diseases* **165**, 464–470.

Geluk, F., Eijk, P.P., van Ham, S.M., Jansen, H.M. and van Alphen, L. (1998). The fimbria gene cluster of nonencapsulated *Haemophilus influenzae*. *Infection and Immunity* **66**, 406–417.

Gilsdorf, J. R. (1998). Antigenic diversity and gene polymorphisms in *Haemophilus influenzae*. *Infection and Immunity* **66**, 5053–5059.

Gilsdorf, J.R. and Ferrieri, P. (1984). Adherence of *Haemophilus influenzae* to human epithelial cells. *Scandinavian Journal of Infectious Diseases* **16**, 271–278.

Gilsdorf, J.R., Forney, L.J. and LiPuma, J. J. (1989a). Reactivity of *Haemophilus influenzae* type b anti-pili antibodies. *Microbial Pathogenesis* **7**, 311–316.

Gilsdorf, J.R., Judd, W.J. and Cinat, M. (1989b). Relationship of *Haemophilus influenzae* type b pilus structure and adherence to human erythrocytes. *Infection and Immunity* **57**, 3259–3260.

Gilsdorf, J.R., Marrs, C.F., McCrea, K.W. and Forney, L.J. (1990a). Cloning, expression, and sequence analysis of the *Haemophilus influenzae* type b strain M43p+ pilin gene. *Infection and Immunity* **58**, 1065–1072.

Gilsdorf, J.R., McCrea, K.W. and Forney, L. (1990b). Conserved and nonconserved epitopes among *Haemophilus influenzae* type b pili. *Infection and Immunity* **58**, 2252–2257.

Gilsdorf, J.R., Chang, H.Y., McCrea, K.W. and Bakaletz, L.O. (1992). Comparison of haemagglutinating pili of *Haemophilus influenzae* type b with similar structures of nontypeable H. influenzae. *Infection and Immunity* **60**, 374–379.

Gilsdorf, J.R., Tucci, M., Forney, L.J., Watson, W., Marrs, C.F. and Hansen, E.J. (1993). Paradoxical effect of pilus expression on binding of antibodies by *Haemophilus influenzae*. *Infection and Immunity* **61**, 3375–3381.

Gilsdorf, J.R., Tucci, M. and Marrs, C.F. (1996). Role of pili in *Haemophilus influenzae* adherence to, and internalization by, respiratory cells. *Pediatric Research* **39**, 343–348.

Gilsdorf, J.R., McCrea, K.W. and Marrs, C.F. (1997). Role of pili in *Haemophilus influenzae* adherence and colonization. *Infection and Immunity* **65**, 2997–3002.

Girardeau, J.P., Bertin, Y. and Callebaut, I. (2000). Conserved structural features in class I major fimbrial subunits (Pilin) in gram-negative bacteria. Molecular basis of classification in seven subfamilies and identification of intrasubfamily sequence signature motifs which might be implicated in quaternary structure. *Journal of Molecular Evolution* **50**, 424–442.

Gratten, M.J., Montgomery, G.G., Gratten, H., Siwi, H., Poli, A. and Koki, G. (1989). Multiple colonization of the upper respiratory tract of Papua New Guinea children with *Haemophilus influenzae* and *Streptococcus pneumoniae*. *Southeast Asian Journal of Tropical Medicine and Public Health* **20**, 501–509.

Guerina, N.G., Langermann, S., Clegg, H.W., Kessler, T.W., Goldman, D.A. and Gilsdorf, J.R. (1982). Adherence of piliated *Haemophilus influenzae* type b to human oropharyngeal cells. *Journal of Infectious Diseases* **146**, 564.

Guerina, N.G., Langermann, S., Schoolnik, G.K., Kessler, T.W. and Goldmann, D.A. (1985). Purification and characterization of *Haemophilus influenzae* pili, and their structural and serological relatedness to *Escherichia coli* P and mannose-sensitive pili. *Journal of Experimental Medicine* **161**, 145–159.

Gutierrez, M.K., Joffe, L.S., Forney, L.J. and Glode, M.P. (1990). Effect of pili on the frequency of bacteremia following challenge with *Haemophilus influenzae* type b (HIB) in the infant rat. *Abstracts of the 30th ICAAC American Society for Microbiology*, Washington, DC **4**, 93.

Hacker, J. (1990). Genetic determinants coding for fimbriae and adhesins of extraintestinal *Escherichia coli*. *Current Topics in Microbiology and Immunology* **151**, 1–27.

Harabuchi, Y., Faden, H., Yamanaka, N., Duffy, L., Wolf, J. and Krystofik, D. (1994). Nasopharyngeal colonization with nontypeable *Haemophilus influenzae* and recurrent otitis media. *Journal of Infectious Diseases* **170**, 862–866.

Holmgren, A., Kuehn, M.J., Branden, C.I. and Hultgren, S.J. (1992). Conserved immunoglobulin-like features in a family of periplasmic pilus chaperones in bacteria. *EMBO Journal* **11**, 1617–1622.

Hultgren, S.J., Abraham, S., Caparon, M., Falk, P., St Geme, J.W. III and Normark, S. (1993). Pilus and nonpilus bacterial adhesins: assembly and function in cell recognition. *Cell* **73**, 887–901.

Kaplan, S.L., Mason, E. O. Jr and Wiedermann, B.L. (1983). Role of adherence in the pathogenesis of *Haemophilus influenzae* type b infection in infant rats. *Infection and Immunity* **42**, 612–617.

Karasic, R.B., Beste, D.J., To, S.C., Doyle, W.J., Wood, S.W., Carter, M.J., To, A.C., Tanpowpong, K., Bluestone, C.D. and Brinton, C.C. Jr (1989). Evaluation of pilus vaccines for prevention of experimental otitis media caused by nontypable *Haemophilus influenzae*. *Pediatric Infectious Disease Journal* **81** (1 Suppl), S62–65.

Ketterer, M.R., Shao, J.Q., Hornick, D.B., Buscher, B., Bandi, V.K. and Apicella, M.A. (1999). Infection of primary human bronchial epithelial cells by *Haemophilus influenzae*: macropinocytosis as a mechanism of airway epithelial cell entry. *Infection and Immunity* **67**, 4161–4170.

Krasan, G.P., Sauer, F.G., Cutter, D., Farley, M.M., Gilsdorf, J.R., Hultgren, S.J. and St Geme, J.W. III (2000). Evidence for donor strand complementation in the biogenesis of *Haemophilus influenzae* haemagglutinating pili. *Molecular Microbiology* **35**, 1335–1347.

Kubiet, M., Ramphal, R., Weber, A. and Smith, A. (2000). Pilus-mediated adherence of *Haemophilus influenzae* to human respiratory mucins. *Infection and Immunity* **68**, 3362–3367.

Kuehn, M.J., Normark, S. and Hultgren, S.J. (1991). Immunoglobulin-like PapD chaparone caps and uncaps interactive surfaces of nascently translocated pilus subunits. *Proceedings of the National Academy of Sciences, USA* **88**, 10586–10590.

Kuehn, M.J., Ogg, D.J., Kihlberg, J., Slonim, L.N., Flemmer, K., Bergfors, T. and Hultgren S.J. (1993). Structural basis of pilus subunit recognition by the PapD chaperone. *Science* **262**, 1234–1241.

Lipsitch, M. and Moxon, E.R. (1997). Virulence and transmissibility of pathogens: what is the relationship? *Trends in Microbiology* **5**, 31–37.

LiPuma, J.J. and Gilsdorf, J.R. (1988). Structural and serological relatedness of *Haemophilus influenzae* type b pili. *Infection and Immunity* **56**, 1051–1056.

Loeb, M R., Connor, E. and Penney, D. (1988). A comparison of the adherence of fimbriated and nonfimbriated *Haemophilus influenzae* type b to human adenoids in organ culture. *Infection and Immunity* **56**, 484–489.

Low, D., Braaten, B. and VanDerWoude, M. (1992). Fimbriae. In *Escherichia coli and Salmonella typhimurium. Cellular and Molecular Biology*, ed. J. L. Ingraham, K. B. Low, B. Magasanik, M. Schaechter and H. E. Umbarger, pp. 146–157. Washington, DC: Society for Microbiology.

McCrea, K.W., Watson, W.J., Gilsdorf, J.R. and Marrs, C.F. (1994). Identification of *hifD* and *hifE* in the pilus gene cluster of *Haemophilus influenzae* type b strain Eagan. *Infection and Immunity* **62**, 4922–4928.

McCrea, K.W., Watson, W.J., Gilsdorf, J.R. and Marrs, C.F. (1997). Identification of two minor subunits in the pilus of *Haemophilus influenzae*. *Journal of Bacteriology* **179**, 4227–4231.

McCrea, K.W., St Sauver, J.M., Marrs, C.F., Clemans, D. and Gilsdorf, J.R. (1998). Immunologic and structural relationships of the minor pilus subunits among *Haemophilus influenzae* isolates. *Infection and Immunity* **66**, 4788–4796.

Melhus, A., Hermansson, A., Forsgren, A. and Prellner, K. (1998). Intra-and interstrain differences of virulence among nontypeable *Haemophilus influenzae* strains. *Acta Pathologica, Microbiologica et Immunologica Scandinavica* **106**, 858–868.

Mhlanga-Mutangadura, T., Morlin, G., Smith, A.L., Eisenstark, A. and Golomb, M. (1998). Evolution of the major pilus gene cluster of *Haemophilus influenzae*. *Journal of Bacteriology* **180**, 4693–4703.

Miyazaki, S., Matsumoto, T., Furuya, N., Tateda, K. and Yamaguchi, K. (1999). The pathogenic role of fimbriae of *Haemophilus influenzae* type b in murine bacteraemia and meningitis. *Journal of Medical Microbiology* **48**, 383–388.

Pichichero, M.E., Loeb, M., Anderson, P. and Smith, D.H. (1982). Do pili play a role in pathogenicity of *Haemophilus influenzae* type B? *Lancet* **2**, 960–962.

Read, R.C., Wilson, R., Rutman, A., Lund, V., Todd, H.C., Brain, A.P.R., Jeffrey, P.K. and Cole, P.J. (1991). Interaction of nontypable *Haemophilus influenzae* with human respiratory mucosa in vitro. *Journal of Infectious Diseases* **163**, 549–558.

Read, T.D., Dowdell, M., Satola, S.W. and Farley, M.M. (1996). Duplication of pilus gene complexes of *Haemophilus influenzae* biogroup aegyptius. *Journal of Bacteriology* **178**, 6564–6570.

Read, T.D., Satola, S. W., Opdyke, J.A. and Farley, M.M. (1998). Copy number of pilus gene clusters in *Haemophilus influenzae* and variation of the *hifE* pilin gene. *Infection and Immunity* **66**, 1622–1631.

Read, T.D., Satala, S.W. and Farley, M.M. (2000). Nucleotide sequence analysis of hypervariable junctions of *Haemophilus influenzae* pilus gene clusters. *Infection and Immunity* **68**, 6896–6902.

Sable, N.S., Connor, E.M., Hall, C.B. and Loeb, M.R. (1985). Variable adherence of fimbriated *Haemophilus influenzae* type b to human cells. *Infection and Immunity* **48**, 119–123.

Sauer, F.G., Futterer, K., Pinkner, J.S., Dodson, K.W., Hultgren, S.J. and Waksman, G. (1999). Structural basis of chaperone function and pilus biogenesis. *Science* **285**, 1058–1061.

Scott, S.S. and Old, D.C. (1981). Mannose-resistant and eluting (MRE) haemagglutinins, fimbriae and surface structure in strains of *Haemophilus*. *FEMS Microbiology Letters* **10**, 235–240.

Sirakova T. K.P., Murwin D., Billy, J., Leake, E., Lim, D., DeMaria, T. and Bakaletz, L. (1994). Role of fimbriae expressed by nontypeable *Haemophilus influenzae* in pathogenesis of and protection against otitis media and relatedness of the fimbrin subunit to outer membrane Protein A. *Infection and Immunity* **62**, 2002–2020.

St Geme, J.W. III and Cutter, D. (1995). Evidence that surface fibrils expressed by *Haemophilus influenzae* type b promote attachment to human epithelial cells. *Molecular Microbiology* **15**, 77–85.

St Geme, J.W. III and Cutter, D. (1996). Influence of pili, fibrils, and capsule on *in vitro* adherence by *Haemophilus influenzae* type b. *Molecular Microbiology* **21**, 21–31.

St Geme, J.W. III, Gilsdorf, J.R. and Falkow, S. (1991). Surface structures and adherence properties of diverse strains of *Haemophilus influenzae* biogroup aegyptius. *Infection and Immunity* **59**, 3366–3371.

St Geme, J.W. III, Falkow, S. and Barenkamp, S.J. (1993). High-molecular-weight proteins of nontypable *Haemophilus influenzae* mediate attachment to human epithelial cells. *Proceedings of the National Academy of Sciences, USA* **90**, 2875–2879.

St Geme, J.W. III, de la Morena, M.L. and Falkow, S. (1994). A *Haemophilus influenzae* IgA protease-like protein promotes intimate interaction with human epithelial cells. *Molecular Microbiology* **14**, 217–233.

St Geme, J.W. III, Cutter, D. and Barenkamp, S.J. (1996a). Characterization of the genetic locus encoding *Haemophilus influenzae* type b surface fibrils. *Journal of Bacteriology* **178**, 6281–6287.

St Geme, J.W. III, Pinkner, J.S. III, Krasan, G.P., Heuser, J., Bullitt, E., Smith, A.L. and Hultgren, S.J. (1996b). *Haemophilus influenzae* pili are composite structures assembled via the HifB chaperone. *Proceedings of the National Academy of Sciences, USA* **93**, 11913–11918.

St Sauver, J., Marrs, C.F., Foxman, B., Somsel, P., Madera, R. and Gilsdorf, J.R. (2000). Risk factors for otitis media and carriage of multiple strains of *Haemophilus influenzae* and *Streptococcus pneumoniae*. *Emerging Infectious Diseases* **6**, 1–9.

Stull, T.L., Mendelman, P.M., Haas, J.E., Schoenborn, M.A., Mack, K.D. and Smith, A. L. (1984). Characterization of *Haemophilus influenzae* type b fimbriae. *Infection and Immunity* **46**, 787–796.

Telen, M.J. (1995). Lutheran antigens, DC44-related antigens, and Lutheran regulatory genes. *Transfusion Clinique et Biologique* **2**, 291–301.

Tosi, M.F., Anderson, D.C., Barrish, J., Mason, E.O. Jr and Kaplan, S.L. (1985). Effect of piliation on interactions of *Haemophilus influenzae* type b with human polymorphonuclear leukocytes. *Infection and Immunity* **47**, 780–785.

Trottier, S., Stenberg, K. and Svanborg-Eden, C. (1989). Turnover of nontypable *Haemophilus influenzae* in the nasopharynges of healthy children. *Journal of Clinical Microbiology* **27**, 2175–2179.

Turk, D.C. (1984). The pathogenicity of *Haemophilus influenzae*. *Journal of Medical Microbiology* **18**, 1–16.

van Alphen, L., Poole, J. and Overbeeke, M. (1986). The anton blood group antigen is the erythrocyte receptor for *Haemophilus influenzae*. *FEMS Microbiology Letters* **37**, 69–71.

van Alphen, L., Poole, J., Geelen, L. and Zanen, H.C. (1987). The erythrocyte and epithelial cell receptors for *Haemophilus influenzae* are expressed independently. *Infection and Immunity* **55**, 2355–2358.

van Alphen, L., Geelen-van den Broek, L., Blaas, L., van Ham, M. and Dankert, J. (1991). Blocking of fimbria-mediated adherence of *Haemophilus influenzae* by sialyl gangliosides. *Infection and Immunity* **59**, 4473–4477.

van Ham, S.M., Mooi, F.R., Sindhunata, M.G., Maris, W.R. and van Alphen, L. (1989). Cloning and expression in *Escherichia coli* of *Haemophilus influenzae* fimbrial genes establishes adherence to oropharyngeal epithelial cells. *EMBO Journal* **8**, 3535–3540.

van Ham, S.M., van Alphen, L., Mooi, F.R. and van Putten, J.P. (1993). Phase variation of *H. influenzae* fimbriae: transcriptional control of two divergent genes through a variable combined promoter region. *Cell* **73**, 1187–1196.

van Ham, S.M., van Alphen, L., Mooi, F.R. and van Putten, J.P. (1994). The fimbrial gene cluster of *Haemophilus influenzae* type b. *Molecular Microbiology* **13**, 673–684.

van Ham, S.M., van Alphen, L., Mooi, F.R. and van Putten, J.P.M (1995). Contribution of the major and minor subunits of fimbria-mediated adherence of *Haemophilus influenzae* to human epithelial cells and erythrocytes. *Infection and Immunity* **63**, 4883–4889.

Virkola, R., Brummer, M., Rauvala, H., van Alphen, L. and Korhonen, T.K. (2000). Interaction of fimbriae of *Haemophilus influenzae* type b with heparin-binding extracellular matrix proteins. *Infection and Immunity* **68**, 5696–5701.

Watson, W.J., Gilsdorf, J.R., Tucci, M.A., McCrea, K.W., Forney, L.J. and Marrs, C.F. (1994). Identification of a gene essential for piliation in *Haemophilus influenzae* type b with homology to the pilus assembly platform genes of Gram-negative bacteria. *Infection and Immunity*, **62**, 468–475.

Weber, A., Harris, K., Lohrke, S., Forney, L. and Smith, A.L. (1991). Inability to express fimbriae results in impaired ability of *Haemophilus influenzae* b to colonize the nasopharynx. *Infection and Immunity* **59**, 4724–4728.

Weiser, J.N. (2000). The generation of diversity by *Haemophilus influenzae*. *Trends in Microbiology* **8**, 433–436

Weiser, J.N., Chong, S.T.H., Greenberg, D. and Fong, W. (1995). Identification and characterization of a cell envelope protein of *Haemophilus influenzae* contributing to phase variation in colony opacity and nasopharyngeal colonization. *Molecular Microbiology* **17**, 555–564.

PART II Effect of adhesion on bacterial structure and function

CHAPTER 7

Transcriptional regulation of meningococcal gene expression upon adhesion to target cells

Muhamed-Kheir Taha

7.1 INTRODUCTION

Initial interactions between pathogenic bacteria and target cells are crucial events in cell infection. Bacteria have developed various regulatory networks to interact with their hosts. These mechanisms tend to favour bacterial survival and multiplication in infectious sites.

In extracellular bacteria such as *Neisseria meningitidis* and *Neisseria gonorrhoeae*, several structures have evolved to permit efficient adhesion to target cells. These two species are exclusive human pathogens. *Neisseria meningitidis*–host cell interactions alternate between asymptomatic carriage and invasive infections. The organism provokes septicaemia and then may invade the subarachnoidal space and causes meningitis. Other localizations are also observed, such as arthritis. Meningococcal infections occur as sporadic or epidemic cases. Factors that provoke meningococcal epidemics have not been fully elucidated. They are thought to be related to both host and bacteria as well as to the environment. Meningococcal strains involved in epidemics are usually different from those isolated from sporadic cases. Using genetic typing approaches, epidemic clones can be clustered into a few groups (clonal complexes). The percentage of healthy carriers in the general population is about 10%; however, very few individuals develop invasive infections. Moreover, meningococcal strains isolated from carriers are highly heterogeneous and differ genetically from invasive strains.

Bacteria first adhere to the epithelium of the nasopharynx and may then be translocated into the bloodstream. To gain entry into the cerebrospinal fluid (CSF), *N. meningitidis* has to adhere to endothelial cells and to cross the blood–CSF barrier. During the septicaemia phase, the organism interacts with components of the host immune system. Meningococci, as well as endotoxin released by the bacteria, are potent inducers of the inflammatory response. The production of cytokines, and particulary tumour necrosis factor (TNF)α, is a key element in this process and in the outcome of meningococcal infection. Subsequently, injury to endothelial cells leads to capillary leakage. Coagulopathy and intravascular thrombosis also result from inappropriate activation of the coagulation system. Bacterium–host cell interaction is therefore essential for *N. meningitidis* pathogenesis and adhesion of the organism to epithelial and endothelial cells are crucial steps in the infection process.

7.2 MENINGOCOCCAL ADHESION: DESCRIPTIVE VIEW

This chapter will focus particularly on the modulation of the expression of bacterial structures during adhesion. Modifications of target cells upon bacterial adhesion will not be extensively described. Bacterial adhesion to target cells often involves a multistep pathway consisting of (i) initial adhesion and (ii) intimate adhesion involving close contact between the organism and the target cell.

7.2.1 Structures involved in initial adhesion

Pili, filamentous structures on the bacterial surface, are composed mainly of protein subunits called pilin. Meningococcal pili are type IV pili and play a key role in mediating interactions with host cells. Non-piliated bacteria are not able to adhere efficiently to target cells. Pilin is encoded by the chromosomal gene *pilE*. It undergoes phase and antigenic variation, phenomena that are implicated in many aspects of meningococcal virulence. Pilin antigenic variation has been shown to modulate adhesiveness to human epithelial cells in both *N. meningitidis* and the closely-related species *N. gonorrhoeae* (Rudel *et al.*, 1992; Virji *et al.*, 1992; Nassif *et al.*, 1993; Jonsson *et al.*, 1994).

Pili make an essential contribution to *N. meningitidis* adhesion (particularly in capsulated meningococci) by allowing initial (localized) adhesion to target cells. Pili are the only structures able to transverse the capsule. The negatively charged meningococcal capsule is expected to hinder bacterium–host cell contact. Indeed, capsule- and/or lipo-oligosaccharide (LOS)-associated sialic acid have been reported to interfere negatively with adhesion of *N. meningitidis* (see below).

Two homologous proteins, PilC1 and PilC2, are also key elements in the structure of pili, since the production of at least one PilC protein is required for pilus assembly. In addition, PilC1 (but not PilC2) modulates adhesiveness, most likely by being the adhesin (Jonsson *et al.*, 1991; Nassif *et al.*, 1994; Rudel *et al.*, 1995). These proteins have been shown to be the pilus tip-located adhesins in *N. gonorrhoeae* (Rudel *et al.*, 1992, 1995). However, PilC proteins have also been reported to be localized in the outer membrane of meningococci (Rahman *et al.*, 1997). In *N. meningitidis*, mutants that do not express PilC1 are non-adhesive, PilC2 alone is not capable of promoting bacterial interaction with cells (Nassif *et al.*, 1994). This suggests that in this organism, unlike *N. gonorrhoeae*, only PilC1 recognizes the eukaryotic receptor. The PilC proteins are subject to phase variation. The mechanism responsible for this event results from changes in the number of nucleotide residues within a tract of poly(G) located in the signal peptide-encoding sequence (Jonsson *et al.*, 1991). Variations of both *pilC* and *pilE* are able to modulate meningococcal pilus-mediated adhesion. Several components of target cells have been proposed to participate in pilus-dependent interactions between bacteria and cells, for example CD46, a complement regulatory glycoprotein. Subsequently, bacterial attachment causes a transient increase in cytosolic free Ca^{2+} in target cells (Källström *et al.*, 1998). Antibodies directed against CD46 and purified CD46 are able to block binding of meningococci to target cells. Moreover, piliated but not non-piliated *Neisseria* are able to bind to transfected Chinese hamster ovary cells expressing human CD46 (Källström *et al.*, 1997). In a recent study, neisserial type IV pili were reported to be required for cortical plaque formation in epithelial cells (Merz *et al.*, 1999). These plaques are enriched with cellular components that are involved in bacterial adhesion to target cells. This recruitment of cellular components in plaques upon pili-mediated initial adhesion may increase the concentration of local receptor(s) at the surface of target cells. Subsequently, other adhesin-receptor pairs could bind efficiently (Merz *et al.*, 1999).

7.2.2 Structures involved in intimate adhesion

During this step, membranes of bacteria and target cells come into close contact and microvilli seem to disappear from the surface of infected cells. Several meningococcal surface structures have been proposed to play a role during this step. Surface adhesins have been reported, for example the Opc outer membrane protein and Opa outer membrane protein families (Stern *et al.*, 1986; Achtman *et al.*, 1988; Aho *et al.*, 1991). Opc and Opa share many features and may confer colonial opacity. The expression of the genes encoding these proteins undergoes phase variation. For *opa* genes, the mechanism responsible for this event results from changes in the number of pentamer repeats, CTCTT, within the signal peptide-encoding sequence (Stern *et al.*, 1986). The expression of *opc* is modified by changes in the number of nucleotide residues within a tract of poly(C) located within the promoter region (Sarkari *et al.*, 1994).

Cellular receptors for Opc and Opa have been reported. The Opa family of neisserial proteins may interact with several members of the CD66 family (Virji *et al.*, 1996; Gray-Owen *et al.*, 1997; Dehio *et al.*, 1998). CD66 (CEA, carcinoembryonic antigen) seems to mediate interactions between Opa-expressing *Neisseria* and human cells such as endothelial cells, polymorpho-nuclear phagocytes, and certain epithelial cells (Virji *et al.*, 1996). However, certain *opa* proteins are also able to use proteoglycans as substrates for bacterial adhesion (Duensing and van Putten, 1997). Opc was reported to interact with integrins on endothelial cells via interaction with serum vitronectin (Virji *et al.*, 1994). Moreover, meningococci producing Opc also bind epithelial cell proteoglycan receptors (de Vries *et al.*, 1998). Opc⁻ and/or Opa⁻ meningococci can be isolated from cases of invasive meningococcal infections, indicating that other structures could also participate in intimate adhesion. Multiple glycolipid-binding adhesins have been reported in pathogenic *Neisseria*; however, further characterization is needed to elucidate their role in the adhesion process (Paruchuri *et al.*, 1990).

7.3 MENINGOCOCCAL ADHESION: MECHANISTIC VIEW

If all surface structures of *N. meningitidis* were present at the same time during the entire period of its interaction with a host cell, this interaction would not be efficient as steric hindrance would obviously prevent intimate adhesion. Capsule and/or LOS-associated sialic acid have been reported to interfere negatively with meningococcal adhesion (Virji *et al.*, 1992; de Vries *et al.*, 1996; Hammerschmidt *et al.*, 1996; Hardy *et al.*, 2000). Indeed, it has been reported that an Opc-expressing *galE* mutant (capsulated but lacking sialic acid on LOS) was able to bind a soluble proteoglycan receptor but was unable to interact efficiently with this receptor on the epithelial cell surface (de Vries *et al.*, 1998). An Opc-expressing *siaD* mutant (non-capsulated but expressing sialic acid on LOS) was unable to interact with the soluble proteoglycan receptor or with this receptor on the epithelial cell surface. *siaA* and *cps* mutants (non-capsulated and lacking sialic acid on LOS) were able to bind the soluble proteoglycan receptor and to interact efficiently with epithelial cell surfaces (de Vries *et al.*, 1998). While capsulated non-piliated strains are not able to adhere efficiently to target cells, non-capsulated meningococci are able to adhere in a pili-independent manner.

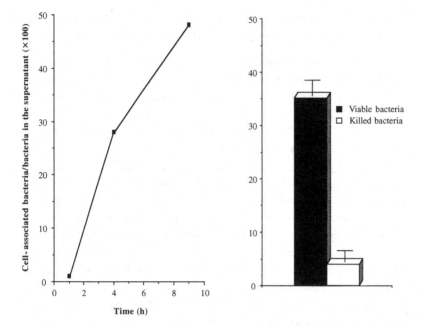

Figure 7.1. *Left*: Extent of adhesion of *N. meningitidis* (strain 8013) to Hec-1-B epithelial cells. Infection was performed with a multiplicity of infection of 1:10 as described by Deghmane *et al*. (2000). At 1 hour, 4 hours and 9 hours post infection, cells were lifted off the plates and the extent of adhesion was measured. *Right*: Extent of adhesion of *N. meningitidis* (strain 8013) to Hec-1-B epithelial cells. Infection was performed as above using live or killed bacteria (glutaraldehyde-fixed bacteria).

Neisseria meningitidis interacts with several cellular barriers during infection (epithelial and endothelial cells). These steps are separated by a dissemination step in the blood stream. It could be hypothesized that modulation of the expression of genes encoding adhesion-involved structures is crucial during this organism's pathophysiological processes.

Contact between capsulated *N. meningitidis* and viable target cells seems to be a signal that promotes effective adhesion. This adhesion requires the presence of pili and PilC1. Indeed, adhesion of the organism to viable epithelial or endothelial cells is more efficient than adhesion to glutaraldehyde-treated cells (Taha *et al*., 1998). Moreover, the adhesion level (the percentage of cell-associated bacteria) of capsulated meningococci seems to increase as bacterium–host cell contact progresses (Fig. 7.1). These data indicate that adhesion could be induced upon cell contact.

7.3.1 Optimal initial adhesion in meningococci necessitates the induction of the expression of *pilC1* but not that of *pilE*

In spite of the fact that PilC1 and PilC2 are highly homologous (80% identity), only PilC1 is involved in meningococcal adhesion. While the C-terminal halves of these two proteins are almost identical, sequence diversity is located in the N-terminal parts of both proteins. The functional difference between PilC1 and PilC2 could be then due to this sequence diversity. Moreover, *pilC1* and *pilC2* are also distinguishable at the transcriptional level, since they are controlled by distinct promoters (Fig. 7.2). The difference between the organization of the regulatory regions of *pilC1* and *pilC2* loci is in a good agreement with the different functions of these proteins. *pilC2* is expressed from a classical −35/−10 promoter, while *pilC1* can be transcribed from three promoters. One corresponds to PC1.1 and is in fact identical with the *pilC2* promoter. Two others (PC1.2 and PC1.3) are located further downstream in a DNA region (size 150 bp) not found in *pilC2*. PC1.3 seems to be the major transcription start point (TSP) for *pilC1*. Transcription from PC1.2 is quite minor. No canonical consensus sequence was found upstream from PC1.3 (Taha *et al.*, 1996, 1998).

The use of reporter genes (such as *lacZ*, which encodes β-galactosidase) has facilitated the monitoring of the expression of meningococcal genes involved in initial adhesion. The expression of *pilC1*, but not that of *pilC2*, seems to be induced early, and temporarily, upon contact with viable cells but not with glutaraldehyde-treated cells (Taha *et al.*, 1998). Furthermore, no induction of *pilC1* expression was observed in the presence of extracts of epithelial cells obtained by sonication. This induction depends on the expression of *pilC1* from the major TSP that is located in the specific fragment in the promoter region of *pilC1* that is absent from *pilC2* (Fig. 7.2; Taha *et al.*, 1996, 1998). The inactivation of this TSP abolished the induction of the expression of *pilC1* upon contact of the bacteria with the host cells and caused a dramatic reduction in adhesion. These data indicate that the up-regulation of *pilC1* expression is at the transcriptional level and is a result of physical contact between the meningococcus and the surface of a viable target cell. The fragment in the promoter region involved in the induction of the transcription of *pilC1* upon contact with target cells has been named CREN for contact regulatory element of *Neisseria*. One obvious mechanism by which this up-regulation might control adhesiveness is by increasing piliation. The expression of *pilE* does not seem to be induced upon contact with cells, and the *pilE* promoter region does not harbour a CREN-like element (Taha *et al.*, 1996; Deghmane *et al.*, 2000). Taken together, these data demon-

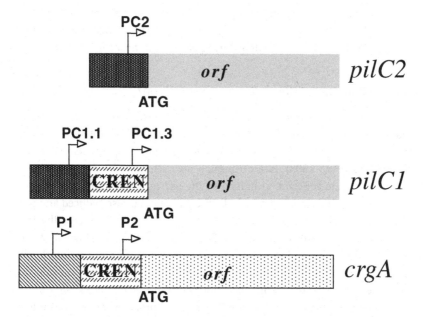

Figure 7.2. Schematic representation of *crgA*, *pilC1* and *pilC2* loci in *N. meningitidis*. The open reading frames (*orf*) and the start codons are indicated. The transcription start points (P1 and P2 for *crgA*, PC1.1 and PC1.3 for *pilC1* and PC2 for *pilC2*) are also indicated by arrows. CREN (*crgA* and *pilC1*) is in a box. The figure is not drawn to scale.

strate that the up-regulation of *pilC1* does not increase meningococcal pilus-mediated adhesion by increasing the number of pili, but suggest that this effect is linked directly to the adhesive function of PilC1. An intriguing finding is the fact that up-regulation of PilC1 does not modify the piliation of meningococci adhering to cells, thus raising the question of the mechanism by which the increasing level of PilC1 up-regulates adhesiveness. One explanation is that a basal level of PilC1 may be sufficient to promote the formation of pili whereas a higher production could be required to localize PilC1 in the pilus and eventually at the tip of the fibre. This would explain why PilC1 of bacteria grown on standard media were reported as being associated with the outer membrane and not with the pili (Rahman *et al.*, 1997), in contrast to a previous report (Rudel *et al.*, 1995).

In meningococci, the CREN element upstream from the *pilC1* gene is conserved in different strains belonging to different genetic lineages but seems to be absent in commensal *Neisseria* strains (Deghmane *et al.*, 2000). As bacterial adhesion to target cells seems to involve several genes, the presence of such a regulatory fragment in the *pilC1* promoter region might suggest that other meningococcal genes may possess this fragment and could be controlled co-ordinately during bacterium–host cell interactions.

7.3.2 Intimate adhesion: how to switch on?

Once initial adhesion has been established, a subsequent negative feedback on the expression of *pilC1* is then observed and the level of expression of *pilC1* declines to its basal level at a late (intimate) step of bacterial adhesion (Taha *et al.*, 1998). *pilE* expression also seems to be down-regulated during intimate adhesion as judged by the *pilE-lacZ* transcriptional fusion. The level of pilin protein (encoded by the *pilE* gene) is also reduced and pili seem to disappear (Fig. 7.3; Pujol *et al.*, 1999). During the intimate adhesion step, structures involved in initial adhesion seem to be modulated at the transcriptional level (down-regulation for *pilE* and negative feedback for *pilC1*). If this is a prerequisite to switch to intimate adhesion, it could be hypothesized that a regulatory protein may be induced upon cell contact and can then act on the expression of *pilE*, *pilC1* and probably other genes interfering with initial adhesion. To be induced by cell contact, the gene encoding such a regulator may be under the control of a CREN-like element. Consistent with this hypothesis is the fact that hybridization experiments using the CREN fragment as a probe suggest that CREN homologues are present in several chromosomal loci (Deghmane *et al.*, 2000).

crgA (contact-regulated gene A) is a gene encoding a protein of 299 amino acid residues that is under the control of a CREN-like element. Promoter mapping of *crgA* detected two TSPs: P1 and P2. They reside within the CREN-like element in the promoter region of *crgA* (Fig. 7.2). As for *pilC1*, *crgA* is also up-regulated by target cell contact by the induction of transcription from P2 (Fig. 7.2; Deghmane *et al.*, 2000). CrgA shares a strong homology (between 47% and 77%) with LysR-type transcriptional regulators (LTTR), possibly the most common type of transcriptional regulator in prokaryotes. LTTRs are involved in very diverse biological functions such as amino acid biosynthesis and regulation of virulence factors (Schell, 1993). CrgA seems to be a good candidate for the control of the switch between initial and intimate adhesion.

Insertional inactivation of *crgA* results in a dramatic reduction of meningococcal adhesion to target cells, particularly at the late stage of meningococcal–host cell interaction (intimate adhesion) (Deghmane *et al.*, 2000). While the wild-type strain was able to establish intimate adhesion, the *crgA* mutant showed a quite different interaction with target cells. Indeed, the *crgA*

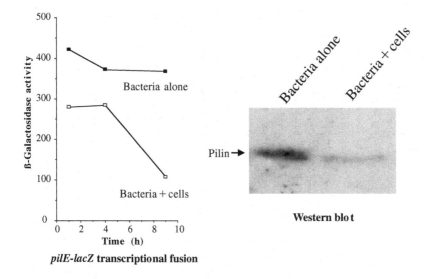

pilE-lacZ transcriptional fusion

Figure 7.3. *pilE* expression is repressed during bacterium–host cell contact. *Left*: Levels of β-galactosidase activity in *N. meningitidis* harbouring a transcriptional *pilE-lacZ* fusion. Hec-1-B epithelial cells were infected with a multiplicity of infection of 1:10 as described by Deghmane *et al.* (2000). At 1 h, 4 h and 9 h post infection, cells were lifted off the plates and the level of β-galactosidase was measured. *Right*: Western blot analysis to detect the level of pilin in *N. meningitidis* during bacterium–host cell contact. Hec-1-B epithelial cells were infected for 9 hours. Cells were lifted off the plates and Western blot analysis was performed using anti-pilin antibodies. Comparison was made with the same number of bacteria that were grown in the absence of epithelial cells.

mutant was unable to adhere intimately to cells. Bacteria were always distant from the cell surface; the bacterial and host cell membranes were never observed to come into direct contact and microvilli were still present at the surface of the cells (Fig. 7.4; Deghmane *et al.* 2000).

7.3.3 CrgA: mode of action

If CrgA is to control the switch from initial to intimate adhesion it should be able to bind to the promoter regions of genes involved in initial and/or intimate adhesion. Indeed, purified CrgA was used in gel mobility shift assays and shown to be able to bind to *pilC1*, *pilE*, *sia* and *crgA* promoters (Deghmane *et al.*, 2000; A. Deghmane and M.-K. Taha, unpublished data). *sia* genes are involved in the biosynthesis of sialic acid-containing capsules.

Alignment of the DNA sequences of these four promoters and analysis

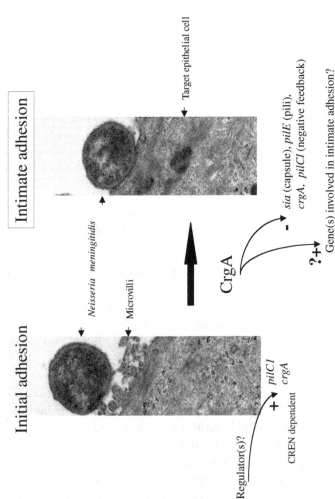

Initial adhesion

Neisseria meningitidis

Microvilli

Regulator(s)?

+   *pilC1*
     crgA

CREN dependent

Intimate adhesion

Target epithelial cell

CrgA

-   *sia* (capsule), *pilE* (pili),
    crgA, *pilC1* (negative feedback)

?+

Gene(s) involved in intimate adhesion?

Figure 7.4. A model for the switch from initial to intimate adhesion. Electron micrographs of Hec-1-B epithelial cells infected for 1 hour (left) or 9 hours (right) by *N. meningitidis*. *left*: One bacterium is seen in close proximity to the cell membrane but is still separated from the cell surface by a space in which microvilli are also observed. *right*: The bacterium is on the cell surface in close contact with the cell (intimate adhesion). Note the disappearance of microvilli. Contact (initial adhesion) with the target cell is a signal that induces the expression of certain bacterial genes (such as *pilC1* and *crgA*) in a CREN-dependent manner. Once induced, CrgA down-regulates pili and capsule and has a negative feedback effect on the expression of *pilC1*. Intimate adhesion could then occur by unmasking of the structures involved. CrgA could also stimulate the expression of gene(s) involved in intimate adhesion. These genes remain to be identified.

by gel mobility shift assays using different promoter fragments enabled the identification of minimal CrgA-binding sequences in these promoters. Interestingly, these latter fragments were found to contain several copies of the T-N11-A motif, which has been suggested to be the recognition site for LTTRs (Schell, 1993). Footprinting analysis using purified CrgA and target promoter regions confirmed the binding site of CrgA (A. Deghmane and M.-K. Taha, unpublished data). The CrgA binding site is located immediately upstream from the CREN element in both *pilC1* and *crgA* (no CREN element is present in the *pilE* and *sia* promoter regions).

CrgA binds to its own promoter and has a negative autoregulatory effect (Deghmane *et al.*, 2000). In a *crgA* mutant, the expression of *pilC1* (as judged by the *pilC1-lacZ* transcriptional fusion) showed an increase in β-galactosidase activity in host cell-associated bacteria, reflecting the induction of *pilC1*. However, in contrast to the wild-type strain, this induction was maintained throughout the adhesion of the *crgA* mutant (no negative feed-back). These results suggest that at least part of the effect of *crgA* inactivation (diminution of adhesion and absence of intimate adhesion) might be due to the absence of the negative feedback regulation on *pilC1* after initial adhesion to target cells. CrgA could, therefore, be a negative regulator of *pilC1* during the intimate phase of bacterial adhesion.

In meningococci, *crgA* seems to be expressed at a very low level in the absence of contact with target cells. However, it is noteworthy that the expression of *crgA*, like that of *pilC1*, is induced upon bacterium–host cell contact. The nature of the signal responsible for the induction of the expression of *crgA* and *pilC1*, as well as the regulatory protein(s) required, remain to be determined. However, this induction depends on the presence of CREN in the promoter region. Once *crgA* is induced, it seems to repress the expression of *pilC1* after the initial phase of induction. The fact that CrgA is able to bind to the *pilC1* promoter immediately upstream from CREN favours this hypothesis. Indeed, the CrgA-binding site is distinct from the sequence necessary for transcriptional stimulation by bacterium–host cell contact (Taha *et al.*, 1998; Deghmane *et al.*, 2000). CrgA may modulate the function of CREN by binding DNA in the near vicinity of the regulatory element.

As CrgA binds to the promoter region of *sia*, and as the capsule is known to hinder bacterium–host cell contact, it is tempting to hypothesize that capsule synthesis is also down-regulated during intimate meningococcus–host cell interaction and, most likely, the expression of *pilE*. The effect of *crgA* on the expression of the capsule of other serogroups (not containing sialic acid, such as serogroup A) remains to be tested.

Trans-acting regulators of the LTTR family that are able to modulate the expression of virulence factors have been reported in several bacteria. One example is *Salmonella typhimurium* SpvR, which regulates the expression of *spv* virulence genes involved in spleen invasion (Caldwell and Gulig, 1991; Coynault *et al.*, 1992). LTTRs are thought to act in a co-inducer-responsive manner in order to permit optimal bacterial survival and adaptation to environmental factors (Schell, 1993). Co-inducers are usually small compounds that are specific to a particular environmental condition. Such a co-inducer remains to be identified in *N. meningitidis*.

7.4 BACTERIUM–HOST CELL CROSS-TALK: THE BACTERIAL SIDE OF THE STORY

Adhesion of *N. meningitidis* to target cells provokes bacterium–host cell cross-talk. A new aspect of this meningococcus–host cell interaction is the fact that the bacterium also undergoes an adaptive response through a signal transduction pathway involving a network of regulators acting in cascade (Fig. 7.4). CrgA could be a member of this network and could be involved in a co-ordinated regulation of bacterial genes such as *pilC1*, *pilE* and *sia*. Other bacterial regulators of this pathway remain to be identified, as well as other targets of CrgA regulation.

Negative feedback of *pilC1* might be necessary for bacterial adhesion to progress further to intimate adhesion. This latter may occur by the unmasking of the structures involved. The fact that CrgA binds to the *sia* gene promoter suggests that CrgA may directly repress capsule synthesis during intimate adhesion. However, capsule diminution may be a consequence of intimate adhesion rather than occurring prior to it. Alternatively, CrgA, once induced, might stimulate the expression of gene(s) involved in intimate adhesion. These genes remain to be identified, and the purification of CrgA should facilitate this identification. According to the second hypothesis, CrgA might act as both a positive and a negative regulator of the transcription of target genes. Regulators of the LTTR family are very diverse. In general, they are positive transcriptional regulators with a negative autoregulatory effect (Schell, 1993). However, several LTTRs can act as repressors or even as repressor-activators, such as the TfdS protein encoded by pJP4 in *Alcaligenes eutrophus* (Kaphammer and Olsen, 1990). The identification of *crgA*, which seems to be required for intimate adhesion, confirms that intimate adhesion is dissociable from initial pilus-mediated meningococcal attachment to target cells (Pujol *et al.*, 1999; Deghmane *et al.*, 2000). The fact that *crgA* is needed for the negative feedback of *pilC1* does not conflict with the observed reduction in adhesion by the *crgA* mutant. Indeed, the reduction in adhesion level is observed after 9 hours of meningococcus–host cell interaction. After 1 hour of infection, the *crg*A mutant and the wild-type strain showed comparable levels of induction of *pilC1* and comparable levels of adhesion to epithelial cells. CrgA would only reduce the level of PilC1 to its initial level. *pilC1* induction may be necessary in the early initial (focal) adhesion but its high level may hinder intimate contact.

Target cells (particularly the cell cortex) seem to undergo several dramatic modifications (morphological and biochemical changes) such as loss of microvilli and carbohydrate-containing surface structures, possibly as a result of signal transduction in the cell. Whether these modifications are induced by *N. meningitidis* or provoked by different physiological states of the target cells remains to be analysed (Deghmane *et al.*, 2000). These different cellular states may be necessary for the bacteria to be able to infect target cells.

Neisseria gonorrhoeae, which is closely related to *N. meningitidis*, has been shown to induce the production of inflammatory cytokines by epithelial cells. This induction is dependent on the activation of the transcriptional factor NF-κB and requires the adhesion of *N. gonorrhoeae* to epithelial cells (Naumann *et al.*, 1997). Moreover, we have recently shown that piliated (adhesive) meningococci, but not non-piliated (not adhesive) meningococci, are able to induce the expression of the TNF-α encoding gene in target cells (Taha, 2000). This induction could be provoked by a pilus-activated signal transduction pathway in target cells. Interestingly, cortical plaque formation in meningococcus-infected epithelial cells also required pilus-mediated adhesion (Merz *et al.*, 1999). Pili may play an inducer role in bacterial adaptation upon cell contact. Complex signalling pathways have evolved in many bacterial systems to permit such an adaptation. Microbial adaptation to the environment may resemble the physiological and genetic differentiations that are usually ascribed to higher organisms (Rosenzweig and Adams, 1994).

In pathogenic *Neisseria*, DNA rearrangements have been shown to allow the appearance of genetically different bacteria. These phenomena would enable the bacteria to escape the host immune response and to modulate bacterial adhesion and invasion to different cellular lines (Nassif and So, 1995). These DNA arrangements appear at a relatively low frequency of 10^{-2} to 10^{-3} per cell per generation. Alternatively, a central and pleiotropic regulatory system could allow these pathogens to adapt in a more co-ordinated and responsive manner to environmental changes. In the latter case, all members of the bacterial population could be involved in this response.

Contact with cells has recently been recognized as a signal for the transcription of bacterial genes (Cornelis, 1997). In uropathogenic *Escherichia coli*, the expression of a gene essential for the response to iron starvation has been shown to be induced by the interaction of P pili with their receptors (Zhang and Normark, 1996). In *Yersinia pseudotuberculosis*, the bacterium–host cell contact modulates the expression of Yop proteins through a type III secretion system (Pettersson *et al.*, 1996).

Epithelial cells are located at the interface between the external environment and the host. Pilus-mediated adhesion could be responsible for targeting the interaction of bacteria to these cells (Abraham *et al.*, 1998). This interaction would be expected to facilitate the passage of bacteria to internal compartments of the host. The activation through signal transduction pathways of bacteria and these strategically located cells is therefore a key element in neisserial pathogenesis.

Adhesion involves a sophisticated sequence of signalling pathways in both the host and the bacterium. Several issues remain to be elucidated, in particular the signal(s) recognized by bacteria upon contact with target cells. Moreover, the identification of other regulatory proteins involved in signalling pathways may help in understanding *Neisseria*–host cell interactions. Indeed, several regulatory loci have been revealed among the neisserial genomic sequences (Parkhill *et al.*, 2000; Tettelin *et al.*, 2000). For example, density-dependent interactions (quorum sensing) may control modulation of gene expression. Extensive adaptation within meningococcal populations may confer upon them many advantages during their interactions with host cells.

ACKNOWLEDGEMENTS

I thank Jean-Michel Alonso for useful discussions and his generous support. The work in the Neisseria Unit is funded by the Institut Pasteur.

REFERENCES

Abraham, S.N., Jonsson, A.B. and Normark, S. (1998). Fimbriae-mediated host-pathogen cross-talk. *Current Opinion in Microbiology* **1**, 75–81.

Achtman, M., Neibert, M., Crowe, B.A., Strittmatter, W., Kusecek, B., Weyse, E., Walsh, M.J., Slawig, B., Morelli, G. *et al.* (1988). Purification and characterization of eight class 5 outer membrane protein variants from a clone of *Neisseria meningitidis* serogroup A. *Journal of Experimental Medicine* **168**, 507–525.

Aho, E.L., Dempsey, J.A., Hobbs, M.M., Klapper, D.G. and Cannon, J.G. (1991).

M.-K. TAHA

Characterization of the opa (class 5) gene family of *Neisseria meningitidis*. *Molecular Microbiology* 5, 1429–1437.

Caldwell, A.L. and Gulig, P.A. (1991). The *Salmonella typhimurium* virulence plasmid encodes a positive regulator of a plasmid-encoded virulence gene. *Journal of Bacteriology* 173, 7176–7185.

Cornelis, G. (1997). Contact with eukaryotic cells: a new signal triggering bacterial gene expression. *Trends in Microbiology* 5, 43–45.

Coynault, C., Robbe-Saule, V., Popoff, M.Y. and Norel, F. (1992). Growth phase and SpvR regulation of transcription of *Salmonella typhimurium spvABC* virulence genes. *Microbial Pathogenesis* 13, 133–143.

de Vries, F.P., van Der Ende, A., van Putten, J.P. and Dankert, J. (1996). Invasion of primary nasopharyngeal epithelial cells by *Neisseria meningitidis* is controlled by phase variation of multiple surface antigens. *Infection and Immunity* 64, 2998–3006.

de Vries, F.P., Cole, R., Dankert, J., Frosch, M. and van Putten, J.P. (1998). *Neisseria meningitidis* producing the Opc adhesin binds epithelial cell proteoglycan receptors. *Molecular Microbiology* 27, 1203–1212.

Deghmane, A., Petit, S., Topilko, A., Pereira, Y., Giorgini, G., Larribe, M. and Taha, M.-K. (2000). Intimate adhesion of *Neisseria meningitidis* to human epithelial cells is under the control of *crgA* gene, a novel LysR-type transcriptional regulator. *EMBO Journal* 19, 1068–1078.

Dehio, C., Gray-Owen, S.D. and Meyer, T.F. (1998). The role of neisserial Opa proteins in interactions with host cells. *Trends in Microbiology* 6, 489–495.

Duensing,, T.D. and van Putten, J.P. (1997). Vitronectin mediates internalization of *Neisseria gonorrhoeae* by Chinese hamster ovary cells. *Infection and Immunity* 65, 964–970.

Gray-Owen, S.D., Dehio, C., Haude, A., Grunert, F. and Meyer, T.F. (1997). CD66 carcinoembryonic antigens mediate interactions between Opa-expressing *Neisseria gonorrhoeae* and human polymorphonuclear phagocytes. *EMBO Journal* 16, 3435–3445.

Hammerschmidt, S., Hilse, R., van Putten, J.P., Gerardy-Schahn, R., Unkmeir, A., Frosch, M. (1996). Modulation of cell surface sialic acid expression in *Neisseria meningitidis* via a transposable genetic element. *EMBO Journal* 15, 192–198.

Hardy, S.J., Christodoulides, M., Weller, R.O. and Heckels, J.E. (2000). Interactions of *Neisseria meningitidis* with cells of the human meninges. *Molecular Microbiology* 36,817–829.

Jonsson, A.B., Nyberg, G. and Normark, S. (1991). Phase variation of gonococcal pili by frameshift mutation in *pilC*, a novel gene for pilus assembly. *EMBO Journal* 10, 477–488.

Jonsson, A.B., Ilver, D., Falk, P., Pepose, J. and Normark, S. (1994). Sequence changes in the pilus subunit lead to tropism variation of *Neisseria gonorrhoeae* to human tissue. *Molecular Microbiology* 13, 403–416.

Källström, H., Liszewski, M.K., Atkinson, J.P. and Jonsson, A.B. (1997). Membrane cofactor protein (MCP or CD46) is a cellular pilus receptor for pathogenic *Neisseria*. *Molecular Microbiology* 25: 639–647.

Källström, H., Islam, M.S., Berggren, P.O. and Jonsson, A.B. (1998). Cell signaling by the type IV pili of pathogenic *Neisseria*. *Journal of Biological Chemistry* 273, 21777–21782.

Kaphammer, B. and Olsen, R. H. (1990). Cloning and characterization of *tfdS*, the repressor-activator gene of *tfdB*, from the 2,4-dichlorophenoxyacetic acid catabolic plasmid pJP4. *Journal of Bacteriology* 172, 5856–5862.

Merz, A.J., Enns, C.A. and So, M. (1999). Type IV pili of pathogenic neisseriae elicit cortical plaque formation in epithelial cells. *Molecular Microbiology* 32, 1316–1332.

Nassif, X. and So, M. (1995). Interactions of pathogenic neisseria with non-phagocytic cells. *Clinical Microbiological Reviews* 8, 376–388.

Nassif, X., Lowy, J., Stenberg, P., O'Gaora, P., Ganji, A. and So, M. (1993). Antigenic variation of pilin regulates adhesion of *Neisseria meningitidis* to human epithelial cells. *Molecular Microbiology* 8, 719–725.

Nassif, X., Beretti, J.C., Lowy, J., Stenberg, P., O'Gaora, P., Pefifer, J., Normark, S. and So, M. (1994). Roles of pilin and PilC in adhesion of *Neisseria meningitidis* to human epithelial and endothelial cells. *Proceedings of the National Academy of Sciences USA* 91, 3769–3773.

Naumann, M., Webler, S., Bartsch, C., Wieland, B. and Meyer, T.F. (1997). *Neisseria gonorrhoeae* epithelial cell interaction leads to the activation of the transcription factor nuclear factor κB and activator protein 1 and the induction of inflammatory cytokines. *Journal of Experimental Medicine* 186, 247–258.

Parkhill, J., Achtman, M., James, K.D., Bentley, S.D., Churcher, C., Klee, S.R., Morelli, G., Basham, D., Brown, D. *et al.* (2000). Complete DNA sequence of a serogroup A strain of *Neisseria meningitidis* Z2491. *Nature* 404, 502–506.

Paruchuri, D.K., Seifert, H.S., Ajioka, R.S., Karlsson, K.A. and So, M. (1990). Identification and characterization of a *Neisseria gonorrhoeae* gene encoding a glycolipid-binding adhesin. *Proceedings of the National Academy of Sciences, USA* 87, 333–337.

Pettersson, J., Nordfelth, R., Dubinina, E., Bergman, T., Gustafsson, M., Magnusson, K.E. and Wolf-Watz, H. (1996). Modulation of virulence factor expression by pathogen target cell contact. *Science* 273, 1231–1233.

Pujol, C., Eugène, E., Marceau, M. and Nassif, X. (1999). The meningococcal PilT protein is required for induction of intimate attachment to epithelial cells following pilus-mediated adhesion. *Proceedings of the National Academy of Sciences, USA* **96**, 4017–4022.

Rahman, M., Källström, H., Normark, S. and Jonsson, A.B. (1997). PilC of pathogenic *Neisseria* is associated with the bacterial cell surface. *Molecular Microbiology* **25**, 11–25.

Rosenzweig, F. and Adams, J. (1994). Microbial adaptation to a changeable environment: cell–cell interactions mediate physiological and genetic differentiation. *BioEssays* **16**, 715–717.

Rudel, T., van Putten, J.P., Gibbs, C.P., Haas, R., and Meyer, T.F. (1992). Interaction of two variable proteins (PilE and PilC) required for pilus-mediated adherence of *Neisseria gonorrhoeae* to human epithelial cells. *Molecular Microbiology* **6**, 3439–3450.

Rudel, T., Scheuerpflug, I. and Meyer, T.F. (1995). *Neisseria* PilC protein identified as type-4 pilus tip-located adhesion. *Nature* **373**, 357–359.

Sarkari, J., Pandit, N., Moxon, E.R. and Achtman, M. (1994). Variable expression of the Opc outer membrane protein in *Neisseria meningitidis* is caused by size variation of a promoter containing poly-cytidine. *Molecular Microbiology* **13**, 207–217.

Schell, M.A. (1993). Molecular biology of the LysR family of transcriptional regulators. *Annual Review of Microbiology* **47**, 597–626.

Stern, A., Brown, M., Nickel, P. and Meyer, T.F. (1986). Opacity genes in *Neisseria gonorrhoeae*: control of phase and antigenic variation. *Cell* **47**, 61–71.

Taha, M.-K. (2000). *Neisseria meningitidis* induces the expression of the TNF-alpha gene in endothelial cells. *Cytokine* **12**, 21–25.

Taha, M.-K., Giorgini, D. and Nassif, X. (1996). The *pilA* regulatory gene modulates the pilus-mediated adhesion of *Neisseria meningitidis* by controling the transcription of *pilC1*. *Molecular Microbiology* **19**, 1073–1084.

Taha, M-.K., Morand, P.C., Pereira, Y., Eugène, E., Giorgini, D., Larribe, M. and Nassif, X. (1998). Pilus-mediated adhesion of *Neisseria meningitidis*: the essential role of cell contact-dependent transcriptional upregulation of the PilC1 protein. *Molecular Microbiology* **28**, 1153–1163.

Tettelin, H., Saunders, N.J., Heidelberg, J., Jeffries, A.C., Nelson, K.E., Eisen, J.A., Ketchum, K.A., Hood, D.W., Peden, J.F. *et al.* (2000). Complete genome sequence of *Neisseria meningitidis* serogroup B strain MC58. *Science* **287**, 1809–1815.

Virji, M., Alexandrescu, C., Ferguson, D.J., Saunders, J.R. and Moxon, E.R. (1992). Variations in the expression of pili: the effect on adherence of *Neisseria meningitidis* to human epithelial and endothelial cells. *Molecular Microbiology* **6**, 1271–1279.

Virji, M., Makepeace, K. and Moxon, E.R. (1994). Distinct mechanisms of interactions of Opc-expressing meningococci at apical and basolateral surfaces of human endothelial cells; the role of integrins in apical interactions. *Molecular Microbiology* **14**, 173–184.

Virji, M., Makepeace, K., Ferguson, D.J. and Watt, S.M. (1996). Carcinoembryonic antigens (CD66) on epithelial cells and neutrophils are receptors for Opa proteins of pathogenic neisseriae. *Molecular Microbiology* **22**, 941–950.

Zhang, J.P. and Normark, S. (1996). Induction of gene expression in *Escherichia coli* after pilus-mediated adherence. *Science* **273**, 1234–1236.

M.-K. TAHA

Induction of protein secretion by *Yersinia enterocolitica* through contact with eukaryotic cells

Dorothy E. Pierson

8.1 INTRODUCTION

Contact-dependent secretion pathways, also known as type III secretion pathways, have been identified in a large number of animal and plant pathogenic bacteria (Van Gijsegem *et al.*, 1993; Cornelis and Van Gijsegem, 2000). The designation 'contact dependent' comes from the observation that proteins are secreted from bacteria after they contact their particular eukaryotic cell target (Ginocchio *et al.*, 1994; Ménard *et al.*, 1994; Rosqvist *et al.*, 1994; Wolff *et al.*, 1998; Vallis *et al.*, 1999; van Dijk *et al.*, 1999). The name type III differentiates this secretion pathway from the four other secretion pathways that have been identified thus far in Gram-negative bacteria (for recent reviews, see Henderson *et al.*, 1998; Wandersman, 1998; Burns, 1999; Stathopoulos *et al.*, 2000; Thanassi and Hultgren, 2000;). Many of the components of the type III secretion pathways of Gram-negative bacteria resemble components of the flagellar assembly apparatus found in these same organisms (Minamino and Macnab, 1999; Bennett and Hughes, 2000).

Yersinia enterocolitica, a pathogen of a variety of mammals, has three apparently independent type III secretion pathways that are involved in virulence. The best studied of these (and probably the best-studied type III secretion pathway in all bacteria), the plasmid-encoded Ysc secretion pathway, is found in all three pathogenic species of *Yersinia* – *Y. enterocolitica*, *Y. pestis* and *Y. pseudotuberculosis* (for a review, see Cornelis and Van Gijsegem, 2000). This contact-dependent secretion pathway is involved in the secretion of 11 proteins, termed Yops, most of which are translocated into host cells, where they alter host cell functions, resulting in the inhibition of antimicrobial activities of phagocytic cells (for a review, see Cornelis, 2000). The action of individual effector Yops on the mammalian cell has been well covered in several recent reviews (e.g. Bliska, 2000; Cornelis, 2000) and will not be discussed here. The other two type III secretion pathways in *Y. enterocolitica*, the Ysa and flagellar type III secretion systems are less well characterized. The Ysa secretion pathway secretes at least eight proteins (Haller *et al.*, 2000). The functions of these secreted proteins are not known individually; however, it is clear that this pathway is required for virulence, possibly in the early stages of infection, as strains containing mutations in this pathway are slightly attenuated for virulence when given orally, but are as virulent as parental strains when given intraperitoneally (Haller *et al.*, 2000). The flagellar type III secretion pathway secretes 14 proteins (Young *et al.*, 1999). One of the targets of this pathway has been identified thus far, a phospholipase, YplA, which contributes to host colonization by *Y. enterocolitica* (Schmiel *et al.*, 1998; Young *et al.*, 1999). Evidence to date suggests these three secretion pathways act independently. The *in vitro* signals that induce secretion differ for each pathway (Portnoy *et al.*, 1981; Straley and Brubaker, 1981; Young *et al.*, 1999; Haller *et al.*, 2000). Mutants in the Ysc secretion pathway are unaltered in secretion of the targets of the Ysa secretion pathway (Haller *et al.*, 2000). Finally, a mutant in one component of the Ysa secretion pathway is unaltered in Yop secretion as well as in phospholipase activity or motility, two targets of the flagellar secretion pathway (Haller *et al.*, 2000).

As described above, type III secretion pathways are also referred to as contact dependent because, in the cases that have been examined, protein secretion occurs when bacteria come in contact with eukaryotic cells. The eukaryotic cell signal sensed by adherent bacteria that lets them know they are attached has not been established. More is known about the bacterial factors that sense this contact as well as the consequences of this contact on gene expression and protein secretion. The Ysa and flagellar secretion pathways have not been demonstrated to be contact dependent. Therefore, this chapter will focus on the Ysc secretion pathway and the consequences of host contact on its expression and activity. Because there are excellent reviews on the function of the secreted proteins, I will not touch on this topic. The work described comes from studies on all three pathogenic species of *Yersinia*. Gene names given are those for *Y. enterocolitica*, the names of their counterparts in *Y. pestis* and *Y. pseudotuberculosis* (where different) are given in parentheses.

Over 20 different proteins make up the Ysc secretion machinery (Hueck, 1998; Cheng and Schneewind, 2000a; Cornelis and Van Gijsegem, 2000). These proteins make up a complex structure crossing the inner and outer membrane of the bacterium, resulting in secretion of components from the cell without a periplasmic intermediate (Van Gijsegem *et al.*, 1993). The Ysc proteins and their location and roles in production of the secretory apparatus are discussed below and reviewed in greater detail elsewhere (Hueck, 1998; Cheng and Schneewind, 2000a; Cornelis and Van Gijsegem, 2000). Several lines of evidence support the hypothesis that secretion and translocation are coupled in *Yersinia* species. A strain that overexpresses a negative regulator of pore size, YopQ (YopK), is blocked for Yop secretion (Hölmstrom *et al.*, 1997). In addition, in the presence of eukaryotic cells, secretion is polarized – it occurs only where bacteria are in contact with cells, when secreted proteins will be translocated into the host cell (Rosqvist *et al.*, 1994; Persson *et al.*, 1995). Thus it appears that, upon contact, most secreted proteins are released only at the site of contact with the host cell, where they are translocated into the host cell. Only upon opening of both channels, the secretion channel and the translocation channel, are Yops released, thus ensuring their contact with their targets inside the eukaryotic cell. Expression of Yops is also induced by contact, ensuring a continuing supply of effectors that target the host cell (Pettersson *et al.*, 1996). *In vitro*, expression and secretion of Yops can be induced by growth at 37 °C in the absence of calcium (referred to as the low calcium response; for a review, see Straley, 1988). Much of what we know about how the bacterium senses the eukaryotic cell, resulting in activation of expression and secretion, comes from studies of mutants that are altered in the expression and secretion of Yops in response to calcium *in vitro* (Portnoy *et al.*, 1983; Goguen *et al.*, 1984; Yother and Goguen, 1985; Straley and Bowmer, 1986).

This chapter will focus primarily on the roles of five major players in the regulation of expression and secretion of Yops upon contact with host cells, YopN (LcrE), LcrG, LcrV, YscM (LcrQ) and YopD. Some of these are also involved in the translocation process, either directly (YopD; Rosqvist *et al.*, 1991a,b; Hartland *et al.*, 1994; Sory and Cornelis, 1994) or indirectly (LcrV; Nilles *et al.*, 1998; Sarker *et al.*, 1998; Pettersson *et al.*, 1999), whereas others are themselves translocated (YopD; Francis and Wolf-Watz, 1998; and YscM; Cambronne *et al.*, 2000). Two of the major players act independently of translocation (YopN; Boland *et al.*, 1996; and LcrG; Nilles *et al.*, 1997).

8.2 THE PLAYERS

8.2.1 The Ysc secretion machinery

As mentioned above, the Ysc secretion apparatus is made up of over 20 proteins that form a complex that traverses the two membranes of the organism. YscC and perhaps YscW are in the outer membrane (Plano and Straley, 1995; Koster *et al.*, 1997), YscD, R, U, V, and probably YscJ, S, and T are in the inner membrane (Michiels *et al.*, 1991; Plano *et al.*, 1991; Allaoui *et al.*, 1994; Plano and Straley, 1995; Payne and Straley, 1998). The locations of some of the other components in this apparatus have not been established. In addition to the complex, there are associated proteins found in the bacterial cytoplasm. These include the putative ATPase that is thought to provide the energy for translocation, YscN (Woestyn *et al.*, 1994), as well as several chaperones that bind to intracytoplasmic Yops (Wattiau and Cornelis, 1993; Wattiau *et al.*, 1994; Day and Plano, 1998, 2000; Iriarte and Cornelis, 1998; Jackson *et al.*, 1998;). Each chaperone binds only one or two specific Yop partners. There are likely to be multiple functions of these chaperone/Yop interactions. Evidence from chaperone mutants suggests that they have a role in preventing degradation of their partner Yops (Frithz-Lindsten *et al.*, 1995). A number of these chaperones aid in secretion of their partner Yops (Cheng *et al.*, 1997) and in some cases appear to help translocation as well (Lee *et al.*, 1998). Finally, it has been suggested that the chaperone binding of Yops inside the bacterial cytoplasm prevents their premature association with the components of the translocation apparatus, which would render them secretion-incompetent (Woestyn *et al.*, 1996).

The Ysc secretion machinery genes are expressed in bacteria grown at 37 °C, the host temperature (Cornelis *et al.*, 1989). Thus the secretion apparatus is in place when bacteria contact host cells. Upon contact, the secretion apparatus opens and the majority of Yops are translocated into the host cell cytoplasm. Several of the Yops, LcrV, and three Ysc proteins (YscO, YscP and YscX) are secreted into the extracellular milieu where either they are directly involved in the translocation process (the translocase components, as well as proposed members of an 'injectisome' that helps introduce proteins into the host cell) or they are involved in the regulation of secretion or translocation in some other manner (Lawton *et al.*, 1963; Payne and Straley, 1998; Lee and Schneewind, 1999; Day and Plano, 2000; Stainier *et al.*, 2000).

8.2.2 The translocation apparatus

Two of the Yops – YopB and YopD – are proposed to form the translocase that allows Yops to enter the eukaryotic cell. Strains containing a mutation in either gene do not inject their Yops inside host cells, as detected by three methods. First, the activity of particular Yops on host cell function (e.g. YopE is cytotoxic) has been measured in *yopB* and *yopD* mutants in association with host cells (Rosqvist *et al.*, 1991a,b; Hartland *et al.*, 1994; Håkansson *et al.*, 1996a,b; Iriarte and Cornelis, 1998). The second method that has been used to detect the translocation of Yops is operon fusions of *yop* genes to the *cya* gene of *Bordetella pertussis* (Sory and Cornelis, 1994; Sory *et al.*, 1995; Boland *et al.*, 1996; Håkansson *et al.*, 1996b; Iriarte and Cornelis, 1998). *Bordetella pertussis* adenylate cyclase, the product of *cya*, is calmodulin activated (Wolff *et al.*, 1980). Thus adenylate cyclase activity is only detected when proteins are inside host cells, in the presence of calmodulin (Sory and Cornelis, 1994). Finally, Yop translocation has been detected by immunofluorescence (Persson *et al.*, 1995; Håkansson *et al.*, 1996a; Skrzypek *et al.*, 1998). YopB is proposed to form a pore; it shares homology with several pore-forming toxins in the RTX family (Håkansson *et al.*, 1993). In addition, YopB has been shown to have contact-dependent cytolytic activity against macrophages and sheep erythrocytes (Håkansson *et al.*, 1996b; Neyt and Cornelis, 1999a). YopD interacts with YopB and has been suggested to be part of the pore (Håkansson *et al.*, 1993; Neyt and Cornelis, 1999a,b). YopD is also translocated into the eukaryotic cell cytoplasm, leading to the suggestion that it may be an extracellular chaperone that helps some or all of the effector Yops to enter the eukaryotic cell through the YopB pore (Francis and Wolf-Watz, 1998). It should be noted that recent data from Schneewind and colleagues examining Yop location by digitonin extraction of eukaryotic cells suggests that YopB is not involved in translocation and that YopD has its effects on translocation from its location inside the bacterial cell (Lee and Schneewind, 1999). These contradictory results are probably due to the different methods used to examine localization of Yops and will need to be examined further to resolve the discrepancies.

8.2.3 The regulators

8.2.3.1 YopN (LcrE)

The YopN protein is found on the bacterial cell surface (Forsberg *et al.*, 1991). It is secreted through the Ysc channel and requires two chaperones, SycN and YscB, for its secretion (Day and Plano, 1998; Jackson *et al.*, 1998). Because of its location and the phenotypes of *yopN* mutants discussed below, YopN is proposed to be the plug for the secretion channel. Additionally, it is proposed to be the surface receptor for eukaryotic cells; the binding of YopN to its putative receptor(s) is thought to unblock the secretion channel when the bacterium has contacted a host cell (Rosqvist *et al.*, 1994). *yopN* mutants express and secrete Yops whether eukaryotic cells are present or not (Forsberg *et al.*, 1991). In the *yopN* mutant, secretion is random; although some Yops are translocated into eukaryotic cells in contact with *yopN* mutants, the majority of Yops are secreted into the medium by these strains

(Rosqvist *et al.*, 1994; Persson *et al.*, 1995; Boland *et al.*, 1996). YopN is thus responsible for the polarized release of Yops from the bacterial cell, which allows their translocation into the host cell. The host molecule(s) with which YopN interacts to signal to the bacterium that it has contacted a host cell are not known. Experiments designed to address this question using a glutathione S-transferase (GST)-YopN fusion to pull out molecules from HeLa cells that bind YopN were unsuccessful due to the instability of the GST-YopN fusion protein (Boyd *et al.*, 1998).

8.2.3.2 LcrG

lcrG mutants have the same phenotype as *yopN* mutants: they constitutively express and secrete Yops at 37 °C (Skrzypek and Straley, 1993). Thus, like YopN, LcrG is a negative regulator of secretion, it prevents Yop expression and release in the absence of host cells. The location of LcrG is less well defined than that of YopN. The majority of LcrG appears to be inside the bacterial cytoplasm, leading to the suggestion that LcrG is the intrabacterial plug for the Ysc secretion channel (Nilles *et al.*, 1997). Some LcrG is also secreted upon contact with eukaryotic cells (Skrzypek and Straley, 1993). The secreted LcrG tends to remain associated with the cell, leading to the alternative suggestion that LcrG is found on the bacterial surface in association with YopN, helping to block the secretion channel there (Skrzypek and Straley, 1993). Experiments using a GST-LcrG fusion to pick out molecules from HeLa cells that bind LcrG pulled out heparin sulphate proteoglycans (Boyd *et al.*, 1998). Heparin does not inhibit expression and secretion of Yops; however, it does affect their translocation into the cell, perhaps suggesting an additional role for LcrG in translocation (Boyd *et al.*, 1998). These results suggest that LcrG may have differing roles inside and outside the cell. In this model, intrabacterial LcrG blocks secretion through the Ysc channel in the absence of host cell contact, whereas surface-exposed LcrG is required for the translocation of Yops into the host cell.

8.2.3.3 LcrV

LcrV (also known as V antigen) appears to have multiple roles in *Yersinia*. LcrV is a secreted effector that has the ability to suppress the immune system. LcrV prevents the production of the pro-inflammatory cytokines tumor necrosis factor (TNF)α and interferon (IFN)γ (Nakajima *et al.*, 1995) and inhibits neutrophil chemotaxis (Welkos *et al.*, 1998). Much of the secreted LcrV is found located on the bacterial cell surface in discrete foci even before contact with eukaryotic cells (Fields *et al.*, 1999; Pettersson *et al.*, 1999). LcrV has been shown to affect the abundance and secretion of YopB

(Nilles *et al.*, 1998), which may explain why *lcrV* mutants are defective in the translocation of Yops into the eukaryotic cell (Nilles *et al.*, 1998; Pettersson *et al.*, 1999; Lee *et al.*, 2000). Antibodies raised against LcrV are protective against *Yersinia* infection (Une and Brubaker, 1984). At least some of this protection is likely to be due to the inhibition of suppression of inflammation that is induced by LcrV. An alternative explanation for protection by these antibodies is that they inhibit Yop translocation. Some, but not all, of these protective antibodies do indeed inhibit Yop translocation into host cells (Fields *et al.*, 1999; Pettersson *et al.*, 1999). In addition to being a secreted effector of the Ysc secretion system, LcrV is found in the cytoplasm (Straley and Brubaker, 1981), where it appears to affect expression and secretion through the Ysc channel via its effects on LcrG (Nilles *et al.*, 1997). LcrV binds LcrG and the two can be cross-linked by a variety of chemical cross-linkers (Nilles *et al.*, 1997). *lcrVlcrG* double mutants have the same phenotype as *lcrG* mutants, they are down-regulated for Yop expression and secretion, indicating that *lcrG* is epistatic to *lcrV* (Skrzypek and Straley, 1995). As expected from its role as a negative regulator, overexpression of LcrG in *Yersinia* results in an inhibition of Yop expression and secretion (Nilles *et al.*, 1998). However, concurrent overexpression of LcrV in the presence of excess LcrG results in a wild-type phenotype: Yops are expressed and secreted in response to contact with cells or in media lacking calcium (Nilles *et al.*, 1998). Taken together, these observations suggest that LcrV has its actions through the inhibition of the negative effects of LcrG. It is proposed that by binding LcrG, LcrV prevents LcrG from binding to the Ysc secretion machinery and blocking secretion.

8.2.3.4 YscM (LcrQ)

Yersinia enterocolitica has two YscM proteins, YscM1 and YscM2 (Stainier *et al.*, 1997). They are homologous to a single protein produced by *Y. pestis* and *Y. pseudotuberculosis*, LcrQ (Rimpiläinen *et al.*, 1992). Overexpression of YscM1 or YscM2 inhibits Yop synthesis and *yscM1 yscM2* double mutants constitutively express *yop* genes (Stainier *et al.*, 1997). However, Yop secretion and translocation are still dependent upon the presence of eukaryotic cells in *yscM1 yscM2* mutants (Stainier *et al.*, 1997). So, although the YscM and LcrQ proteins are negative regulators like YopN and LcrG, the YscM and LcrQ proteins have their effects inside the cell. Operon fusions of *yopH* to the *cat* gene demonstrate that the effect of the YscM proteins on *yop* expression is at the level of transcription (Stainier *et al.*, 1997). This control is indirect, *yscM1* alone does not prevent transcription, it requires other genes on the virulence plasmid (Stainier *et al.*, 1997). There is no evidence that the YscM and

LcrQ proteins bind DNA and they are thought to act through their effects on some other protein (Stainier et al., 1997; Cambronne et al., 2000). The YscM and LcrQ proteins are secreted from bacteria when they contact eukaryotic cells (Stainier et al., 1997). Digitonin extraction experiments suggest that YscM1 and LcrQ are translocated into the eukaryotic cell cytoplasm, but YscM2 is not translocated (Cambronne et al., 2000). The secretion of LcrQ is rapid: secretion can be detected within three minutes of a shift to medium without calcium and no LcrQ can be detected inside the bacterium by five minutes after the shift (Pettersson et al., 1996). The secretion of the YscM and LcrQ proteins leads to rapid expression of yop genes (Pettersson et al., 1996; Stainier et al., 1997). This secretion of a negative regulator by the Ysc secretion machinery is similar to the secretion of the anti-sigma factors that negatively regulate flagellar gene expression by the flagellar apparatus in a variety of bacteria, which results in an increase in flagellar gene expression (Hughes et al., 1993; Kutsukake, 1994). The ultimate result of secretion control of the negative regulator of gene expression in Yersinia is that Yops are down-regulated until bacteria contact host cells, when they are needed to carry out their anti-host activities.

8.2.3.5 YopD

As mentioned above, YopD appears to be part of the translocase that brings the Yop effectors into the eukaryotic cell (Håkansson et al., 1993; Neyt and Cornelis, 1999a,b) and has been proposed to be the extracellular chaperone that aids in this process (Francis and Wolf-Watz, 1998). In addition to its role in translocation, YopD has a role in the regulation of yop gene expression. The role of YopD in regulation was first noticed in an examination of phenotypes of mutants that were defective in production of the intracellular chaperone required for the stabilization and secretion of YopD, SycD (LcrH) (Wattiau et al., 1994). sycD mutants constitutively express Yops (Price and Straley, 1989; Bergman et al., 1991), raising the possibility that SycD or YopD are involved in yop gene regulation. yopD mutants also constitutively express Yops, although as expected, since YopD is important in formation of the translocase, secreted Yops are not translocated into the eukaryotic cell (Williams and Straley, 1998). The high-level expression of Yops occurs even though LcrQ, the negative regulator of gene expression, is present in the cytoplasm. As mentioned above, when LcrQ is overexpressed from a heterologous promoter in a wild-type strain, it down-regulates yop expression (Stainier et al., 1997). However, yop gene expression is constitutive when LcrQ is overexpressed in a yopD mutant (Williams and Straley, 1998). Thus, although this negative regulator is present at high levels it did not downregulate *yop* gene expression in the *yopD* mutant, suggesting that YopD is required for the negative regulatory activity of YscM (LcrQ).

8.2.4 Other components

8.2.4.1 Regulation of gene expression

As described above, *yop* gene expression is induced in bacteria at 37 °C in the presence of eukaryotic cells (or absence of calcium *in vitro*). The regulation of *yop* gene expression by temperature is controlled by a positive activator, VirF (LcrF) (Cornelis *et al.*, 1989). The activity of VirF is, in turn, controlled by a regulator of chromatin structure, YmoA (Cornelis *et al.*, 1991). VirF is a member of the AraC family of transcriptional activators (Gallegos *et al.*, 1997; Cornelis *et al.*, 1989). The *virF* gene itself is temperature regulated; *virF* gene transcription is induced at 37 °C (Cornelis *et al.*, 1989). Overexpression of *virF* at 28 °C under control of a heterologous promoter is insufficient to induce *yop* gene transcription a temperature shift to 37 °C is required. This temperature dependence is controlled by the histone-like protein YmoA (Lambert de Rouvroit *et al.*, 1992). In a *ymoA* mutant, *yop* gene transcription is activated by overexpression of VirF. Expression of *yop* genes at 37 °C thus requires chromatin structure changes that allow VirF to bind and activate transcription (Cornelis, 1993).

8.2.4.2 Control of translocation

In *Salmonella enterica*, *Shigella flexneri* and enterotoxigenic *Escherichia coli* (ETEC), cell surface appendages can be observed by electron microscopy in bacteria producing type III secretion machinery but not in mutants that are lacking a type III secretion system (Knutton *et al.*, 1998; Kubori *et al.*, 1998, 2000; Tamano *et al.*, 2000). These cell surface appendages show some similarities to flagella, in particular in the region of attachment to the bacterial cell surface (Kubori *et al.*, 1998). As it has been shown that some components of the type III secretion pathways resemble the flagellar assembly machinery, the similarities in appearance of the macromolecular structures might be expected. It has been proposed that these cell surface appendages are the injectisomes that introduce the type III effector proteins from these bacteria into their host cells (Knutton *et al.*, 1998; Kubori *et al.*, 1998; Tamano *et al.*, 2000). Electron microscopic analysis of the surface of *Yersinia* species has not revealed similar macromolecular structures (Stainier *et al.*, 2000). However, several proteins in *Yersinia* are predicted to be part of an injectisome structure due to their surface location and/or the translocation-defect phenotypes seen in mutants lacking these proteins. Potential components of an injectisome would include some of the previously described proteins that affect translocation, including LcrV, LcrG, YopB, and YopD. Additionally, TyeA and YscP in particular have been proposed to be components of a cell surface-localized injectisome structure in *Yersinia* (Iriarte *et al.*, 1998; Stainier *et al.*, 2000). *tyeA* mutants are defective in translocation of a subset of Yop effectors (Iriarte *et al.*, 1998; Cheng and Schneewind, 2000b). Interestingly, only the Yops that have intrabacterial chaperones are affected by a *tyeA* mutant, suggesting that one role of TyeA might be in removing the chaperone from the protein being secreted and translocated (Iriarte *et al.*, 1998). The TyeA protein binds both YopN and YopD, leading to the suggestion that TyeA may serve as a tether between these two molecules, allowing the coupling of secretion (induced by the release of the YopN block upon contact) and translocation (aided by the YopD extracellular chaperone/translocase component) (Iriarte *et al.*, 1998; Cheng and Schneewind, 2000b). YscP is required for Yop secretion (Stainier *et al.*, 2000). The designation of YscP as a member of the injectisome comes from the observations that the protein is located on the bacterial cell surface and that it can be removed by simple mechanical shearing, a procedure that removes extracellular organelles such as flagella and pili (Stainier *et al.*, 2000). Due to their extracellular location, two components of the secretion machinery, YscO and YscX, have also been proposed to be members of the injectisome (Iriarte and Cornelis, 1999; Day and Plano, 2000), but this possibility has not been well investigated.

8.3 THE MODEL

As shown in Figure 8.1, bacteria that have just entered the host produce the Ysc secretion machinery and low levels of the Yop effectors, as well as the main players in the control of secretion described above. The expression of this system is positively regulated by VirF at the host temperature of 37 °C. No secretion occurs through the Ysc secretion channel because YopN blocks secretion from the outside of the cell and LcrG blocks secretion from the inside. Low levels of LcrV are found on the bacterial surface in discrete foci; this LcrV is probably associated with the secretion channel. TyeA is also found on the surface; because it is able to bind YopN it is presumed to be in association with this protein. Inside the host, the VirF activator can bind to *yop* genes and induce their transcription only if the negative regulatory signal is removed. This negative regulatory signal is controlled by YscM1, YscM2 and YopD. As there is no evidence that these molecules bind DNA, they must

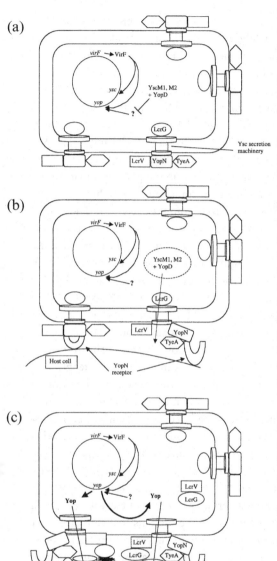

Figure 8.1. Model for induction of protein secretion by *Yersinia* upon contact with host cells. This process is described in greater detail in the text. (a) Before coming into close contact with a eukaryotic cell, *Yersinia* inside the host express the Ysc secretion machinery but proteins are secreted only at a low level due to the presence of YopN and LcrG, which plug the machinery. (b) Upon contact, sensed by YopN, the secretion pore opens and the negative regulators YscM1 and YscM2 are released from the cell. This release results in a lifting of the repression, and the *yop* and *lcrV* genes are expressed. (c) LcrV protein binds LcrG, removing it from the Ysc machinery, further opening the secretion channel. Proteins involved in injection of Yops into the host cell are secreted and perhaps assembled into a macromolecular injectisome structure on the bacterial cell surface. Yops are secreted directly from the bacterium into the host cell through this injectisome structure. HSP, heparin sulphate proteoglycans.

work through another molecule(s) to repress *yop* gene expression. This other molecule (or molecules ('?' in Fig. 8.1]) may be an activator like VirF, in which case YscM1, YscM2 and YopD negatively regulate its activity as shown in Fig. 8.1. Alternatively, this other molecule could be a repressor which is activated by the YscM1, YscM2, YopD negative regulatory system. Bacteria that come into intimate contact with a host cell through the action of adhesins such as YadA, Inv, Ail and pH6 antigen (Isberg and Falkow, 1985; Miller and Falkow, 1988; Bliska *et al.*, 1993; Yang and Isberg, 1993; Yang *et al.*, 1996) sense this contact through the YopN protein attached to the external portion of the Ysc secretion machinery (Fig. 8.1b). This binding of YopN to its receptor leads to opening of the channel that has contacted the cell through YopN, and the negative regulator YscM is released from the bacterium. The level of secretion from the channel at this point is low because the LcrG protein is still blocking secretion from the inside of the bacterium. The removal of the negative regulator leads to the up-regulation of *yop* and *lcrV* gene expression. Increased levels of LcrV bind LcrG and remove it from its location blocking the inside of the channel (Fig. 8.1c). LcrV also aids in YopB secretion. The secretion of YopB and YopD results in the formation of the pore through which the effector Yops are translocated. Small amounts of LcrG are secreted, aiding in translocation in some unknown manner. This translocation is dependent upon the binding of heparin sulphate proteoglycans. Additionally YscO, YscP and YscX (not shown) are released into the extracellular space, where they may be involved in forming an injectisome. The full opening of the secretion channel and formation of the injectisome and translocation pore results in full secretion and translocation of the YscM proteins such that *yop* and *lcrV* gene expression is elevated to even higher levels. Ultimately, high levels of Yops are introduced into the eukaryotic cell cytoplasm, where they poison normal cell functions, and secreted LcrV suppresses the immune system. The bacterium senses its contact with a host cell and only upon this contact does it release the proteins that can have their effects on host cell function. This tight coupling of gene expression, secretion and translocation is useful to the bacterium as it allows the organism to secrete the Yop effector molecules when the target of the molecule (inside the eukaryotic cell) is readily available.

REFERENCES

Allaoui, A., Woestyn, S., Sluiters, C. and Cornelis, G.R. (1994). YscU, a *Yersinia enterocolitica* inner membrane protein involved in Yop secretion. *Journal of Bacteriology* **176**, 4534–4542.

D. E. PIERSON

Bennett, J.C.Q. and Hughes, C. (2000). From flagellum assembly to virulence: the extended family of type III export chaperones. *Trends in Microbiology* 8, 202–204.

Bergman, T., Håkansson, S., Forsberg, Å., Norlander, L., Macellaro, A., Backman, A., Bolin, I. and Wolf-Watz, H. (1991). Analysis of the V antigen *lcrGVH-yopBD* operon of *Yersinia pseudotuberculosis*: evidence for a regulatory role of LcrH and LcrV. *Journal of Bacteriology* 173, 1607–1616.

Bliska, J.B. (2000). Yop effectors of *Yersinia* spp. and actin rearrangements. *Trends in Microbiology* 8, 205–208.

Bliska, J.B., Copass, M.C. and Falkow, S. (1993). The *Yersinia pseudotuberculosis* adhesin *yadA* mediates intimate bacterial attachment to and entry into HEp-2 cells. *Infection and Immunity* 61, 3914–3921.

Boland, A., Sory, M.-P., Iriarte, M., Kerbourch, C., Wattiau, P. and Cornelis, G. R. (1996). Status of YopM and YopN in the *Yersinia* Yop virulon: YopM of *Yersinia enterocolitica* is internalized inside the cytosol of PU5–1.8 macrophages by the YopB, D, N delivery apparatus. *EMBO Journal* 15, 5191–5201.

Boyd, A.P., Sory, M.-P., Iriarte, M. and Cornelis, G.R. (1998). Heparin interferes with translocation of Yop proteins into HeLa cells and binds to LcrG, a regulatory component of the *Yersinia* Yop apparatus. *Molecular Microbiology* 27, 425–436.

Burns, D.L. (1999). Biochemistry of type IV secretion. *Current Opinion in Microbiology* 2, 25–29.

Cambronne, E.D., Cheng, L.W. and Schneewind, O. (2000). LcrQ/YscM1, regulators of the *Yersinia* yop virulon, are injected into host cells by a chaperone-dependent mechanism. *Molecular Microbiology* 37, 263–273.

Cheng, L.W. and Schneewind, O. (2000a). Type III machines of Gram-negative bacteria: delivering the goods. *Trends in Microbiology* 8, 214–220.

Cheng, L.W. and Schneewind, O. (2000b). *Yersinia enterocolitica* TyeA, an intracellular regulator of the type III machinery is required for specific targeting of YopE, YopH, YopM and YopN into the cytosol of eukaryotic cells. Journal of Bacteriology 182, 3183–3190.

Cheng, L.W., Anderson, D.M. and Schneewind, O. (1997). Two independent type III secretion mechanisms for YopE in *Yersinia enterocolitica*. *Molecular Microbiology* 24, 757–765.

Cornelis, G.R. (1993). Role of the transcriptional activator VirF and the histone-like protein YmoA in the thermoregulation of virulence functions in *Yersiniae*. *Zentralblatt für Bakteriologie* 278, 149–164.

Cornelis, G.R. (2000). Molecular and cell biology aspects of plague. *Proceedings of the National Academy of Sciences, USA* 97, 8778–8783.

Cornelis, G.R. and Van Gijsegem, E. (2000). Assembly and function of type III secretion systems. *Annual Review of Microbiology* 54, 735–774.

Cornelis, G.R., Sluiters, C., Lambert de Rouvroit, D. and Michiels, T. (1989). Homology between VirF, the transcriptional activator of the *Yersinia* virulence regulon, and AraC, the *Escherichia coli* arabinose operon regulator. *Journal of Bacteriology* **171**, 254–262.

Cornelis, G.R., Sluiters, C., Delor, I., Geib, D., Kaniga, K., Lambert de Rouvroit, C., Sory, M.-P., Vanooteghem, J.-C. and Michiels, T. (1991). *ymoA*, a *Yersinia enterocolitica* chromosomal gene modulating the expression of virulence functions. *Molecular Microbiology* **5**, 1023–1034.

Day, J.B. and Plano, G.V. (1998). A complex composed of SycN and YscB functions as a specific chaperon for YopN in *Yersinia pestis*. *Molecular Microbiology* **30**, 777–788.

Day, J.B. and Plano, G.V. (2000). The *Yersinia pestis* YscY protein directly binds YscX, a secreted component of the type III secretion machinery. *Journal of Bacteriology* **182**, 1834–1843.

Fields, K.A., Nilles, M.L., Cowan, C. and Straley, S.C. (1999). Virulence role of V antigen of *Yersinia pestis* at the bacterial surface. *Infection and Immunity* **67**, 5395–5408.

Forsberg, Å., Viitanen, A.-M., Skurnik, M. and Wolf-Watz, H. (1991). The surface-located YopN protein is involved in calcium signal transduction in *Yersinia pseudotuberculosis*. *Molecular Microbiology* **5**, 977–986.

Francis, M.S. and Wolf-Watz, H. (1998). YopD of *Yersinia pseudotuberculosis* is translocated into the cytosol of HeLa epithelial cells: evidence of a structural domain necessary for translocation. *Molecular Microbiology* **29**, 799–813.

Frithz-Lindsten, E., Rosqvist, R., Johansson, L. and Forsberg, Å. (1995). The chaperone-like protein YerA of *Yersinia pseudotuberculosis* stabilizes YopE in the cytoplasm but is dispensable for targeting to the secretion loci. *Molecular Microbiology* **16**, 635–647.

Gallegos, M.T., Schleif, R., Bairoch, A., Hofmann, K. and Ramos, J.L. (1997). AraC/XylS family of transcriptional regulators. *Microbiology and Molecular Biology Reviews* **61**, 393–410.

Ginocchio, C.C., Olmsted, S.B., Wells, C.L. and Galán, J.E. (1994). Contact with epithelial cells induces the formation of surface appendages on *Salmonella typhimurium*. *Cell* **76**, 714–724.

Goguen, J.D., Yother, J. and Straley, S.C. (1984). Genetic analysis of the low calcium response in *Yersinia pestis* Mu d1(Ap *lac*) insertion mutants. *Journal of Bacteriology* **160**, 842–848.

Håkansson, S., Bergman, T., Vanooteghem, J.C., Cornelis, G. and Wolf-Watz, H. (1993). YopB and YopD constitute a novel class of *Yersinia* Yop proteins. *Infection and Immunity* **61**, 71–80.

Håkansson, S., Galyov, E.E., Rosqvist, R. and Wolf-Watz, H. (1996a). The *Yersinia*

YpkA Ser/Thr kinase is translocated and subsequently targeted to the inner surface of the HeLa cell plasma membrane. *Molecular Microbiology* **20**, 593–603.

Håkansson, S., Schesser, K., Persson, C., Galyov, E.E., Rosqvist, R., Homble, F. and Wolf-Watz, H. (1996b). The YopB protein of *Yersinia pseudotuberculosis* is essential for the translocation of Yop effector proteins across the target cell plasma membrane and displays a contact-dependent membrane disrupting activity. *EMBO Journal* **15**, 5812–5823.

Haller, J.C., Carlson, S., Pederson, K.J. and Pierson, D.E. (2000). A chromosomally-encoded type III secretion pathway in *Yersinia enterocolitica* is important in virulence. *Molecular Microbiology* **36**, 1436–1446.

Hartland, E.L., Green, S.P., Philipps, W.A. and Robins-Browne, R.M. (1994). Essential role of YopD in inhibition of the respiratory burst of macrophages by *Yersinia enterocolitica*. *Infection and Immunity* **62**, 4445–4453.

Henderson, I.R., Navarro-Garcia, F. and Nataro, J.P. (1998). The great escape: structure and function of autotransporter proteins. *Trends in Microbiology* **6**, 370–378.

Hölmstrom, A., Pettersson, J., Rosqvist, R., Håkansson, S., Tafazoli, F., Fällman, M., Magnusson, K.-E., Wolf-Watz, H. and Forsberg, Å. (1997). YopK of *Yersinia pseudotuberculosis* controls translocation of Yop effectors across the eukaryotic cell membrane. *Molecular Microbiology* **24**, 73–91.

Hueck, C.J. (1998). Type III protein secretion systems in bacterial pathogens of animals and plants. *Microbiology and Molecular Biology Review* **62**, 379–433.

Hughes, K.T., Gillen, K.L., Semon, M.J. and Karlinsey, J.E. (1993). Sensing structural intermediates in bacterial flagellar assembly by export of a negative regulator. *Science* **262**, 1277–1280.

Iriarte, M. and Cornelis, G.R. (1998). YopT, a new *Yersinia* Yop effector protein, affects the cytoskeleton of host cells. *Molecular Microbiology* **29**, 915–929.

Iriarte, M. and Cornelis, G.R. (1999). Identification of SycN, YscX and YscY, three new elements of the *Yersinia yop* virulon. *Journal of Bacteriology* **181**, 675–680.

Iriarte, M., Sory, M.-P., Boland, A., Boyd, A.P., Mills, S.D., Lambermont, I. and Cornelis, G.R. (1998). TyeA, a protein involved in control of Yop release and translocation of *Yersinia* Yop effectors. *EMBO Journal* **17**, 1907–1918.

Isberg, R.R. and Falkow, S. (1985). A single genetic locus encoded by *Yersinia pseudotuberculosis* permits invasion of cultured animal cells by *Escherichia coli* K-12. *Nature* **317**, 262–264.

Jackson, M.W., Day, J.B. and Plano, G.V. (1998). YscB of *Yersinia pestis* functions as a specific chaperone for YopN. *Journal of Bacteriology* **180**, 4912–4921.

Knutton, S., Rosenshine, I., Pallen, M.J., Nisan, I., Neves, B.C., Bain, C., Wolff,

C., Dougan, G. and Frankel, G. (1998). A novel EspA-associated surface organelle of enteropathogenic *Escherichia coli* involved in protein translocation into epithelial cells. *EMBO Journal* 17, 2166–2176.

Koster, M., Bitter, W., de Cock, H., Allaoui, A., Cornelis, G. R. and Tommassen, J. (1997). The outer membrane component, YscC, of the Yop secretion machinery of *Yersinia enterocolitica* forms a ring-shaped multimeric complex. *Molecular Microbiology* 26, 789–797.

Kubori, R., Matsushima, Y., Nakamura, D., Uralil, J., Lara-Tejero, M., Sukhan, A., Galán, J.E. and Aizawa, S.-I. (1998). Supramolecular structure of the *Salmonella typhimurium* type III protein secretion system. *Science* 280, 602–605.

Kubori, R., Sukhan, A., Aizawa, S.-I. and Galán, J.E. (2000). Molecular characterization and assembly of the needle complex of the *Salmonella typhimurium* type III protein secretion system. *Proceedings of the National Academy of Sciences, USA* 97, 10225–10230.

Kutsukake, K. (1994). Excretion of the anti-sigma factor through a flagellar substructure couples flagellar gene expression with flagellar assembly in *Salmonella typhimurium*. *Molecular and General Genetics* 243, 605–612.

Lambert de Rouvroit, C., Sluiters, C. and Cornelis, G.R. (1992). Role of the transcriptional activator, VirF, and temperature in the expression of the pYV plasmid genes of *Yersinia enterocolitica*. *Molecular Microbiology* 6, 395–409.

Lawton, W.D., Erdman, R.I. and Surgalla, M.J. (1963). Biosynthesis and purification of V and W antigens in *Pasteurella pestis*. *Journal of Immunology* 91, 179–184.

Lee, V.T. and Schneewind, O. (1999). Type III machines of pathogenic yersiniae secrete virulence factors into the extracellular milieu. *Molecular Microbiology* 31, 1619–1629.

Lee, V.T., Anderson, D.M. and Schneewind, O. (1998). Targeting of *Yersinia* Yop proteins into the cytosol of HeLa cells: one-step translocation of YopE across bacterial and eukaryotic membranes is dependent on SycE chaperone. *Molecular Microbiology* 28, 593–601.

Lee, V.T., Tam, C. and Schneewind, O. (2000). LcrV, a substrate for *Yersinia enterocolitica* type III secretion, is required for toxin targeting into the cytosol of HeLa cells. *Journal of Biological Chemistry* 275, 36869–36875.

Ménard, R., Sansonetti, P. and Parsot, C. (1994). The secretion of the *Shigella flexneri* Ipa invasins is activated by epithelial cells and controlled by IpaB and IpaD. *EMBO Journal* 13, 5293–5302.

Michiels, T., Vanooteghem, J.C., Lambert de Rouvroit, C., China, B., Gustin, A., Boudry, P. and Cornelis, G.R. (1991). Analysis of *virC*, an operon involved in the secretion of Yop proteins by *Yersinia enterocolitica*. *Journal of Bacteriology* 173, 4994–5009.

Miller, V.L. and Falkow, S. (1988). Evidence for two genetic loci in *Yersinia enterocolitica* that can promote invasion of epithelial cells. *Infection and Immunity* 56, 1242–1248.

Minamino, R. and Macnab, R.M. (1999). Components of the *Salmonella* flagellar export apparatus and classification of export substrates. *Journal of Bacteriology* 181, 1388–1394.

Nakajima, R., Motin, V.L. and Brubaker, R.R. (1995). Suppression of cytokines in mice by protein A–V antigen fusion peptide and restoration of synthesis by active immunization. *Infection and Immunity* 63, 3021–3029.

Neyt, C. and Cornelis, G.R. (1999a). Insertion of a Yop translocation pore into the macrophage plasma membrane by *Yersinia enterocolitica*: requirement for translocators YopB and YopD, but not LcrG. *Molecular Microbiology* 33, 971–981.

Neyt, C. and Cornelis, G.R. (1999b). Role of SycD, the chaperone of the *Yersinia* Yop translocators YopB and YopD. *Molecular Microbiology* 31, 143–156.

Nilles, M.L., Williams, A.W., Skrzypek, E. and Straley, S.C. (1997). *Yersinia pestis* LcrV forms a stable complex with LcrG and may have a secretion-related regulatory role in the low-Ca^{2+} response. *Journal of Bacteriology* 179, 1307–1316.

Nilles, M.L., Fields, K.A. and Straley, S.C. (1998). The V antigen of *Yersinia pestis* regulates Yop vectorial targeting as well as Yop secretion through effects on YopB and LcrG. *Journal of Bacteriology* 180, 3410–3420.

Payne, P.L. and Straley, S.C. (1998). YscO of *Yersinia pestis* is a mobile core component of the Yop secretion system. *Journal of Bacteriology* 180, 3882–3890.

Persson, C., Nordfelth, R., Hölmstrom, A., Håkansson, S., Rosqvist, R. and Wolf-Watz, H. (1995). Cell-surface-bound *Yersinia* translocate the protein tyrosine phosphatase YopH by a polarized mechanism into the target cell. *Molecular Microbiology* 18, 135–150.

Pettersson, J., Nordfelth, R., Dubinina, E., Bergman, T., Gustafsson, M., Magnusson, K. E. and Wolf-Watz, H. (1996). Modulation of virulence factor expression by pathogen target cell contact. *Science* 273, 1231–1233.

Pettersson, J., Hölmstrom, A., Hill, J., Leary, S., Frithz-Lindsten, E., von Euler-Matell, A., Carlsson, W., Titball, R., Forsberg, Å. and Wolf-Watz, H. (1999). The V-antigen of *Yersinia* is surface exposed before target cell contact and involved in virulence protein translocation. *Molecular Microbiology* 32, 961–976.

Plano, G.V. and Straley, S.C. (1995). Mutations in *yscC*, *yscD*, and *yscG* prevent high-level expression and secretion of V antigen and Yops in *Yersinia pestis*. *Journal of Bacteriology* 177, 3843–3854.

Plano, G.V., Barve, S.S. and Straley, S.C. (1991). LcrD, a membrane-bound regulator of the *Yersinia pestis* low-calcium response. *Journal of Bacteriology* 173, 7293–7303.

Portnoy, D.A., Moseley, S.L. and Falkow, S. (1981). Characterization of plasmids and plasmid-associated determinants of *Yersinia enterocolitica* pathogenesis. *Infection and Immunity* **31**, 775–782.

Portnoy, D.A., Blank, H.F., Kingsbury, D.T. and Falkow, S. (1983). Genetic analysis of essential plasmid determinants of pathogenicity in *Yersinia pestis*. *Journal of Infectious Diseases* **148**: 297–304.

Price, S.B. and Straley, S.C. (1989). *lcrH*, a gene necessary for virulence of *Yersinia pestis* and for the normal response of *Y. pestis* to ATP and calcium. *Infection and Immunity* **57**, 1491–1498.

Rimpiläinen, M., Forsberg, Å. and Wolf-Watz, H. (1992). A novel protein, LcrQ, involved in the low-calcium response of *Yersinia pseudotuberculosis* shows extensive homology to YopH. *Journal of Bacteriology* **174**, 3355–3363.

Rosqvist, R., Forsberg, Å. and Wolf-Watz, H. (1991a). Intracellular targeting of the *Yersinia* YopE cytotoxin in mammalian cells induces actin microfilament disruption. *Infection and Immunity* **59**, 4562–4569.

Rosqvist, R., Forsberg, Å. and Wolf-Watz, H. (1991b). Microinjection of the *Yersinia* YopE cytotoxin in mammalian cells induces actin microfilament disruption. *Biochemical Society Transactions* **19**, 1131–1132.

Rosqvist, R., Magnusson, K.-E. and Wolf-Watz, H. (1994). Target cell contact triggers expression and polarized transfer of *Yersinia* YopE cytotoxin into mammalian cells. *EMBO Journal* **13**, 964–972.

Sarker, M.R., Neyt, C., Stainier, I. and Cornelis, G.R. (1998). The *Yersinia* Yop virulon: LcrV is required for extrusion of the translocators YopB and YopD. *Journal of Bacteriology* **180**, 1207–1214.

Schmiel, D.H., Wagar, E., Karamanou, L., Weeks, D. and Miller, V.L. (1998). Phospholipase A of *Yersinia enterocolitica* contributes to pathogenesis in a mouse model. *Infection and Immunity* **66**, 3941–3951.

Skrzypek, E. and Straley, S.C. (1993). LcrG, a secreted protein involved in negative regulation of the low-calcium response in *Yersinia pestis*. *Journal of Bacteriology* **175**, 3520–3528.

Skrzypek, E. and Straley, S.C. (1995). Differential effects of deletions in *lcrV* on secretion of V antigen, regulation of the low-Ca^{2+} response, and virulence of *Yersinia pestis*. *Journal of Bacteriology* **177**, 2530–2542.

Skrzypek, E., Cowan, C. and Straley, S.C. (1998). Targeting of the *Yersinia pestis* YopM protein into HeLa cells and intracellular trafficking to the nucleus. *Molecular Microbiology* **30**, 1051–1065.

Sory, M.-P. and Cornelis, G.R. (1994). Translocation of a hybrid YopE-adenylate cyclase from *Yersinia enterocolitica* into HeLa cells. *Molecular Microbiology* **14**, 583–594.

Sory, M.-P., Boland, A., Lambermont, I. and Cornelis, G.R. (1995). Identification of the YopE and YopH domains required for secretion and internalization into the cytosol of macrophages, using the *cyaA* gene fusion approach. *Proceedings of the National Academy of Sciences, USA* **92**, 11998–12002.

Stainier, I., Iriarte, M. and Cornelis, G.R. (1997). YscM1 and YscM2, two *Yersinia enterocolitica* proteins causing downregulation of *yop* transcription. *Molecular Microbiology* **26**, 833–843.

Stainier, I., Bleves, S., Josenhans, C., Karmani, L., Kerbourch, C., Tötemeyer, S., Boyd, A. and Cornelis, G.R. (2000). YscP, a *Yersinia* protein required for Yop secretion that is surface exposed, and released in low Ca^{2+}. *Molecular Microbiology* **37**, 1005–1018.

Stathopoulos, C., Hendrixson, D.R., Thanassi, D.G., Hultgren, S.J., St Geme, J.W. III and Curtiss, R. III (2000). Secretion of virulence determinants by the general secretory pathway in Gram-negative pathogens: an evolving story. *Microbes and Infection* **2**, 1061–1072.

Straley, S.C. (1988). The plasmid-encoded outer-membrane proteins of *Yersinia pestis*. *Review of Infectious Diseases* **10**, S323–S326.

Straley, S.C. and Bowmer, W.S. (1986). Virulence genes regulated at the transcriptional level by Ca^{2+} in *Yersinia pestis* include structural genes for outer membrane proteins. *Infection and Immunity* **51**, 445–454.

Straley, S.C. and Brubaker, R.R. (1981). Cytoplasmic and membrane proteins of yersiniae cultivated under conditions simulating mammalian intracellular environment. *Proceedings of the National Academy of Sciences, USA* **78**, 1224–1228.

Tamano, K., Aizawa, S.-I., Katayama, E., Nonaka, T., Imajoh-Ohmi, S., Kuwae, A., Nagai, S. and Sasakawa, C. (2000). Supramolecular structure of the *Shigella* type III secretion machinery: the needle part is changeable in length and essential for delivery of effectors. *EMBO Journal* **19**, 3876–3887.

Thanassi, D.G. and Hultgren, S.J. (2000). Multiple pathways allow protein secretion across the bacterial outer membrane. *Current Opinion in Cell Biology* **12**, 420–430.

Une, T. and Brubaker, R.R. (1984). Roles of V antigen in promoting virulence and immunity in yersiniae. *Journal of Immunology* **133**, 2226–2230.

Vallis, A.J., Yahr, T.L., Barbieri, J.T. and Frank, D.W. (1999). Regulation of ExoS production and secretion by *Pseudomonas aeruginosa* in response to tissue culture conditions. *Infection and Immunity* **67**, 914–920.

van Dijk, K., Fouts, D.E., Rehn, A.H., Hill, A.R., Collmer, A. and Alfano, J.R. (1999). The Avr (effector) proteins HrmA (HopPsyA) and AvrPto are secreted in culture from *Pseudomonas syringae* pathovars via the Hrp (type III) protein secretion system in a temperature- and pH-sensitive manner. *Journal of Bacteriology* **181**, 4790–4797.

Van Gijsegem, F., Genin, S. and Boucher, C. (1993). Conservation of secretion pathways for pathogenicity determinants of plant and animal bacteria. *Trends in Microbiology* **1**, 175–180.

Wandersman, C. (1998). Protein and peptide secretion by ABC exporters. *Research in Microbiology* **149**, 163–170.

Wattiau, P. and Cornelis, G.R. (1993). SycE, a chaperone-like protein of *Yersinia enterocolitica* involved in the secretion of YopE. *Molecular Microbiology* **8**, 123–131.

Wattiau, P., Bernier, B., Deslee, P., Michiels, T. and Cornelis, G.R. (1994). Individual chaperones required for Yop secretion by *Yersinia*. *Proceedings of the National Academy of Sciences, USA* **91**, 10493–10497.

Welkos, S., Friedlander, A., McDowell, D., Weeks, J. and Tobery, S. (1998). V antigen of *Yersinia pestis* inhibits neutrophil chemotaxis. *Microbial Pathogenesis* **24**, 185–196.

Williams, A.W. and Straley, S.C. (1998). YopD of *Yersinia pestis* plays a role in negative regulation of the low-calcium response in addition to its role in translocation of Yops. *Journal of Bacteriology* **180**, 350–358.

Woestyn, S., Allaoui, A., Wattiau, P. and Cornelis, G.R. (1994). YscN, the putative energizer of the *Yersinia* Yop secretion machinery. *Journal of Bacteriology* **176**, 1561–1569.

Woestyn, S., Sory, M.-P., Boland, A., Lequenne, O. and Cornelis, G.R. (1996). The cytosolic SycE and SycH chaperones of *Yersinia* protect the region of YopE and YopH involved in translocation across eukaryotic cell membranes. *Molecular Microbiology* **20**, 1261–1271.

Wolff, C., Nisan, I., Hanski, E., Frankel, G. and Rosenshine, I. (1998). Protein translocation into host epithelial cells by infecting enteropathogenic *Escherichia coli*. *Molecular Microbiology* **28**, 143–155.

Wolff, J., Cook, G.H., Goldhammer, A.R. and Berkowtiz, S.A. (1980). Calmodulin activates prokaryotic adenylate cyclase. *Proceedings of the National Academy of Sciences, USA* **77**, 3841–3844.

Yang, Y. and Isberg, R.R. (1993). Cellular internalization in the absence of invasin expression is promoted by the *Yersinia pseudotuberculosis yadA* product. *Infection and Immunity* **61**, 3907–3913.

Yang, Y., Merriam, J.J., Mueller, J.P. and Isberg, R.R. (1996). The *psa* locus is responsible for thermoinducible binding of *Yersinia pseudotuberculosis* to cultured cells. *Infection and Immunity* **64**, 2483–2489.

Yother, J. and Goguen, J.D. (1985). Isolation and characterization of Ca^{2+}-blind mutants of *Yersinia pestis*. *Journal of Bacteriology* **164**, 704–711.

Young, G.M., Schmiel, D.H. and Miller, V.L. (1999). A new pathway for the secretion of virulence factors by bacteria: the flagellar export apparatus functions as a protein-secretion system. *Proceedings of the National Academy of Sciences, USA* **96**, 6456–6461.

CHAPTER 9

Functional modulation of pathogenic bacteria upon contact with host target cells

Andreas U. Kresse, Frank Ebel and Carlos A. Guzmán

9.1 INTRODUCTION

(203)

The list of prokaryotic microorganisms able to cause illness or death in animals and plants is long. In fact, infectious diseases constitute the most common cause of human morbidity and death in the world. Our fight against these pathogens, however, started to become a serious scientific business only in our very recent history. The experience gathered during the past century has already demonstrated that even the initial hopes of the antibiotic era, namely to save uncountable lives by eradicating or limiting bacterial infections, turned out to be idealistic dreams of a scientific generation rather than deliverable promises. We now realize that the concept of either a pathogen-free world or the existence of a universally active therapeutic tool belongs to the realms of science fiction.

The treatment of infected patients has been rendered complicated by the constant emergence of multidrug-resistant strains. This is further impaired by the appearance of novel infectious agents that are not treatable by conventional therapies. Although it is expected that, as a result of the combined efforts of pharmaceutical companies, we will see new antibiotics coming onto the market, it has become clear that this strategy to combat pathogenic bacteria is insufficient and has limited value. In fact, a single point mutation in a gene from a member of the microbial community can destroy the outcome of many years of research and development. The identification of novel molecular targets for therapeutic intervention, in combination with high throughput screening for novel active compounds, will have a key role in the discovery of new tools to address these problems. However, it is essential that we first of all gain a better understanding of microbial pathogenesis, in order to identify the critical bottlenecks during the infection process. The fine modulation of pathogenesis-related functions and structures during infection constitutes a critical physiological event for infectious agents, providing a golden opportunity to devise novel strategies to attack pathogens, and thereby counteracting their attempts to colonize host target tissues.

9.2 THE NEED FOR A TIGHT MODULATION OF BACTERIAL VIRULENCE

Different portals of entry into the human body are exploited by different pathogens. On the one hand, some bacteria cause infections of the skin or lung epithelium that may progress to the underlying tissues. However, pre-existing injury or host predisposition are usually a prerequisite. An example for this group of opportunistic pathogens is *Pseudomonas aeruginosa*, which is naturally present in the environment but triggers disease in immuno-compromised individuals or patients suffering from cystic fibrosis. On the other hand, a large number of pathogens target the intestinal tract as this rep-resents the largest interface between the host and its external environment. To reach the relatively friendly environment of their target areas in the gut, bacteria have to survive harsh conditions in the stomach, overcome protec-tive barriers (e.g. bile salts) and compete with the commensal flora. A large inoculum, enclosure within and protection by food particles, and the pres-ence of alkaline substances that may temporarily increase the pH in the stomach are factors that facilitate an infection by otherwise-susceptible bac-teria. Interestingly, some pathogens have developed extremely sophisticated strategies to tolerate low pH. Examples of such highly infectious pathogens are enterohaemorrhagic *Escherichia coli* (EHEC) and *Shigella* spp., for which an incredibly small inoculum (10–100 bacteria) is sufficient for them to reach their targets and promote disease, overcoming the non-specific host clearance mechanisms.

The pathway from the external environment to a specific habitat within the host can be long, dangerous and plagued by stress factors. In order to survive, a microorganism has to adapt itself to rapidly changing environmen-tal conditions. Certain virulence factors may be necessary to ensure survival of a pathogen within a particular habitat, but it has been shown that their untimely expression may have a devastating effect on the pathogen (Akerley and Miller, 1996). Virulence factors are often regulated at the transcriptional level, but also they may be directly activated or else assembled from stored components to form functional complexes. In all cases, this takes place in response to a variety of environmental signals and is orchestrated by an extremely complex cascade of regulatory factors (Mekalanos, 1992; Finlay and Falkow, 1997).

The precise regulation of the pathogenic repertoire is essential for most bacterial pathogens and especially those that transit across several, and completely different, habitats on their way to and through the host. A finely tuned regulation of protein expression requires the ability to sense external stimuli in order to identify the actual environment in which the pathogen finds itself. The major means by which this is accomplished is through the presence of response regulators that are located in the bacterial envelope and can sense a wide variety of signals (e.g. temperature, oxygen concentration, osmolarity, ion and nutrient gradients, and cell density) and respond accordingly by activating the expression of appropriate genes. The larger the variety of environments a given pathogen has to encounter, the greater is its need to be able to recognize its actual location. A pathogen like *Helicobacter*, which is specialized to colonize gastric epithelial cells, may achieve a clear identification of its habitat by sensing temperature and pH, whereas an organism like *Salmonella typhi*, which breaches the gut mucosal barrier spreading systemically via infected macrophages, is faced with a variety of situations and habitats, thereby needing to integrate multiple signals to accurately identify the actual environment in order to activate the appropriate set of genes.

9.3 TYPE III SECRETION SYSTEMS ARE COMPLEX STRUCTURES WHICH ARE MODULATED DURING INFECTION

The bacterial flagellum enables motility and spread of the organism. However, this organelle is generated at the expense of restricted energy and nutrient resources and under certain conditions the possession of an active flagellum may even be harmful to the organism. Hence, flagella can be excellent targets for the host immune system or they may impair bacterial adhesion to host cells. Therefore, in many pathogens, the expression of flagella is tightly regulated, although the means by which such organisms downregulate the production of, or even shed, these organelles, are currently poorly understood. However, the situation is more clear for *Caulobacter crescentus*, in which an active cell cycle-controlled degradation of a flagellar motor protein has been reported that triggers the subsequent release of the organelle (Jenal and Shapiro, 1996).

The so-called type III secretion system, which is found in many Gram-negative bacteria such as EHEC, enteropathogenic *E. coli* (EPEC), *Shigella*, *Yersinia* and *Salmonella*, facilitates the export of proteins in a one-step mechanism from the bacterial cytosol to either the surrounding medium (secretion) or into the cytoplasm of a cell of the infected host (translocation). Proteins of the type III secretion system show striking homologies to components of the flagella basal body, and both systems share a similar structural organization (Stephens and Shapiro, 1996; Kubori *et al.*, 1998; Blocker *et al.*, 1999). Because of this, it has been proposed that the flagellum basal body and the pathogenic injection apparatus are designated 'type III systems a and b', respectively (Stephens and Shapiro, 1996). However, for the sake of convenience we will use the term 'type III secretion system' only for the injection apparatus of pathogenic bacteria.

The core of the type III secretion machinery in the bacterial envelope appears to be similar in most pathogens. In contrast, the surface-exposed portions can differ significantly and may comprise specialized structures that reflect different strategies to gain intimate contact with the target host cell surface (Ebel *et al.*, 1998; Knutton *et al.*, 1998; Kubori *et al.*, 1998). Like the flagellum, type III secretion systems are assembled and disassembled in a tightly regulated manner, thus providing an excellent model to study the functional and structural modulation of pathogenic bacteria upon contact with target cells.

9.4 THE TYPE III SECRETION MACHINERY OF EHEC IS INVOLVED IN THE ATTACHMENT TO, AND MODIFICATION OF, HOST CELLS

Our work in the last few years has focused on the assembly, structure and function of the type III secretion system of EHEC, which belongs to the larger group of Shiga toxin-producing *E. coli* (STEC). This microorganism can cause food-borne epidemics of gastrointestinal illnesses that range from mild diarrhoea to severe diseases such as haemorrhagic colitis and haemolytic-uremic syndrome (Boyce *et al.*, 1995; Nataro and Kaper, 1998). The primary source of EHEC infections in humans is livestock, particularly contaminated meat or dairy products (Renwick *et al.* 1994; Gallien *et al.*, 1997). Interestingly, animals carrying EHEC are often asymptomatic, but it is so far unclear whether EHEC actually might be a commensal in these hosts. Although contaminated food is the major source of infection, EHEC are also able to survive for a long time in the environment, for example lakes or drinking water reservoirs (Swerdlow *et al.*, 1992; Keene *et al.*, 1994). They can also survive at a low pH and have a low infective dose – between 10 and 100 viable cells (Gorden and Small, 1993). Treatment of EHEC infections is complicated, as the administration of antibiotics can actually worsen the clinical symptoms due to the induction of the massive release of Shiga toxins. Therefore an improved knowledge of the function and regulation of the type III secretion system, which is essential for this pathogen to establish an infection in the gut, may facilitate the discovery of alternative therapeutic approaches.

After their passage through the stomach, EHEC must quickly adhere to the intestinal epithelium to avoid being shed. The colonization of the duodenum by EHEC resembles the colonization of the small intestine by the closely related EPEC, both infections being associated with histopathological changes known as 'attaching and effacing' (A/E) lesions (Moon et al., 1983; Tzipori et al., 1995). Bacteria penetrate the mucus layer in an unknown manner, establish direct contact with the epithelial cells, induce an effacement of the microvilli, attach intimately, and trigger a locally restricted reorganization of the cytoskeleton (Donnenberg et al., 1997). This massive recruitment of cellular cytoskeletal elements leads to the formation of protruding pedestal-like structures on which the organisms reside (see Plate 9.1).

9.5 THE LINKAGE BETWEEN THE TYPE III SECRETION MACHINERY AND THE FORMATION OF A/E LESIONS

Most of the factors required to produce A/E lesions are encoded by a large chromosomal locus called LEE (locus of enterocyte effacement). This pathogenicity island comprises (i) the genes for a type III secretion system, (ii) the outer membrane protein intimin (eae), (iii) the secreted proteins EspA, EspD, EspB, and EspF, and (iv) the translocated intimin receptor Tir (for STEC, this was originally designated EspE). Tir is injected (translocated) into eukaryotic cells by bacteria via their type III-system and inserts itself into the cellular membrane, where it binds to intimin (Deibel et al., 1998). Evidence suggests that Tir is also directly involved in the reorganization of filamentous actin, thereby linking the attached pathogen directly to the cytoskeleton (Goosney et al., 2000). Genetic experiments in which the LEE element was transferred to E. coli K12 showed that the LEE$_{EPEC}$ is sufficient to confer the capacity to cause A/E lesions (McDaniel and Kaper, 1997), whereas the LEE$_{EHEC}$ is not (Elliott et al., 1999). This might in part be due to the lack of a so far unknown essential accessory gene located outside of the LEE$_{EHEC}$ or an altered transcriptional regulation of factors encoded by the pathogenicity island.

Analysis of the LEE-encoded proteins secreted under different in vitro conditions revealed that they are not (or only poorly) expressed after growth at low temperatures or in bacterial media (e.g. Luria broth). In contrast, supernatants collected after bacterial growth in cell culture media at 37 °C contain nearly exclusively the three LEE-encoded proteins EspA, EspB and EspD (Jarvis et al., 1995; Ebel et al., 1996). We and others have demonstrated (by microscopy) that the EspA protein assembles into thread-like (Fig. 9.1 and 9.2) surface structures (Ebel et al., 1998; Knutton et al., 1998). Similar

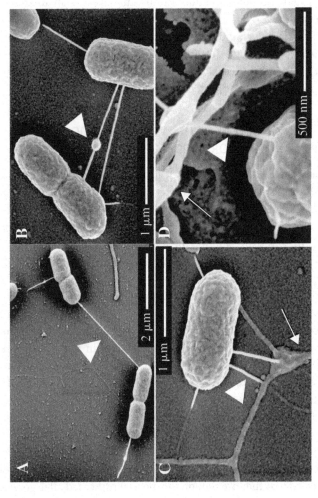

Fig. 9.1. Scanning electron microscopic analysis of surface appendages from the EHEC strain EDL933. Surface appendages contribute to the initial contact between bacteria and eukaryotic cells. Bacterial cells are interconnected via rigid filamentous appendages (A to C, see arrowheads). Some filaments show round thickenings (B). The initial contact between bacteria and eukaryotic cells is established via microvilli (D, see arrow) and the filamentous appendages (D, see arrowhead).

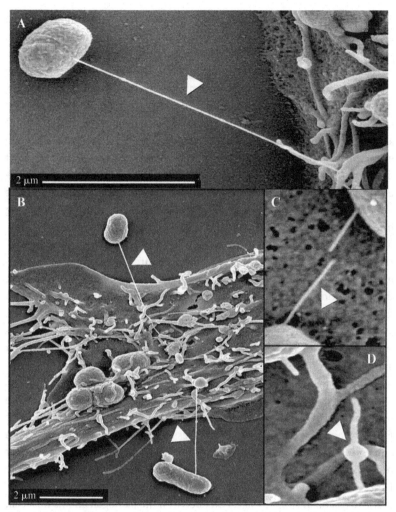

Fig. 9.2. Scanning electron microscopic analysis of EDL933 infected cells. Some of the bacterial filamentous appendages involved in bacteria–host cell contact via microvilli can be more than 2 μm in length (A and B, see arrowheads). However, shorter filaments in direct contact with the cellular surface (C) or thick filaments with round enlargements (D) are also apparent.

type III-dependent filaments have so far been described only for the plant pathogen *Pseudomonas syringae* (Roine *et al.*, 1997). Using isogenic deletion mutants, it was shown that the EspA-filaments are essential for the translocation of the effector proteins EspB and Tir, whereas they are not required for the *in vitro* secretion (Ebel *et al.*, 1998; Knutton *et al.*, 1998; Wolff *et al.*,

1998). EspD was shown to be inserted into the host cell membrane and a deletion in the *espD* gene led to the formation of shorter, or even abrogated, EspA-filaments, which were not functional with respect to protein translocation (Knutton *et al.*, 1998; Kresse *et al.*, 1999). Both EspB and EspD share structural features with some pore-forming proteins, and a haemolytic activity has been demonstrated for these proteins (Wawara *et al.*, 1999; Shaw *et al.*, 2001). All of these data can be integrated into a model in which (i) EspB and EspD are located at the tip of the EspA filament, (ii) they insert into the eukaryotic membrane, and (iii) they form a pore that represents the most distal part of the injection apparatus and the channel that allows translocation into the host cell. Interestingly, translocation of the EspB protein has also been shown, which might therefore serve different functions during infection (Wolff *et al.*, 1998).

For EPEC (as described in greater detail in Chapter 12) the initial attachment to host cells is, at least in *in vitro* tissue culture models of infection, facilitated by type IV pili. These so-called bundle-forming pili (Bfp) also support the formation of microcolonies by cross-linking individual bacteria into a three-dimensional network (Donnenberg *et al.*, 1992; Rosenshine *et al.*, 1996). EHEC do not possess Bfp (Gunzburg *et al.*, 1995), which certainly contributes to the longer incubation time that is needed to initiate attachment and the formation of microcolonies. The absence of additional adhesive structures in EHEC enabled us to hypothesize and then demonstrate that the EspA-filaments are bi-functional, being not only an important part of the injection apparatus, but also playing a crucial role in the initial binding of EHEC to host cells (Ebel *et al.*, 1998; Kresse *et al.*, 2000). Mutations in *espA* and *espD* both abrogated attachment of EHEC to HeLa cells, whereas similar mutations in *espB* reduced, but did not abolish, binding, suggesting that the EspD located at the tip of the filament may embed itself into the target membrane to establish a first physical link between the pathogen and its host cell.

9.6 THE ESP OPERON AND ITS TRANSCRIPTIONAL REGULATION

The genes encoding the putative components of the functional EHEC filament, EspA, B, and D, are clustered and part of an operon (Beltrametti *et al.*, 1999). This transcriptional linkage enables co-regulation of components that build up a structure that is believed to be the 'injection needle' of the translocation machinery. We have analysed the stimuli that control expression of the *esp* operon using a fusion of *lacZ* to the *esp* promoter, which is localized 94 bp upstream from the *espA* gene. Consistent with results obtained at the level of protein expression, we found that the *esp* promoter is activated during growth at 37 °C in tissue culture medium (Beltrametti *et al.*, 1999). These conditions might somehow mimic an environmental cue, but Hepes, which is commonly used to buffer these media, also triggers the induction of the *esp* promoter in a pH-independent manner (Beltrametti *et al.*, 1999). Type III secretion is often referred to as a contact-induced translocation process. Consistently, we found that the physical contact to target cells is an important stimulus that leads to a 5- to 10-fold induction of the *esp* promoter (Beltrametti *et al.*, 1999).

Further studies revealed additional environmental stimuli that influenced the activation of the *esp* promoter. The potential role in regulation of micronutrients that are available in the intestine was also analysed. Ca^{2+} was found to strongly induce the *esp* promoter, supporting an earlier notion that calcium is essential for signal transduction events triggered by EHEC to induce the rearrangement of cytoskeletal proteins (Ismaili *et al.*, 1995). A recent study, however, provided evidence that the formation of A/E lesions is not accompanied by Ca^{2+} influxes (Bain *et al.*, 1998). Nevertheless, the results of the transcriptional studies are in agreement with the general role played by Ca^{2+} in the regulation of virulence genes from other pathogenic microorganisms (Mekalanos, 1992; Puente *et al.*, 1996; Kenny *et al.*, 1997). Similar activation levels were observed when media were supplemented with manganese instead of calcium ions (Beltrametti *et al.*, 1999). Regulation of virulence factors by Mn^{2+} has already been shown for unrelated pathogens such as *Streptococcus* spp. and *Yersinia* spp. (Dintilhac *et al.*, 1997; Bearden *et al.*, 1998). In all these cases, surface proteins were affected, suggesting common underlying processes, although the precise role of Mn^{2+} in the induction of the *esp* promoter remains to be investigated in more detail.

Surprisingly, when temperature was tested as a potential physiological parameter, a relatively weak effect on the activity of the *esp* promoter was observed. However, when combined with high osmolarity, a significant increment in promoter activation was achieved (Beltrametti *et al.*, 1999). Although EHEC can be confronted with any of these stimuli outside the host, the combination of 37 °C and high osmotic pressure represents an excellent indication that bacteria have reached their target site within the host intestine. Interestingly, a similar global activation pattern in response to the tested stimuli was observed also for the *pas* gene (Beltrametti *et al.*, 2000), strengthening the concept of common regulatory cascades for the expression of virulence factors in EHEC. In fact, the Pas protein seems to play a key role in pathogenesis, since it is apparently required for the secretion of Esp proteins (Kresse *et al.*, 1998).

9.7 REGULATORY PROTEINS THAT CONTROL THE *esp* OPERON

In EPEC the expression of the *esp* genes as well as those coding for the adhesin intimin and other virulence factors (e.g. Bfp pilus) are controlled by the *per* locus, which is located on the 60 MDa plasmid pMAR2. The presence of this plasmid is required in EPEC to achieve full virulence (Gomez-Duarte and Kaper, 1995). So far, a similar regulator has not been identified in EHEC, but the presence of a megaplasmid, which is different from pMAR2, results in a significant increment in the activation of the *esp* operon (Beltrametti *et al.*, 1999). This suggests a regulatory pattern in which the activation of the *esp* promoter is fine-tuned by a product(s) encoded by this megaplasmid.

Which other regulatory factors are involved in the expression of the proteins encoded by LEE that contribute to the initial attachment? The results obtained with an *rpoS* mutant of EHEC and the high degree of homology between putative consensus sequences located upstream from the *esp* operon and the σ^S-dependent promoter of *osmE* (Conter *et al.*, 1997) led to the hypothesis that *esp* transcription is σ^S dependent. However, activation of the *esp* promoter preferentially occurs during the exponential phase of growth. On the other hand, the role of σ^S is more complex than that of other alternative sigma factors, since it plays a role under various inducing conditions of slow growth, such as those observed during the stationary phase or osmotic shock (Hengge-Aronis *et al.*, 1993; Hengge-Aronis, 1996). Although the basal expression levels of the reporter were strongly reduced in the *rpoS* mutant, osmoinduction was still preserved. Interestingly, Tanaka *et al.* (1993) showed that several promoters can be recognized by both Eσ^{70} and Eσ^S RNA polymerase holoenzymes. Therefore, σ^S-independent transcription of the *esp* promoter may be directly dependent on Eσ^{70}. The *esp* promoter also exhibits homology with the *bfpA* promoter of EPEC (Beltrametti *et al.*, 1999), which has been suggested to be σ^{70} dependent. It is intriguing that promoters driving the expression of different proteins involved in the synthesis of surface appendages that are required for the initial bacterial attachment have common motifs.

The H-NS protein is involved in the regulation of many genes activated by environmental signals (Atlung and Ingmer, 1997). Our initial studies demonstrated that the levels of transcription of the *esp* promoter are significantly increased in an *hns* mutant (Beltrametti *et al.*, 1999). The presence of this regulator usually results in 2- to 20-fold repression, which is stronger when H-NS both acts at the promoter level and affects the expression of positive regulators (Atlung and Ingmer, 1997). Therefore, in the *hns* mutant, the observed influence of H-NS in activation of the *esp* promoter can be explained by (i) hyperexpression of the σ^S factor, which is repressed by H-NS (Atlung and Ingmer, 1997), and (ii) a direct effect on the promoter itself, since putative H-NS-binding regions have been identified. Subsequent studies from Ogierman *et al.* (2000) resulted in the characterization of a LEE gene coding for an EHEC H-NS homologue, Ler, which is not only involved in the expression of proteins that are essential for the A/E lesions (e.g. the type III secretion system, Esp proteins, Tir and intimin), but also other genes not required for A/E lesions, which are located within and outside LEE (Elliott *et al.*, 2000). In EPEC, Ler is believed to be positively regulated by the megaplasmid-encoded Per, thereby explaining one of the main differences between EPEC and EHEC in the regulatory cascades leading to the expression of LEE-encoded virulence factors (Elliott *et al.*, 2000; Ogierman *et al.*, 2000). This dual regulatory cascade resembles the fine-tuning of the type III secretion system and secreted proteins by VirF and VirB in *Shigella flexneri* (Adler *et al.*, 1989; Dorman and Porter, 1998).

9.8 SWITCHING PROPERTIES IN THE RIGHT PLACE AT THE RIGHT TIME

After their passage through the stomach, EHEC encounter their preferred habitat within the host and the expression of LEE-encoded pathogenic factors is activated. The biosynthesis of EspA filaments is necessary to enable bacterial attachment to the host tissues. The luminal surface of the intestinal epithelial cells is highly organized and microvilli are the major architectural element. They can represent a morphological barrier for the formation of actin pedestals, thereby hindering the colonization process. EspA filaments may enable EHEC both to establish distant contact with epithelial cells and to promote a subsequent alteration of the cellular surface by the injection of effector proteins. This notion is in agreement with the recent finding that type III-mediated haemolysis of red blood cells by EHEC and EPEC does not require an intimate physical contact (Shaw *et al.*, 2001). This is in contrast to the *Shigella* and *Yersinia* systems in which such intimate contact has first to be established by centrifugation (Håkansson *et al.*, 1996; Blocker *et al.*, 1999). The fact that EPEC are able to induce haemolysis at a distance, suggests that the translocation apparatus is fully functional at this stage, although translocation into red blood cells has not yet been demonstrated.

A very interesting finding is that the EspA filaments disappear when bacteria are in intimate contact with their target cells (Knutton *et al.*, 1998). How their disassembly takes place is unknown, but it is tempting to speculate that these structures may be retractable, as has been reported for type IV pili (Wall and Kaiser, 1999). The ability to actively reduce the length of the EspA filaments, once the initial contact has been made, would (i) reduce the distance to the target cell surface and lead to an intimate contact, (ii) enable the bacteria to recycle a certain percentage of their secreted proteins, and (iii) minimize epitope presentation to the host defence system by removing surface structures that are not needed any more. The phenotypic changes in this organelle are accompanied by a sequential up- and down-regulation of the *esp* promoter. This shows the tight time frame in which the EspA filaments are formed to fulfil the requirements of initial adherence, protein translocation and intimate attachment, and their subsequent disappearance to minimize the exposition of potential antigens and the additional energetic cost associated with their synthesis. However, it is important to keep in mind that the colonization of the intestinal epithelium by EHEC is a dynamic process characterized by the formation and growth of microcolonies. Consequently, bacterial multiplication will provide the need for establishing new translocation machinery in daughter cells. However, due to the already intimate attachment, the required quantities of Esp proteins might be much smaller during this phase than in the initial infection.

Interestingly, it has been shown that intimin also disappears in a later phase of infection when actin pedestals have already been formed (Knutton *et al.*, 1997) and that bacterial *de novo* protein synthesis is not required to maintain these structures. Taken together, these data suggest a co-ordinated induction and subsequent stepwise down-regulation of LEE-encoded products during the formation of the A/E lesion, indicating that bacteria are able to sense the morphological changes that they are promoting in the infected host cells.

9.9 CONCLUSIONS

We are at the dawn of an era in which we will gain a global understanding of the modulation of pathogenic factors and functions during infection. During the complex cross-talk between pathogens and target cells, the expression of the relevant genes is tightly controlled and fine-tuned. The type III secretion apparatus provides an excellent model to study this vital physiological process in pathogenic bacteria. The mechanisms for the assembly and disassembly of the type III machinery are still an enigma, and even the organization and functional modulation of EspA filaments are not yet fully understood. Further studies are needed to investigate the stability of the type III apparatus and its potential substructures. It would also be important to elucidate whether the disappearance of EspA filaments is linked to a disassembly of the secretion apparatus or if the intimately attached bacteria simply do not need elongated filaments.

The emerging picture shows us that the products encoded by the LEE are integrated into extremely complex regulatory networks. However, the co-ordinated expression of two type III secretion systems in *Salmonella* provides an example of an even more complex system. *Salmonella* pathogenicity island 1 (SPI1) is required for the invasion of epithelial cells, whereas SPI2 is essential for the intracellular survival in macrophages (Hensel, 2000). These two pathogenicity islands seem to be expressed in a highly organized manner to avoid unproductive interference of different type III-mediated processes. It has been shown that mutations in SPI2 substantially reduce expression of SPI1 genes (Deiwick *et al.*, 1998) and that genes encoding secreted effector proteins are not necessarily located on the same island as their secretion system (Mirold *et al.*, 1999). Different gene clusters coding for pathogenic functions have been independently acquired by horizontal gene transfer and are located at different topologies within the genome. These genes have been finally integrated into global regulatory networks that orchestrate their harmonious expression and integration during the complex dynamics of the infection process. The clear understanding of the regulatory cascades and their interplay during infection remains as a major challenge for future research in the field of bacterial pathogenesis. The fulfilment of this task would enable us not only to understand the pathophysiology of bacterial infections but also to identify new molecular targets for therapeutic interventions.

ACKNOWLEDGEMENTS

We thank M. Rohde for providing the electron microscopy pictures presented in this work.

REFERENCES

Adler, B., Sasakawa, C., Tobe, T., Makino, K., Komatsu, K. and Yoshikawa, M. (1989). A dual transcriptional activation system for the 230 kb plasmid genes coding for virulence-associated antigens of *Shigella flexneri*. *Molecular Microbiology* **3**, 627–635.

Akerley, B.J. and Miller, J.F. (1996). Understanding signal transduction during bacterial infection. *Trends in Microbiology* **4**, 141–146.

Atlung, T. and Ingmer, H. (1997). H-NS: a modulator of environmentally regulated gene expression. *Molecular Microbiology* **24**, 7–17.

Bain, C., Keller, R., Collington, G.K., Trabulsi, L.R. and Knutton, S. (1998). Increased levels of intracellular calcium are not required for the formation of attaching and effacing lesions by enteropathogenic and enterohaemorrhagic *Escherichia coli*. *Infection and Immunity* **66**, 3900–3908.

Bearden, S.W., Staggs, T.M. and Perry, R.D. (1998). An ABC transporter system of *Yersinia pestis* allows utilization of chelated iron by *Escherichia coli* SAB11. *Journal of Bacteriology* **180**, 1135–1147.

Beltrametti, F., Kresse, A.U. and Guzmán, C.A. (1999). Transcriptional regulation of the *esp* genes of enterohemorrhagic *Escherichia coli*. *Journal of Bacteriology* **181**, 3409–3418.

Beltrametti, F., Kresse, A.U. and Guzmán, C.A. (2000). Transcriptional regulation of the *pas* gene of enterohemorrhagic *Escherichia coli*. *FEMS Microbiology Letters*, **184**, 119–125.

Blocker, A., Gounon, P., Larquet, E., Niebuhr, K., Cabiaux, V., Parsot, C. and Sansonetti, P. (1999). The tripartite type III secreton of *Shigella flexneri* inserts IpaB and IpaC into host membranes. *Journal of Cellular Biology* **147**, 683–693.

Boyce, T.G., Swerdlow, D.L. and Griffin, P.M. (1995). *Escherichia coli* O157:H7 and the hemolytic-uremic syndrome. *New England Journal of Medicine* **333**, 364–368.

Conter, A., Menchon, C. and Gutierrez, C. (1997). Role of DNA supercoiling and RpoS sigma factor in the osmotic and growth phase-dependent induction of the gene *osmE* of *Escherichia coli* K12. *Journal of Molecular Biology* **273**, 75–83.

Deibel, C., Kramer, S., Chakraborty, T. and Ebel, F. (1998). EspE, a novel secreted protein of attaching and effacing bacteria, is directly translocated into infected host cells, where it appears as a tyrosine-phosphorylated 90 kDa protein. *Molecular Microbiology* **28**, 463–474.

Deiwick, J., Nikolaus, T., Shea, J.E., Gleeson, C., Holden, D.W. and Hensel, M. (1998). Mutations in *Salmonella* pathogenicity island 2 (SPI2) genes affecting transcription of SPI1 genes and resistance to antimicrobial agents. *Journal of Bacteriology* **180**, 4775–4780.

Dintilhac, A., Alloing, G., Granadel, C. and Claverys, J. (1997). Competence and virulence of *Streptococcus pneumoniae*: Adc and PsaA mutants exhibit a requirement for Zn and Mn resulting from inactivation of putative ABC metal permeases. *Molecular Microbiology* **25**, 727–739.

Donnenberg, M.S., Giròn, J.A., Nataro, J.P. and Kaper, J.B. (1992). A plasmid-encoded type IV fimbrial gene of enteropathogenic *Escherichia coli* associated with localized adherence. *Molecular Microbiology* **6**, 3427–3437.

Donnenberg, M.S., Kaper, J.B. and Finlay, B.B. (1997). Interaction between enteropathogenic *Escherichia coli* and host epithelial cells. *Trends in Microbiology* **5**, 109–114.

Dorman, C.J. and Porter, M.E. (1998). The *Shigella* virulence gene regulatory cascade: a paradigm of bacterial gene control mechanisms. *Molecular Microbiology* **29**, 677–684.

Ebel, F., Deibel, C., Kresse, A.U., Guzmán, C.A. and Chakraborty, T. (1996). Temperature- and medium-dependent secretion of proteins by Shiga toxin-producing *Escherichia coli*. *Infection and Immunity* **64**, 4472–4479.

Ebel, F., Podzadel, T., Rohde, M., Kresse, A.U., Krämer, S., Deibel, C., Guzmán, C.A. and Chakraborty, T. (1998). Initial binding of Shiga toxin-producing *Escherichia coli* to host cells and subsequent induction of actin rearrangements depend on filamentous EspA-containing surface appendages. *Molecular Microbiology* **30**,147–161.

Elliott, S.J., Yu, J. and Kaper, J.B. (1999). The cloned locus of enterocyte effacement from enterohemorrhagic *Escherichia coli* O157:H7 is unable to confer the attaching and effacing phenotype upon *E. coli* K-12. *Infection and Immunity* **67**, 4260–4263.

Elliott, S.J., Sperandio, V., Giron, J.A., Shin, S., Mellies J.L., Wainwright, L., Hutcheson, S.W., McDaniel, T.K. and Kaper, J.B. (2000). The locus of enterocyte effacement (LEE)-encoded regulator controls expression of both LEE- and non-LEE-encoded virulence factors in enteropathogenic and enterohemorrhagic *Escherichia coli*. *Infection and Immunity* **68**, 6115–6126.

Finlay, B.B. and Falkow, S. (1997). Common themes in microbial pathogenicity revisited. *Microbiology and Molecular Biology Reviews* **61**, 136–139.

Gallien, P., Richter, H., Klie, H., Timm, M., Karch, H., Perlberg, K.W., Steinruck, H., Riemer, S., Djuren, M. and Protz, D. (1997). Detection of STEC and epidemiological investigations in surrounding of a HUS patient. *Berliner und Muenchner Tieraerztliche Wochenschrift* **110**, 342–346.

Gomez-Duarte, O. and Kaper, J.B. (1995). A plasmid-encoded regulatory region activates chromosomal *eaeA* expression in enteropathogenic *Escherichia coli*. *Infection and Immunity* **63**, 1767–1776.

Goosney, D.L., DeVinney, R., Pfuetzner, R.A., Frey, E.A., Strynadka, N.C. and Finlay, B.B. (2000). Enteropathogenic *E. coli* translocated intimin receptor Tir interacts directly with alpha-actinin. *Current Biology* **10**, 735–738.

Gorden, J. and Small, P.L.C. (1993). Acid resistance in enteric bacteria. *Infection and Immunity* **61**, 364–367.

Gunzburg, S.T., Tornieporth, N.G. and Riley, L.W. (1995). Identification of enteropathogenic *Escherichia coli* by PCR-based detection of the bundle-forming pilus gene. *Journal of Clinical Microbiology* **33**, 1375–1377.

Håkansson, S., Schesser, K., Persson, C., Galyov, E.E., Rosqvist, R. and Wolf-Watz, H. (1996). The YopB protein of *Yersinia pseudotuberculosis* is essential for the translocation of Yop effector proteins across the target cell membrane and displays a contact-dependent membrane disrupting activity. *EMBO Journal* **15**, 5812–5823.

Hengge-Aronis, R. (1996). Back to the log phase: σ^S as a global regulator in the osmotic control of gene expression in *Escherichia coli*. *Molecular Microbiology* **21**, 887–893.

Hengge-Aronis, R., Lange, R., Henneberg, N. and Fisher, D. (1993). Osmotic regulation of *rpoS*-dependent genes in *Escherichia coli*. *Journal of Bacteriology* **175**, 259–265.

Hensel, M. (2000). *Salmonella* pathogenicity island 2. *Molecular Microbiology* **36**, 1015–1023.

Ismaili, A., Philpott, D.J., Dytoc, M.T. and Sherman, P.M. (1995). Signal transduction responses following adhesion of verocytotoxin-producing *Escherichia coli*. *Infection and Immunity* **63**, 3316–3326.

Jarvis, K.G., Giròn, J.A., Jerse, A.E., McDaniel, T.K., Donnenberg, M.S. and Kaper, J.B. (1995). Enteropathogenic *Escherichia coli* contains a putative type III secretion system necessary for the export of proteins involved in attaching and effacing lesion formation. *Proceedings of the National Academy of Sciences, USA* **92**, 7996–8000.

Jenal, U. and Shapiro, L. (1996). Cell cycle-controlled proteolysis of a flagellar motor protein that is asymmetrically distributed in the *Caulobacter* predivisional cell. *EMBO Journal* **15**, 2393–2406.

Keene, W.E., McAnulty, J.M. and Hoesly, F.C. (1994). A swimming-associated outbreak of hemorrhagic colitis caused by *Escherichia coli* O157:H7 and *Shigella sonnei*. *New England Journal of Medicine* **331**, 579–584.

Kenny, B., De Vinney, R., Stein, M., Reinsheid, D.J., Frey, E.A. and Finlay, B.B. (1997). Enteropathogenic *Escherichia coli* (EPEC) transfer its receptor for intimate adherence to mammalian cells. *Cell* **91**, 511–520.

Knutton, S., Adu-Bobie, J., Bain, C., Phillips, A.D., Dougan, G. and Frankel, G. (1997). Down regulation of intimin expression during attaching and effacing *Escherichia coli* adhesion. *Infection and Immunity* **65**, 1644–1652.

Knutton S., Rosenshine, I., Pallen, M.J., Nisan, I., Neves, B.C., Bain, C., Wolff, C., Dougan, G. and Frankel, G. (1998). A novel EspA-associated surface organelle of enteropathogenic *Escherichia coli* involved in protein translocation into epithelial cells. *EMBO Journal* **17**, 2166–2176.

Kresse, A.U., Schulze, K., Deibel, C., Ebel, F., Rohde, M., Chakraborty, T. and Guzmán, C.A. (1998). Pas, a novel protein required for protein secretion and attaching and effacing activities of enterohemorrhagic *Escherichia coli*. *Journal of Bacteriology* **180**, 4370–4379.

Kresse A.U., Rohde, M. and Guzmán, C.A. (1999). The EspD protein of enterohemorrhagic *Escherichia coli* is required for the formation of bacterial surface

A. U. KRESSE *ET AL.*

appendages and is incorporated in the cytoplasmic membranes of target cells. *Infection and Immunity* 67, 4834–4842.

Kresse, A.U., Beltrametti, F., Müller, A., Ebel, F. and Guzmán, C.A. (2000). Characterization of SepL of enterohemorrhagic *Escherichia coli*. *Journal of Bacteriology* 182, 6490–6498.

Kubori, T., Matsushima, Y., Nakamura, D., Uralil, J., Lara-Tejero, M., Sukhan, A., Galan, J.E. and Aizawa, S.I. (1998). Supramolecular structure of the *Salmonella typhimurium* type III secretion system, *Science* 280, 602–605.

McDaniel, T.K. and Kaper, J.B. (1997). A cloned pathogenicity island from enteropathogenic *Escherichia coli* confers the attaching and effacing phenotype on *E. coli* K-12. *Molecular Microbiology* 23, 399–407.

Mekalanos, J.J. (1992). Environmental signals controlling expression of virulence determinants in bacteria. *Journal of Bacteriology* 174, 1–7.

Mirold, S., Rabsch, W., Rohde, M., Stender, S., Tschäpe, H., Rüssmann, H., Jgwe, E. and Hardt, W.D. (1999). Isolation of a temperent bacteriophage encoding the type III effector protein SopE from an epidemic *Salmonella typhimurium* strain. *Proceedings of the National Academy of Sciences, USA* 96, 9845–9850.

Moon, H.V., Whipp, S.C., Argenzio, R.A., Levine, M.M. and Ginnella, R.A. (1983). Attaching and effacing activities of rabbit and human enteropathogenic *Escherichia coli* in pig and rabbit intestines. *Infection and Immunity* 41, 1340–1351.

Nataro, J.P. and Kaper, J.B. (1998). Diarrheagenic *Escherichia coli*. *Clinical Microbiology Reviews* 11, 142–201.

Ogierman, M.A., Paton, A.W. and Paton, J.C. (2000). Up-regulation of both intimin and *eae*-independent adherence of Shiga toxigenic *Escherichia coli* O157 by *ler* and phenotypic impact of a naturally occurring *ler* mutation. *Infection and Immunity* 68, 5344–5353.

Puente, J.L., Bieber, D., Ramer, S.W., Murray, W. and Schoolnik, G.K. (1996). The bundle-forming pili of enteropathogenic *Escherichia coli*: transcriptional regulation by environmental signals. *Molecular Microbiology* 20, 87–100.

Renwick, S.A., Wilson, J.B., Clarke, R.C., Lior, H., Borczyck, A., Spika, J.S., Rahn, K., McFadden, K., Brouwer, A. and Copps, A. (1994). Evidence of direct transmission of *Escherichia coli* O157:H7 infection between calves and a human — Ontario. *Canada Communicable Disease Report* 20, 73–75.

Roine, E., Wei, W., Yuan, J., Nurmiaho-Lassila, E.-L., Kalkkinen, N., Romantschuk, M. and He, S. Y. (1997). Hrp pilus: an *hrp*-dependent bacterial surface appendage produced by *Pseudomonas syringae* pv. *tomato* DC3000. *Proceedings of the National Academy of Sciences, USA* 94, 3459–3464.

Rosenshine, I., Ruschkowski, S., Stein, M., Reinscheid, D.J., Mills, S.D. and Finlay, B.B. (1996). A pathogenic bacterium triggers epithelial signals to

form a functional bacterial receptor that mediates actin pseudopod formation. *EMBO Journal* **15**, 2613–2624.

Shaw, R.K., Daniell, S., Ebel, F., Frankel, G. and Knutton, S. (2001). EspA
filament-mediated protein translocation into red blood cells. *Cellular
Microbiology* **3**, 213–222.

Stephens, C. and Shapiro, L. (1996). Bacterial pathogenesis: delivering of the
payload. *Current Biology* **6**, 927–930.

Swerdlow, D.L., Woodruff, B.A., Brady, R.C., Griffin, P.M., Tippen, S., Donnell,
H.D., Jr., Geldreich, E., Payne, B.J., Meyer, A., Jr, Wells, J.G. *et al.* (1992). A
waterborne outbreak in Missouri of *Escherichia coli* O157:H7 associated with
bloody diarrhea and death. *Annals of Internal Medicine* **117**, 812–819.

Tanaka, K., Takayanagi, Y., Fujita, N., Ishihama, A. and Takahashi, H. (1993).
Heterogeneity of the principal σ factor in *Escherichia coli*: the *rpoS* gene
product σ^{38} is a secondary principal σ factor of RNA polymerase in station-
ary-phase. *Proceedings of the National Academy of Sciences, USA*, **90**,
3511–3515.

Tzipori, S., Gunzer, F., Donnenberg, M.S., de Montigny, L., Kaper, J.B. and
Donohue Rolfe, A. (1995). The role of the *eaeA* gene in diarrhea and neuro-
logical complications in a gnotobiotic piglet model of enterohemorrhagic
Escherichia coli infection. *Infection and Immunity* **63**, 3621–3627.

Wall, D. and Kaiser, D. (1999). Type IV pili and cell motility. *Molecular
Microbiology* **32**, 1–10.

Wawara, J., Finlay, B.B. and Kenny, B. (1999). Type III secretion-dependent hemolytic activity of enteropathogenic *Escherichia coli*. *Infection and Immunity* **67**,
5538–5540.

Wolff, C., Nisan, I., Hanski, E., Frankel, G. and Rosenshine, I. (1998). Protein
translocation into host epithelial cells by infecting enteropathogenic
Escherichia coli. *Molecular Microbiology* **28**, 143–155.

PART III Consequences of bacterial adhesion for the host

CHAPTER 10

Adhesion, signal transduction and mucosal inflammation

Catharina Svanborg, Goran Bergsten, Hans Fischer, Björn Frendéus, Gabriela Godaly, Erika Gustafsson, Long Hang, Maria Hedlund, Ann-Charl0tte Lundstedt, Martin Samuelsson, Patrik Samuelsson, Majlis Svensson and Björn Wullt

10.1 INTRODUCTION

Mucosal surfaces continue where the skin leaves off, completing the boundaries between the environment and internal tissues. It is to the mucosal barrier that pathogens first attach, and the ensuing molecular interactions determine whether the pathogens will cause infection or if health will be maintained.

The different cell populations in the mucosal barrier are equipped to sense and respond to the molecular contents in the lumen and to translate this molecular information into signals that can reach local or distant tissue sites. Inflammatory cascades are often the first to be activated by bacteria or viruses, and explain many aspects of acute disease. It is not the presence of microbes in the tissues that makes us sick. Vast numbers of bacteria may colonize different sites in the body, and viruses may persist for long periods of time, while the host remains perfectly healthy. It is the host response to the infecting agents that causes the symptoms and tissue damage. Inflammatory mediators thus provide a direct link between the microbe and the host in disease pathogenesis (for a review, see Svanborg et al., 1999).

It is commonly accepted that the most virulent bacteria cause the most severe acute symptoms and long-term effects, but the mechanisms used to trigger acute disease manifestations are less clear. Pathogens elicit the strongest host response, even though members of the indigenous microflora contain many of the same molecules. This is quite puzzling, but at least two factors may explain the difference. First, pathogens may have a broader repertoire of host-activating molecules than do the commensals or they may express molecular variants with a better specificity for the host response pathways. Second, the pathogens may present the host-activating molecules more efficiently to the cells in the mucosa (Smith and Linggood, 1972; Svanborg et al., 1999; Hedlund et al., 2001). Finally, it has been suggested that commensals inactivate the host response in order to enhance their own survival (Neish et al. 2000).

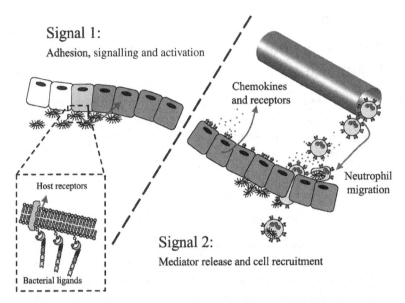

Signal 1:

Adhesion, signalling and activation

Chemokines
and receptors

Host receptors

Neutrophil
migration

Bacterial ligands

Signal 2:

Mediator release and cell recruitment

Figure 10.1. The 'two-signal model' for host response induction by attaching bacteria. Signal 1: Bacterial attachment to epithelial cell receptors triggers cell activation through specific transmembrane signalling pathways. Patients with a dysfunctional first signal have no response, leading to an asymptomatic carrier state.

Signal 2: Cell activation causes inflammation through the activation of the second signal. Chemokines are secreted and inflammatory cells are recruited to the mucosa. Adherence and the fimbrial receptor specificity direct the cellular infiltrate through the repertoire of mediators that is produced as a consequence. In the patients with a well-functioning inflammatory response the cells first migrate into and then exit from the tissues, and in the process infection is cleared. These inividuals may develop symptoms of acute disease but no tissue pathology. Defective inflammation leads to tissue damage and renal scarring, owing to poor bacterial clearance and tissue destruction by cells trapped in the tissues (see Plate 10.2).

Bacterial attachment serves a number of basic functions. Early studies emphasized the importance of attachment for bacterial persistence at sites of colonization or infection (Freter, 1969; Smith and Linggood, 1972; Gibbons, 1973). The attached state was thought to promote nutrient uptake and to allow bacteria to multiply (Zobell, 1943). The effect on persistence, however, does not explain why adherence is linked to the severity of infection. Smith and Linggood (1972) showed that adherence promotes the delivery of toxins, and we have demonstrated that pathogens use adherence as a direct tissue attack mechanism, to initiate the local host response to infection (Svanborg-Edén *et al.*, 1976; Linder *et al.*, 1988; Svanborg *et al.*, 1999).

Figure 10.1 illustrates the 'two-signal' model for host response induction by attaching bacteria. Attachment of bacterial adhesins to cell surface receptors triggers the first signal. Specific signal transduction pathways are activated, and host response-activating virulence factors are delivered to the tissues. This interaction determines whether bacteria will remain surface associated, cross the mucosal barrier through or between the cells, or export their virulence products to modify the behaviour of the cells. The second signal involves the release of pro-inflammatory mediators by the activated cells. Chemokines recruit inflammatory cells and chemokine receptors direct their interaction with the mucosal barrier. The inflammation is amplified by the recruited cells and their mediators determine subsequent steps in the inflammatory process.

The effects of bacterial adhesion depend on the host. In the 'high responders' both the first and the second signal are very active. If the resulting inflammatory response is fully functional, the patient may develop transient symptoms and clear the infection, but if the response is dysfunctional, clearance is impaired and chronic tissue pathology may follow (see below).

The 'low responders', on the other hand, have a suppressed 'first signal', and do not mount an inflammatory response or a defence. This allows bacteria to establish without causing inflammation, and the patients become asymptomatic carriers, due to their refractoriness to activation by attaching bacteria.

This chapter describes how bacterial attachment influences the signal transduction pathways involved in cell activation and directs pathogenesis through the inflammatory response, and indirectly the resistance to mucosal infection. We will use P fimbriated *Escherichia coli* urinary tract infection (UTI), as a model to illustrate the link between bacterial attachment and cell activation. Many mechanisms are common to the different fimbrial types on uropathogens, and to mucosal pathogens infecting other sites.

10.2 THE 'FIRST SIGNAL': ATTACHMENT – MECHANISMS AND CONSEQUENCES FOR CELL ACTIVATION

Adherence is a virulence factor for uropathogenic *E. coli*. Strains causing severe disease (acute pyelonephritis, urosepsis) are more highly adhesive than strains causing less severe forms of UTI (asymptomatic bacteriuria, ABU) or in the faecal flora of healthy controls (Fig. 10.2). In epidemiological studies, P fimbriae show the strongest association with acute disease severity, with at least 90% of acute pyelonephritis isolates but fewer than 20% of ABU strains expressing this phenotype (Svanborg-Edén *et al.* 1976; Svanborg-Edén and Hansson, 1978; Leffler and Svanborg-Edén, 1981; Plos *et al.*, 1995).

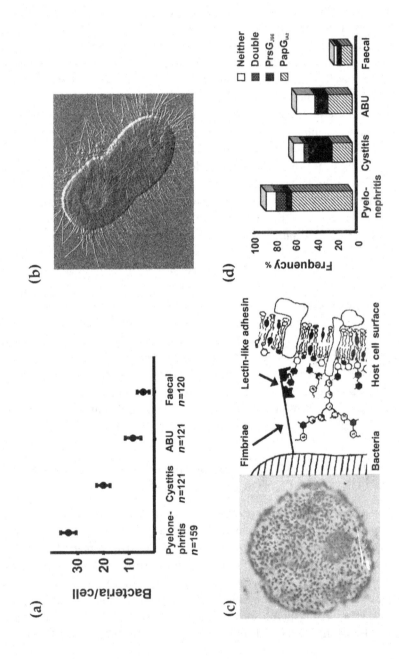

(a)

Bacteria/cell

30

20

10

Pyelone-   Cystitis   ABU   Faecal
phritis   *n*=121   *n*=121   *n*=120
n=159

(b)

(c)

Fimbriae   Lectin-like adhesin

Bacteria   Host cell surface

(d)

Frequency %

100

80

60

40

20

0

Pyelo-   Cystitis   ABU   Faecal
nephritis

☐ Neither
▨ Double
■ PrsG$_{J96}$
▨ PapG$_{IA2}$

Figure 10.2. Bacterial adherence: mechanisms and epidemiological correlates. (a) Adherence to uro-epithelial cells *in vitro* is associated with the severity of UTI. Isolates from patients with acute pyelonephritis adhere most efficiently, acute cystitis strains show intermediate adherence and ABU strains are poorly adhesive, as are isolates from the faecal flora of healthy children (Svanborg-Edén et al., 1976, 1978).

(b) Fimbriae are the bacterial surface organelles that mediate adherence to different cell types (Duguid et al., 1955; Brinton, 1965; Svanborg-Edén et al., 1978; Korhonen et al., 1982), including uro-epithelial cells. Transmission electron micrograph of *E. coli* 83972 *pap*+ after platinum shadowing.

(c) Molecular mechanism of P fimbriae-mediated adherence to uro-epithelial cells. (*Left panel*) P fimbriae-mediated attachment to a squamous epithelial cell from human urine. (*Right panel*) P fimbriae recognize specific receptor epitopes composed of oligosaccharide sequences in the globo series of glycosphingolipids. The fimbrial adhesin is shown at the tip of the fimbrial rod, extending towards the epithelial cell surface. P fimbriae bind specifically to Galα1-4Galβ-containing receptor epitopes. (Data from Leffler and Svanborg-Edén, 1980. Used with permission from the *Swedish Medical Journal*.)

(d) *papG* adhesin sequences in relation to disease severity. The columns describe the proportion of *pap*+ strains in each diagnostic group, detected by probes encompassing the entire *pap* operon. The shaded areas indicate the proportion of strains carrying each G adhesin, as detected by polymerase chain reaction with *papG*-specific primers. The *papG*_{IA2} sequences predominated among the acute pyelonephritis strains, many of which carried multiple *pap* copies. The *prsG*_{J96} sequences were relatively more common in acute cystitis, but were found only in about 25% of these strains. The *papG*_{J96} adhesin, which has been used as a model by several groups (Hull et al., 1981; Normark et al., 1983), is atypical and rare among disease isolates. (Data from Plos et al., 1990; Plos, 1992; Johanson et al., 1993.)

10.2.1 Receptor specificity of P fimbriae

For many years, adherence was regarded as a non-specific process, involving simply charge and hydrophobicity. We were unable to reconcile our findings with this concept, as uropathogens showed differential adherence depending on the individual host and tissue where the cells had been obtained. We demonstrated that adherence is a specific process involving bacterial surface ligands and host cell receptors, and that fimbriae-associated lectins use cell surface glycoconjugate receptors to target sites where they initiate infection (C. Svanborg, oral presentation, 1979; Leffler and Svanborg-Edén, 1980; Svanborg-Edén et al., 1981).

P fimbriae attach to epithelial cells through specific receptors composed of Galα1–4Galβ-oligosaccharide sequences in the globoseries of glycosphingolipids (GSLs) (Leffler and Svanborg-Edén, 1980). The receptor specificity of P fimbriae has been extensively characterized. P fimbriated bacteria fail to adhere to cells from host individuals of blood group P (Källenius et al., 1980; Leffler and Svanborg-Edén 1980), who lack these receptors, but bind to receptor-coated inert surfaces, showing that the receptors are necessary, and sufficient for attachment. Furthermore, in vivo attachment is inhibited by soluble receptor analogues (Svanborg-Edén et al., 1982).

10.2.2 P fimbriae, colonization and host response induction

P fimbriae augment the virulence of uropathogenic E. coli at different stages during the pathogenesis of UTIs. Prior to infection of the urinary tract, P fimbriated strains establish in the intestinal flora of UTI-prone patients and spread more efficiently than other bacteria to the urinary tract (Wold et al., 1988; Plos et al., 1995). P fimbriae enhance the establishment of bacteriuria and the cytokine response to E. coli in the murine urinary tract (Hagberg et al., 1983; Hedges et al., 1991; Hedlund et al., 1996; Godaly et al., 1998; Frendéus et al., 2001; Wullt et al., 2001).

The role of P fimbriae in bacterial persistence and host response induction has recently been studied in the human urinary tract (Wullt et al., 2000, 2001). As carriage of avirulent bacteria may prevent symptomatic superinfections, we have developed a protocol for deliberate colonization of the human urinary tract with an avirulent ABU strain, and have used P fimbriated transformants of this strain for comparison. The pap^+ transformant expressed P fimbriae, as shown by electron microscopy and receptor-specific haemagglutination assays (Fig. 10.3). Patients with recurrent UTI were invited to participate, and were inoculated intravesically with 10^5

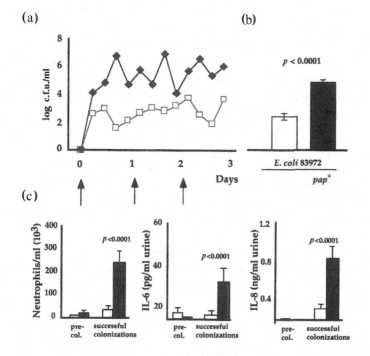

Figure 10.3. Role of P fimbriae in bacterial persistence and host response induction in the human urinary tract. ABU has been shown to prevent symptomatic UTI (Hanson *et al.*, 1972; Lindberg, 1975). In the patients who do not develop ABU, a protocol for deliberate colonization has been developed. An avirulent ABU strain was used (□) and it was transformed with the *pap* sequences in an attempt to enhance bacterial establishment (◆). The *pap*[+] transformant expressed P fimbriae, as shown by electron microscopy. (a) Patients with recurrent UTI were inoculated intravesically with 10[5] c.f.u./ml of either strain on three consecutive days (denoted by arrows). The figure illustrates the kinetics of bacterial establishment for the P fimbriated transformant and the non-fimbriated ABU strain. (b) The columns show cumulative bacterial counts from the first three days. Open columns, non-fimbriated ABU strain; shaded columns, fimbriated transformant. (c) Host response induction, quantified by urine neutrophil counts and IL-6 and IL-8 concentrations in the urine. The P fimbriated transformant was shown to trigger the local host responses more efficiently than the ABU host strain. (Data from Wullt *et al.*, 2000, 2001.)

colony-forming units (c.f.u.) of the ABU strain or the P fimbriated transformant per millilitre. The P fimbriated transformants established bacteriuria more rapidly than did the ABU host strain, and triggered the local host responses more efficiently, as shown by the neutrophil, interleukin (IL)-6 and IL-8 measurements (Fig. 10.3d). These results demonstrate that P fimbriae fulfil the molecular Koch postulates as a colonization enhancer and host response inducer in the human urinary tract.

10.2.3 Mechanisms of cell activation

Two general mechanisms have been proposed to explain the link between attachment and inflammation. First, the coupling of fimbriae to their receptors activates receptor-determined signalling pathways in the host cell (de Man *et al.*, 1989; Hedges *et al.*, 1992; Hedlund *et al.*, 1996, 1998, 1999, 2001). Second, fimbriated bacteria deliver to the tissues other virulence factors that themselves lack specific targeting mechanisms, and the ensuing tissue responses are due mainly to the delivered virulence factor with attachment as a facilitator.

10.2.3.1 Fimbriae, transmembrane signalling and cell activation

P fimbriae are directly involved in host cell activation. Coupling of fimbriated bacteria or isolated fimbriae to their receptors elicits a cytokine response in the target cell (Plate 10.1).

10.2.3.1.1 P fimbriae and transmembrane signalling through ceramide

Membrane sphingolipids play an important role in signal transduction. While the carbohydrate head groups are essential for ligand recognition, the membrane domain determines signalling and cell activation. After binding of P fimbriae to their Galα1–4Galβ receptor epitopes, ceramide is released (Hedlund *et al.*, 1996) and downstream signalling involves Ser/Thr kinases, consistent with the ceramide signalling pathway being activated.

Other exogenous ligands such as vitamin D3, IL-1, tumour necrosis factor (TNF) or interferon (IFN)-γ (Kim *et al.*, 1991) release ceramide from sphingomyelin and the released ceramide activates CAPP (ceramide-activated protein phosphatase), a heterotrimeric protein phosphatase, and Ser/Thr specific protein kinases (Schütze *et al.*, 1992). P fimbriae differ from these activators by binding directly to ceramide-anchored receptors and triggering the release of ceramide from the glycosphingolipid (Hedlund *et al.*, 1996, 1998).

Receptor fragmentation into ceramide and free oligosaccharide may represent a highly efficient strategy of the host defence (Hedlund *et al.*, 1996, 1998). Release of the carbohydrate receptor makes the bacteria 'lose their grip', and the host cells probably remain refractory to further adherence until new receptors are expressed. Soluble receptors may competitively inhibit further attachment as shown by the prevention of experimental UTI by receptor analogues *in vivo* (Svanborg-Edén *et al.*, 1982). Inflammation is the second and more obvious advantage to the host defence, as released cera-mide may activate mediators involved in the recruitment of inflammatory cells that clear the infection.

10.2.3.1.2 P fimbriae recruit Toll-like receptor 4 as a co-receptor

Recent studies have demonstrated that P fimbriae recruit Toll-like receptor 4 (TLR4) as co-receptor in cell activation (Svanborg *et al.*, 2001; Frendéus *et al.*, 2001). The involvement of TLR4 was first suggested from results of experimental UTI in TLR4-deficient mice (Shahin *et al.*, 1987; Svanborg-Edén *et al.*, 1987), carrying a mutation in TLR4 that renders them unrespon-sive to lipopolysaccharide (LPS) (Poltorak *et al.*, 1998). Recombinant P fimbriated strains triggered inflammation in C3H/HeN mice but responses were virtually absent in the TLR4-deficient mice. The possible relevance to human disease was shown by expression in uro-epithelial cells of several TLR mRNAs, including TLR4, and by the up-regulation of TLR4 following exposure to the P fimbriated strains. By confocal microscopy, TLR4 was shown to co-localize with the GSL receptors in caveoli. We proposed that P fimbriae bind to the receptor GSLs and recruit TLR4 for signal transduction (Fig. 10.4) (Frendéus *et al.*, 2001).

LPS is thought to be the principal component of Gram-negative bacteria that alerts the host to systemic infection. LPS-binding protein transfers LPS to a binding site on CD14, and cell activation occurs through the recruitment of TLRs (Ulevitch and Tobias, 1995) (Fig. 10.4). Most epithelial cells are CD14-negative, however, and do not become activated by free LPS but they respond to whole Gram-negative bacteria and secrete cytokines and other pro-inflammatory mediators (Hedges *et al.*, 1992; Hedlund *et al.*, 1996, 1999). LPS might still play a role in this response, if delivered to CD14-independent signalling pathways.

The possible involvement of LPS in P fimbrial responses was investi-gated by mutational inactivation of lipid A. The *msbB* gene encodes an acyl transferase that couples myristic acid to the lipid IV_A precursor (Somerville *et al.*, 1996). Bacteria expressing the mutated lipid A lose their toxicity for CD14-positive cells. We used mutant strains expressing normal ($msbB^+$) or mutated ($msbB^-$) lipidA as hosts for recombinant plasmids carrying the *pap*

P fimbriae

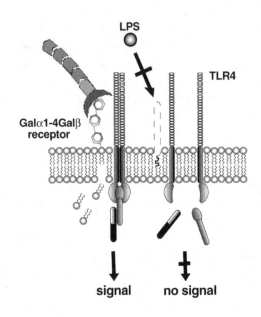

LPS

TLR4

Galα1-4Galβ
receptor

signal   no signal

(b)

Type 1 fimbriae

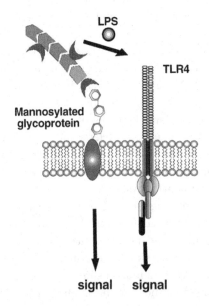

LPS

TLR4

Mannosylated
glycoprotein

signal   signal

sequences encoding P fimbriae. The recombinant strains expressing P fimbriae in the two LPS backgrounds were then used for cell activation. The P fimbriated strains elicited strong cellular responses, but the non-fimbriated controls were poor host response inducers, demonstrating the importance of the fimbriae for cell activation in this model (Hedlund *et al.*, 1999; Frendéus *et al.*, 2001). Responses to the P fimbriated transformants were independent of the *msbB* mutation. Furthermore, known inhibitors of LPS responses in CD14-positive cells showed no inhibitory effect.

The results suggest that the TLR4-dependent signalling pathway allows P fimbriae to overcome the LPS refractoriness of the CD14-negative epithelial cells, and that P fimbriae recruit TLR4 by molecular mechanisms different from those involved in cell activation by LPS in CD14-positive cells.

10.2.3.2 Difference in LPS delivery between P and type 1 fimbriated *Escherichia coli*

Type 1 fimbriae recognize mannosylated glycoprotein receptors on a variety of cells (Fig. 10.4). Uroplakins are one candidate on epithelial cells, and the Tamm–Horsfall protein may bind type 1 fimbriae in the lumen, but the exact structure of the uro-epithelial cell receptors is not known (Orskov *et al.*, 1980; Wu *et al.*, 1996). Like P fimbriae, type 1 fimbriae mediate attachment and activate the production of inflammatory mediators in the target cell

Figure 10.4. Difference in TLR4 recruitment and LPS delivery between P and type 1 fimbriae. (a) The uro-epithelial cells are CD14-negative. Still, their response to P fimbriae is TLR4 dependent, but with no detectable influence of LPS. Addition of LPS inhibitors such as polymyxin B or BPI did not reduce the cellular responses to P fimbriated *E. coli*. Also, mutational inactivation of the *msbB* sequences that encode lipidA myristoylation and reduce the responses to Gram-negative bacteria in CD14-positive cells had no effect on responses to P fimbriated bacteria (Hedlund et al., 1999; Frendéus *et al.*, 2000). The results suggest that P fimbriae overcome the LPS refractoriness of the epithelial cells and activate them through mechanisms that converge at the level of TLR4. Interestingly, both the LPS receptor CD14 and the glycosphingolipid receptor for P fimbriae are lipid anchored, lack a transmembrane signalling domain and may thus activate TLR4 via similar mechanisms. (b) The epithelial cell responses to type 1 fimbriated *E. coli* are partially TLR4 and LPS dependent. Type 1 fimbriated bacteria deliver a weak LPS/TLR4 signal accounting for about 25% of the cellular response. The remaining response is lectin-dependent, requiring the binding to mannosylated glycoprotein receptors (Hedlund *et al.*, 2001). Both responses are inhibited by α-methyl-D-mannoside, which blocks the binding of type 1 fimbriae to their receptors, demonstrating the importance of adhesion for the LPS-dependent as well as the lectin-dependent signal.

(Fig. 10.4). They contribute to virulence in the murine urinary tract, as shown by the loss of virulence in adhesin-negative *fim*H gene knockouts (Connell *et al.*, 1996).

Type 1 fimbriae use different mechanisms for cell activation. They trigger both a lectin-dependent/TLR4-independent and an LPS/TLR4-dependent signal (Fig. 10.5). Mutational inactivation of lipid A caused about a 20% reduction in the host response, and polymyxin B and bacterial permeability-inducing (BPI) protein gave a similar reduction of cellular responses to type 1 fimbriated bacteria. The type 1 fimbriated bacteria elicited a strong inflammatory response in TLR4-proficient mice, but the responses were slower and lower in the TLR4-deficient mice. Still, the type 1 fimbriated strains caused a significant and fimbriae-dependent response also in these mice, which was not seen for the P fimbriae. These results demonstrate that type 1 fimbriated *E. coli* trigger both an LPS-dependent and a lectin-dependent cytokine response (Hedlund *et al.*, 2001).

These studies illustrate that bacterial fimbriae target LPS and other microbial molecules to the host cells and that they activate host responses via LPS-dependent and LPS-independent mechanisms. LPS is released from the surface of Gram-negative bacteria as membrane 'blebs' containing most of the outer membrane components. Host cells probably see such membrane blebs rather than LPS monomers. It is not just a question of the molecules involved and their concentration at the cell surface, but of the molecular context into which these molecules are delivered to the cell surface. Fimbriae-mediated attachment provides an interesting example where the delivery mechanism itself determines the pathway of cell activation and where an additional microbial product may contribute to this process if delivered to the appropriate target.

10.2.3.3 Host receptor variation

There exist great individual differences in the expression of receptors for P fimbriae, and they influence the susceptibility to infection. Patients prone to UTI show a higher density of epithelial cell receptors (Stamey and Sexton, 1975). Furthermore individuals of blood-group P_1 run an increased risk of developing recurrent pyelonephritis (Lomberg *et al.*, 1986), and the A_1P_1 blood group predisposes to infection with bacteria recognizing the globo A receptor (Lindstedt *et al.*, 1991). One may predict that a host lacking receptors would be resistant to UTI, but there are too few receptor-negative individuals to enable investigation of this hypothesis.

10.2.4 Summary of the first signal

Mucosal responses to infectious agents depend on the delivery of microbial components and on the activation of signalling pathways in the responding cell(s). Fimbriae-mediated attachment is an important mechanism for tissue targeting of Gram-negative bacteria, and enhances mucosal inflammation. While fimbriae themselves can trigger cell activation, they also dock the bacteria to sites where receptors are expressed, and may initiate the host response. Furthermore, attachment is the delivery mechanism for other virulence-associated molecules, causing, for example, membrane perturbation, invasion or apoptosis.

10.3 THE SECOND SIGNAL: EPITHELIAL CELL MEDIATORS AND THE INFLAMMATORY RESPONSE

10.3.1 Epithelial cell cytokine production

For a long time epithelial cells were regarded mainly as mechanical building blocks in the mucosal barrier, with the sole duty to secrete immunoglobulin (Ig) A. We proposed that the link between attachment and disease severity could be explained if the bacteria were able to elicit a response in the epithelial cells. We proposed that epithelial cells sense external danger, and showed that uro-epithelial cells produce an array of mediators that transmit signals across the mucosal barrier to adjacent cells or underlying tissues (for a review, see Svanborg *et al.*, 1999). More specifically, bacteria elicit mucosal cytokine production. By enhancing epithelial cell cytokine responses, bacterial adherence provides a direct link between adherence, inflammation and disease severity (de Man *et al.*, 1989; Hedges *et al.*, 1990, 1991; Linder *et al.*, 1991; Agace *et al.*, 1993a; Svensson *et al.*, 1994).

10.3.1.1 Attachment and the chemokine repertoire

Attachment influences the magnitude of the chemokine response as well as the uro-epithelial chemokine repertoire (Fig. 10.5). The type 1 fimbriated bacteria stimulated chemokines fairly selective for neutrophils, including IL-8 and GRO-α, while P fimbriated bacteria stimulated a response that should favour the recruitment of lymphocytes and monocytes, in addition to neutrophils.

The *in vitro* data on the chemokine repertoire were confirmed by clinical studies. The CXC chemokines IL-8, IP-10, GRO-α and ENA-78 were found in the urine of most patients with febrile UTI, as were the CC chemokines MCP-1, MIP-1α and RANTES. The CXC or CC chemokine responses

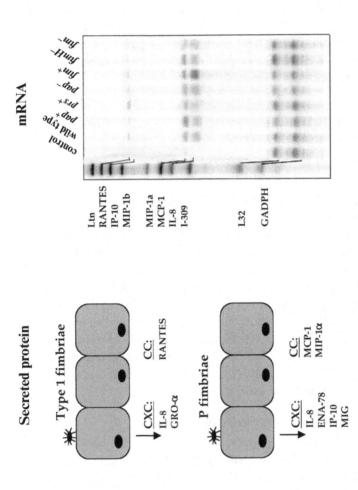

Figure 10.5. Fimbriae influence the chemokine repertoire of uro-epithelial cells. A498 kidney cells were challenged with bacteria expressing type 1 or P fimbriae and the chemokine response was quantified in culture supernatants, or by RNA display. The left panel shows the classes of chemokines produced by the cells. The gel shows mRNA species in control cells after bacterial challenge (data after 12 hours). ENA, epithelial cell-derived neutrophil-activating protein; GRO, growth-related oncogene; GADPH, glyceraldehyde-3-phosphate dehydrogenase; IL, interleukin; IP, interferon-γ-inducible protein; MCP, monocyte chemoattractant protein; MIG, monokine inducible by interferon-γ; MIP, macrophage inflammatory protein; RANTES, regulated upon activation, normal T cell expressed and secreted.

of the patients were also shown to differ with the *pap* genotype of the infecting strain. Patients infected with the *pap*⁺ *E. coli* isolates had higher MCP-1 responses, and patients infected with *prs*⁺ strains had higher RANTES responses (Otto *et al.*, 1999; G. Godaly *et al.*, unpublished data).

10.3.1.2 Infection up-regulates chemokine receptor expression

Chemokines mediate their biological effects by binding to chemokine receptors belonging to the large serpentine receptor family with seven transmembrane loops, linked to a G-protein for signal transduction (Baggiolini *et al.*, 1997; Damaj *et al.*, 1996; Laudanna *et al.*, 1998). Human uro-epithelial cells express both CXCR1 and CXCR2 (Godaly *et al.*, 2000). Thus the molecular basis for CXCR-dependent neutrophil–epithelial cell interactions is in place along the mucosal lining of the human urinary tract. Experimental infection caused a rapid increase in the expression of the murine IL-8 receptor by epithelial cells in kidney and bladders *in vivo*. CXCR1, but not CXCR2, was shown to account for the increased binding of IL-8 to infected cells and for the increased neutrophil migration across infected cell layers *in vitro*. The epithelial cells thus present IL-8 to incoming neutrophils and lead them through the cell layer into the lumen (Fig. 10.5).

10.3.2 Neutrophil recruitment to the urinary tract

The recruitment of neutrophils to the urinary tract results in so-called 'pyuria'. The different steps involved in neutrophil migration to the urinary tract (Agace *et al.*, 1993a,b; Godaly *et al.*, 1997) have been identified, using the Transwell model system (Parkos *et al.*, 1991; McCormick *et al.*, 1993). Bacteria first stimulate the epithelial cells to secrete chemokines, and a gradient is established. Neutrophils leave the bloodstream, migrate through the tissues and cross the epithelial barrier into the lumen, and IL-8 appears to be the main force that drives neutrophil migration across epithelial cell layers. In the mouse UTI model, MIP-2 was shown to be one IL-8 equivalent, directing neutrophil migration through the lamina propria and across the epithelium (Hang *et al.*, 1999).

10.3.2.1 Dysfunctional neutrophil migration in the mIL-8Rh knockout mice.

The *in vivo* relevance of epithelial CXCR expression was studied in IL-8Rh knockout (KO) mice. While no single rodent homologue for human IL-8 has been identified (Cochran *et al.*, 1983; Tekamp-Olson *et al.*, 1990) several CXC and CC chemokines are produced in response to UTI *in vivo* (Hang *et* al., 1999). These chemokines converge on the single murine IL-8 receptor homologue, which was deleted in the mIL-8Rh KO mouse (Cacalano *et al.*, 1994). A dysfunctional neutrophil response was found in the IL-8R KO mice as compared with the Balb/c controls. The IL-8R mutant mice had an intact mucosal chemokine response to infection and, by immunohistochemistry, neutrophils were shown to accumulate in bladder and kidney tissue (Godaly *et al.*, 2000; Hang *et al.*, 2000). The neutrophils were unable to cross the epithelium into the lumen, resulting in low urine neutrophil numbers at all times (Plate 10.2). The influx of neutrophils showed that the chemotactic gradient was intact despite the lack of IL-8 receptors, and demonstrated that the epithelium formed a virtually impermeable barrier when the chemokine receptor was absent.

10.3.2.2 Innate defences are important in the urinary tract

Specific immune functions and innate host defences have converged at mucosal sites (for a review, see Kagnoff and Kiyono, 1996). While specific immunity can be particularly effective in preventing systemic infections by invading pathogens, innate immune mechanisms provide the most potent defence against UTI, and neutrophils are important local effectors of the mucosal defence (Shahin *et al.*, 1987; Haraoka *et al.*, 1999). The work was based on earlier observations in C3H/HeJ (lps^d, lps^d) mice that lack a neutrophil response to UTI, and fail to clear bacteria from the tissues (Shahin *et al.*, 1987; Svanborg-Edén *et al.*, 1987). To prove that neutrophils were the critical effector cells, C3H/HeN mice were depleted of neutrophils with the granulocyte-specific antibody RB6-8C5. As a consequence, bacterial clearance from kidneys and bladders was drastically impaired (Haraoka *et al.*, 1999). The results demonstrate that neutrophils are essential for bacterial clearance from the urinary tract and that the neutrophil recruitment deficiency in C3H/HeJ mice explains their susceptibility to Gram-negative mucosal infection.

10.3.2.3 The IL-8R deficiency confers susceptibility to acute pyelonephritis and renal scarring

Even though the IL-8 receptor KO mice recruited neutrophils into the tissues, they were highly susceptible to infection (Plate 10.2b) (Frendéus *et al.*, 2000). The mutant mice were unable to clear the bacteria from kidney and bladder tissues and eventually developed bacteraemia and symptoms of systemic disease. This is the first case of an animal model where disruption of a single gene causes a disease closely resembling acute pyelonephritis in humans, as well as long-term tissue pathology with the characteristics of renal scarring. These results emphasize the role of innate immunity, neutrophils and IL-8 receptor function in resistance to acute urinary tract infection and in renal tissue integrity.

10.3.2.4 Neutrophils recognize type 1 but not P fimbriated bacteria

Fimbriae influence the interactions of *E. coli* with neutrophils and the susceptibility to phagocytosis and killing. Type 1 fimbriae bind neutrophils, utilizing mannosylated glycoproteins as receptors, elicit an oxidative burst, and bacteria are taken into the cells and killed depending on other factors such as the capsule, outer membrane proteins LPS and P fimbriae bind poorly to human neutrophils, owing to the paucity of GSL receptors on these cells (Svanborg-Edén *et al.*, 1984; Tewari *et al.*, 1994), and the bacteria are phagocytosed quite poorly (Öhman *et al.*, 1982; Goetz, 1989).

Taken together with the chemokine response profile, the results suggest that while bacteria with type 1 fimbriae mainly trigger a cellular response leading to their own destruction, bacteria with P fimbriae enhance their virulence by activating a cellular infiltrate that they can evade.

10.3.3 Summary of the second signal

Epithelial cells secrete inflammatory mediators that set the stage for subsequent recruitment of inflammatory and lymphoid cells. Attachment regulates this response by influencing the chemokine repertoire and chemokine receptor expression.

Neutrophils are essential for the antibacterial defence, and IL-8 and IL-8 receptors are crucial in order for neutrophils to cross the epithelial barriers. In the absence of functional chemokine receptors, mice become highly susceptible to infection and develop systemic disease resembling acute pyelonephritis in humans (Frendéus *et al.*, 2000). Neutrophils accumulate in the tissues and the animals develop renal scarring (Hang *et al.*, 2000). These findings illustrate the need to regulate neutrophil migration across epithelial barriers and explain why resident cells have developed mechanisms for differential expression of chemokines and their receptors.

10.4 LOW CXCR1 EXPRESSION IN CHILDREN WITH ACUTE PYELONEPHRITIS

The phenotypes of the mIL-8Rh KO mice suggested that susceptibility to UTI in humans might reflect poor IL-8 receptor expression. This was, indeed, shown to be the case (Frendéus *et al.*, 2000). Children with recurrent

UTI and acute pyelonephritis were compared with age-matched children without a history of UTI. Samples for CXCR1 quantification were obtained twice during an infection-free interval, with one year in between. CXCR1 expression was examined by fluorescence-activated cell sorting (FACS) using monoclonal antibodies, and mRNA levels were determined by RNA display. The children with recurrent UTI and acute pyelonephritis were shown to have reduced CXCR1 expression, and their CXCR1 mRNA levels were low as compared with the controls (Plate 10.3). Ongoing studies have suggested that the UTI-prone children have sequence heterogeneity in the CXCR1 promoter region as compared to controls.

These results suggest that the susceptibility to attaching bacteria can be regulated by mutations influencing the expression of molecules involved in host defence. The results illustrate how attachment triggers mucosal inflammation and determines the disease through control of the host response.

REFERENCES

Agace, W., Hedges, S., Andersson, U., Andersson, J., Ceska, M. and Svanborg, C. (1993a). Selective cytokine production by epithelial cells following exposure to *Escherichia coli*. *Infection and Immunity* **61**, 602–609.

Agace, W., Hedges, S.R., Ceska, M. and Svanborg, C. (1993b). Interleukin-8 and the neutrophil response to mucosal gram-negative infection. *Journal of Clinical Investigation* **92**, 780–785.

Baggiolini, M., Dewald, B., Moser, B. (1997). Human chemokines: an update. *Annual Review of Immunology* **15**, 675–705.

Beutler, B. (2000). Tlr4: central component of the sole mammalian LPS sensor. *Current Opinion in Immunology* **12**, 20–26.

Brinton, C. (1965). The structure, function, synthesis, and genetic control of bacterial pili, and a molecular model of DNA and RNA transport in Gram-negative bacteria. *Transactions of the New York Academy of Sciences* **27**, 1003–1054.

Cacalano, G., Le, J., Kikly, K., Ryan, A.M., Pitts-Meek, S., Hultgren, B., Wood, W.I. and Moore, M.W. (1994). Neutrophil and B cell expansion in mice that lack the murine IL-8 receptor homolog. *Science* **265**, 682–684.

Cochran, B.H., Reffel, A.C. and Stiles, C.D. (1983). Molecular cloning of gene sequences regulated by platelet-derived growth factor. *Cell* **33**, 939–947.

Connell, I., Agace, W., Klemm, P., Schembri, M., Marild, S. and Svanborg, C. (1996). Type 1 fimbrial expression enhances *Escherichia coli* virulence for the urinary tract. *Proceedings of the National Academy of Sciences, USA* **93**, 9827–9832.

Plates

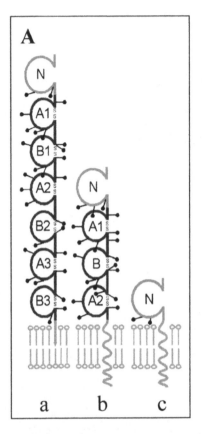

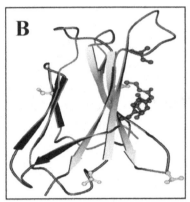

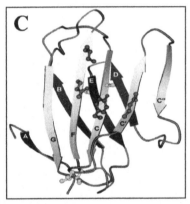

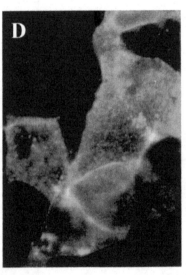

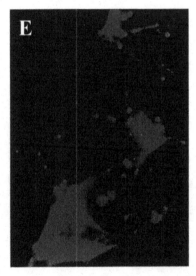

Plate 2.1. CEACAM structure and interactions with meningococci. (A) The domain structure of several CEA family members of the immunoglobulin superfamily (adapted from the CEA website: http://www.ukl.uni-freiburg.de/immz/cea/). The family characteristically contains a single N-terminal IgV-like domain and, in addition, most members contain several 1gC2-like domains (A1, B1, etc.). The proteins are heavily glycoslylated and the positions of glycans are shown as ball and sticks. CEA (a) is anchored via a GPI extension whereas CEACAM1 (b) and CEACAM3 (c) contain hydrophobic transmembrane domains and their isoforms may contain a long or a short cytoplasmic tail. The long cytoplasmic domains contain ITAM/ITIM motifs.

(B–C) Ribbon diagrams (B, side view; C, front view) of the N-domain of CEACAM1 showing the amino acid residues that were mutated for studies on the receptor binding domain (Virji *et al.*, 1999, 2000). The CEACAM1 N-domain, modelled after the immunoglobulin fold, is predicted to consist of nine antiparallel beta strands joined by loop regions and arranged in two sheets (faces), ABED and C″C′CFG (CFG for brevity). The mutations in residues were specifically introduced at sites predicted to be exposed on the molecule and therefore likely to be involved in cellular interactions. Beta strands are represented by arrows. Side chains on the CFG face that have been mutated to alanine are shown in ball-and-stick representation in red: Ile-91 and Tyr-34, blue: Glu-89 and 44, green: Ser-32 and Val-39. The side chains associated with confirmed N-linked glycosylation sites in CEA are shown in yellow (Beauchemin *et al.*, 1999; Virji *et al.*, 1999)

(D–E) Immunofluorescence micrographs showing the adherence of capsulate meningococci expressing Opa proteins to COS cells expressing high levels of CEACAM1. COS cells were transfected with cDNA encoding CEACAM1 and labelled with the N-domain-reactive monoclonal antibody YTH71:3 (D). The monolayers were overlaid with capsulate meningococci and adherent bacteria were labelled with antibodies directed against capsular polysaccharide (E). (D) and (E) represent a single field.

A colour version of these plates is available for download from www.cambridge.org/9780521126755

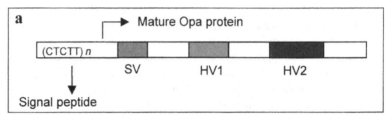

a

(CTCTT) *n*

Mature Opa protein

SV   HV1   HV2

Signal peptide

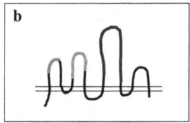

b

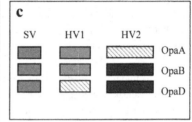

c

SV	HV1	HV2	
			OpaA
			OpaB
			OpaD

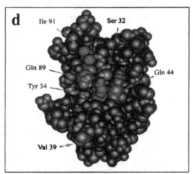

d

Ile 91   Ser 32

Gln 89

Tyr 34   Gln 44

Val 39

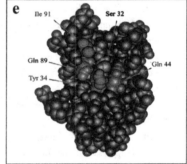

e

Ile 91   Ser 32

Gln 89

Tyr 34   Gln 44

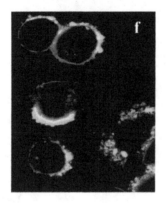

f

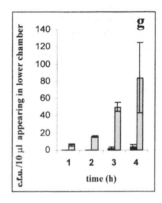

g

c.f.u./10 µl appearing in lower chamber

140
120
100
80
60
40
20
0

1   2   3   4

time (h)

Plate 2.2. Structure of meningococcal Opa proteins, receptor domains targeted by *N. meningitidis* and *H. influenzae* and consequences of interactions via the receptor. (a) Overall organization of *opa* genes of *N. meningitidis*, *N. gonorrhoeae* and commensal neisseriae. The repeats (CTCTT) in the signal peptide coding region are responsible for phase variation of individual *opa* genes by a slipped-strand mispairing mechanism.

(b) A representation of predicted secondary structure of Opa proteins (Malorny *et al.*, 1998) showing the surface-exposed variable domains. (c) Opa proteins of the serogroup A strain C751 OpaA and OpaB share the HV1 structure whereas OpaB and OpaD share the HV2 structure. HV2 of OpaA and HV1 of OpaD are unique. The SV structures are identical in these cases.

(d–e) Van der Waals surface representation of the model (front views) of the N-domain of CEACAM1 showing the critical residues that are required for the interactions of *N. meningitidis* (d) and *H. influenzae* (e) on the CFG face. The colour scheme is the same as shown in the ribbon model (B and C, Plate 2.1). The amino acid residues exerting the greatest influence on binding of *N. meningitidis* and *H. influenzae* are shown in red. Ile-91 lies towards the top of the CFG face but in close proximity to Tyr-34, which lies in the centre of the face. The other amino acid residues that appear to determine receptor tropism of distinct strains and variants also lie in close proximity (Virji *et al.*, 1999, 2000). The adhesiotopes are overlapping such that *N. meningitidis* and *H. influenzae* may compete for binding to the receptor (Virji *et al.*, 2000). The binding regions for the two organisms are located at the exposed protein CFG region that is also implicated in homotypic interactions of CEACAMs.

(f) Cytoskeletal rearrangement in CHO cells stably transfected with CEACAM1 on interaction with *H. influenzae*. Similar observations were made with *N. meningitidis*. Confocal imaging shows F-actin localization (phalloidin–rhodamine stain) underneath bacterially induced receptor caps (bacteria labelled with anti-LPS-fluorescein stain).

(g) Distinct rates of transmigration of *N. meningitidis* (black bars) and *H. influenzae* (grey bars) through polarized Caco-2 cells. Although the primary interactions of both bacteria with the target cells is mediated via CEACAM receptors, the final outcome is distinct. Both bacteria are able to cross intact monolayers without disruption of the monolayer since the electrical resistance remains unaltered at >4 hours postinfection. Confocal imaging suggests that, whereas *H. influenzae* may transmigrate by a paracellular route, *N. meningitidis* may utilize a transcellular route (D.J. Hill, J. Griffith and M. Virji, unpublished data).

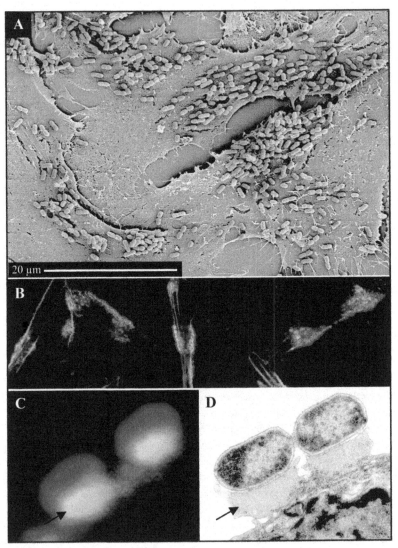

Plate 9.1. Formation of pedestals following infection of eukaryotic cells by EHEC. Analysis of HeLa cells infected with the EDL933 strain for 4 hours by scanning electron microscopy (A), immunofluorescence (B and C) and transmission electron microscopy (D). Anti-O157 antibodies (red) and phalloidin (green) were used for detecting bacteria and polymerized actin, respectively (B and C). Like EPEC, EHEC EDL933 is able to induce pedestal formation (see arrows in panels C and D).

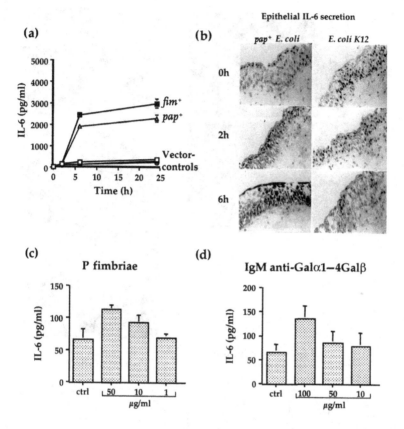

Epithelial IL-6 secretion

Plate 10.1. Fimbriae enhance epithelial cell cytokine responses. (a) Cytokine response of human kidney cells exposed to recombinant *E. coli* strains expressing P (*pap⁺*) or type 1 (*fim⁺*) fimbriae. The non-fimbriated host strain was *E. coli* HB101 or *E. coli* AAEC, and the vector controls carried the respective plasmids without the insert. Transformation with the *pap* or *fim* sequences changed the unfimbriated strain to a host response inducer (Hedlund *et al.*, 1998, 1999, 2001). (b) Cytokine response of human urinary tract biopsies exposed to recombinant *E. coli* strains differing in P fimbrial expression. The biopsies were obtained from patients undergoing urinary tract surgery, and were infected *in vitro* from the apical side with the *pap⁺* or *fim⁺* recombinant *E. coli* strain. The P fimbriated strain triggered a more rapid and stronger epithelial IL-6 response than did the non-fimbriated control (Hang *et al.*, unpublished data). (c) Purified P fimbriae stimulate epithelial cytokine responses.

(d) IgM antibodies to the globo series GSLs activate epithelial cell cytokine responses. The results suggest that ligand–receptor interaction is sufficient for cell activation, even though the responses were low. The results suggest that GSLs can sense their environment and directly translate this molecular information into cell-activating signals. ctrl, control.

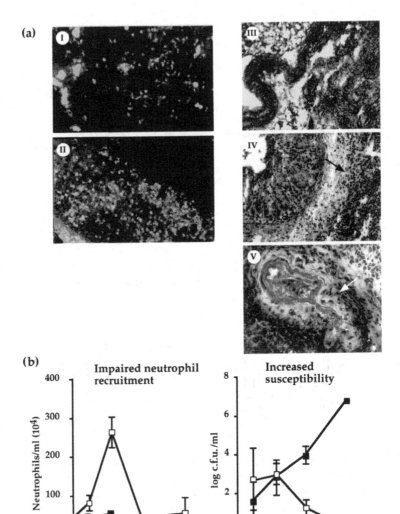

(a)

(b)

Impaired neutrophil recruitment

Increased susceptibility

Neutrophils/ml (10⁴)

log c.f.u./ml

2h   6h   24h   7d

2h   6h   24h   7d

log time

Plate 10.2. IL-8 receptor KO mice develop acute pyelonephritis and renal scarring. The mice were inoculated intravesically with uropathogenic *E. coli*. Urine samples for bacterial and neutrophil counts were obtained after 2, 6 and 24 hours, and tissues were obtained after sacrifice. Neutrophils were visualized by immunohistochemistry, and bacteria by viable counts on tissue homogenates. (The figures are published with the permission of the University of Chicago Press (from Hang *et al.* (2000), *Journal of Infectious Diseases*, **182**, 1738–1748, © 2000 University of Chicago Press) and the *Journal of Experimental Medicine* (Frendéus *et al.*, 2000).)

(a) Neutrophils were recruited into the tissues but were unable to cross the epithelial barrier, owing to the lack of chemokine receptors. The cells accumulated under the epithelium in the cortex and medulla (I + II) and after about five weeks renal tissues were destroyed, leading to renal scarring with fibrosis around the blood vessels and under the epithelial lining (III–V). (b) Control mice (—□—) developed pyuria, but within a few hours after intravesical inoculation, KO mice (—■—) had no or few neutrophils in the urine. Control mice cleared infection within 3 to 7 days and did not develop detectable symptoms, but mIL-8Rh KO mice failed to clear the infection. Bacterial numbers increased, and the animals developed bacteraemia and symptoms of acute infection.

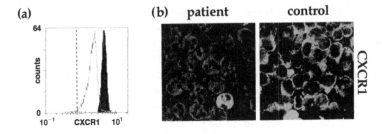

(a)

(b) patient   control

(c)

	Number	Age range	CXCR1 expression range*	Mean CXCR1 expression
Patients	15	1–12	(–4.96) – (–0.02)	(–1.47)
Controls	18	4–13	(–0.31) – 1.32	0.42

* $P < 0.001$.

Plate 10.3. Low chemokine receptor expression in children with acute pylonephritis, as compared with age-matched controls. Children with recurrent UTI and acute pyelonephritis were prospectively enrolled in a study of chemokine receptor expression and were compared with age-matched controls that had no experience of UTI. Peripheral blood neutrophils were isolated and the surface expression of chemokine receptors was detected by (a) flow cytometry (shaded column, controls; open column, study groups) or (b) confocal microscopy. (c) Neutrophils were obtained from children with recurrent UTI and acute pyelonephritis or from healthy age matched controls. CXCR1 expression was quantified by flow cytometry, using specific antibodies. Results are expressed as the difference between standard and patient or control (*). As can be seen from the table, the UTI-prone children had low CXCR1 expression. (Data overlap, but are not identical, with those published by Frendéus et al. (2000) and are published with the permission of the Journal of Experimental Biology.)

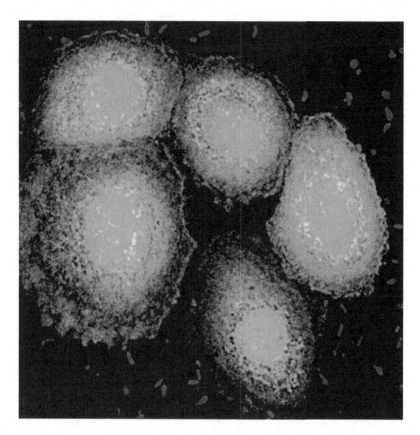

Plate 13.1. Fluoresence microscopy image of a 0.2 µm optical section through gingival epithelial cells (green) exposed to *P. gingivalis* (red) for 15 minutes. Internalized *P. gingivalis* (orange/yellow) accumulate in the perinuclear region. (Image provided by Drs Izutsu and Belton, University of Washington. Reproduced with permission of Blackwell Science Ltd, from Belton *et al.*, 1999.)

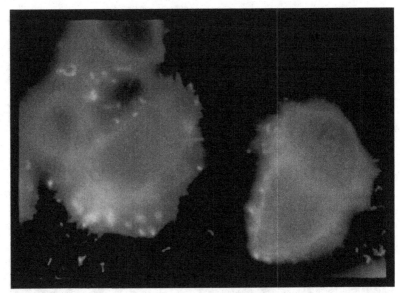

Plate 13.2. Immunofluorescence micrograph of *A. actinomycetemcomitans* (green) within KB cells (red) connected by intercellular protrusions. Internalized bacteria appear yellow. (Image provided by Drs Fives-Taylor and Lippman, University of Vermont. Reproduced, with permission of American Society for Microbiology, from Meyer *et al.*, 1996.)

Damaj, B.B., McColl, S.R., Mahana, W., Crouch, M.F. and Naccache, P.H. (1996). Physical association of Gi2alpha with interleukin-8 receptors. *Journal of Biological Chemistry* **271**, 12783–12789.

de Man, P., van Kooten, C., Aarden, L., Engberg, I., Linder, H. and Svanborg-Edén, C. (1989). Interleukin-6 induced at mucosal surfaces by Gram-negative bacterial infection. *Infection and Immunity* **57**, 3383–3388.

Duguid, J., Smith, I., Dempster, G. and Edmunds, P. (1955). Non-flagellar filamentous appendages ('fimbriae') and haemagglutinating activity in *Bacterium coli*. *Journal of Pathology and Bacteriology* **70**, 335–348.

Frendéus, B., Godaly, G., Hang, L., Karpman, D. and Svanborg, C. (2000). IL-8R deficiency confers susceptibility to acute experimental pyelonephritis and may have a human counterpart. *Journal of Experimental Medicine* **192**, 881–890

Frendéus, B., Wachtler, C., Hedlund, M., Fischer, H., Samuelsson, P., Svensson, M. and Svanborg, C. (2001). *Escherichia coli* P fimbriae utilize the TLR4 receptor pathway for cell activation. *Molecular Microbiology* **40**, 37–51.

Freter, R. (1969). Studies on the mechanism of action of intestinal antibody in experimental cholera. *Texas Reports on Biology and Medicine* **27**, 299–316.

Gibbons, R. (1973). Bacterial adherence in infection and immunity. *Reviews in Microbiology* **4**, 49–60.

Godaly, G., Proudfoot, A.E., Offord, R.E., Svanborg, C. and Agace, W.W. (1997). Role of epithelial interleukin-8 (IL-8) and neutrophil IL-8 receptor A in *Escherichia coli*-induced transuroepithelial neutrophil migration. *Infection and Immunity* **65**, 3451–3456.

Godaly, G., Frendéus, B., Proudfoot, A., Svensson, M., Klemm, P. and Svanborg, C. (1998). Role of fimbriae-mediated adherence for neutrophil migration across *Escherichia coli*-infected epithelial cell layers. *Molecular Microbiology* **30**, 725–735.

Godaly, G., Hang, L., Frendéus, B. and Svanborg, C. (2000). Transepithelial neutrophil migration is CXCR1 dependent *in vitro* and is defective in IL-8 receptor knock out mice. *Journal of Immunology* **165**, 5287–5294.

Goetz, M.B. (1989). Priming of polymorphonuclear neutrophilic leukocyte oxidative activity by type 1 pili from Escherichia coli. *Journal of Infectious Diseases* **159**, 533–542.

Hagberg, L., Engberg, I., Freter, R., Lam, J., Olling, S. and Svanborg-Edén, C. (1983). Ascending, unobstructed urinary tract infection in mice caused by pyelonephritogenic *Escherichia coli* of human origin. *Infection and Immunity* **40**, 273–283.

Hang, L., Haraoka, M., Agace, W.W., Leffler, H., Burdick, M., Strieter, R. and Svanborg, C. (1999). Macrophage inflammatory protein-2 is required for neutrophil passage across the epithelial barrier of the infected urinary tract. *Journal of Immunology* 162, 3037–3044.

Hang, L., Frendéus, B., Godaly, G. and Svanborg, C. (2000). Il-8 receptor KO mice have subepithelial neutrophil entrapment and renal scarring following acute pyelonephritis. *Journal of Infectious Diseases* 182, 1738–1748.

Hanson, L.A., Holmgren, J., Jodal, U., Lincoln, K. and Lindberg, U. (1972). Asymptomatic bacteriuria – a serious disease? *British Medical Journal*, 2, 530.

Haraoka, M., Hang, L., Frendéus, B., Godaly, G., Burdick, M., Strieter, R. and Svanborg, C. (1999). Neutrophil recruitment and resistance to urinary tract infection. *Journal of Infectious Diseases* 180, 1220–1229.

Hedges, S., de Man, P., Linder, H., Kooten, C. V. and Svanborg-Edén, C. (1990). Interleukin-6 is secreted by epithelial cells in response to Gram-negative bacterial challenge. In *Advances in Mucosal Immunology*, Proceedings of the Fifth International Congress of Mucosal Immunology, ed. T.T. MacDonald, S. Challacombe, P. Bland, C. Stokes, A. Heatley and A. Mowat, pp. 144–148. London: Kluwer.

Hedges, S., Anderson, P., Lidin-Janson, G., de Man, P. and Svanborg, C. (1991). Interleukin-6 response to deliberate colonization of the human urinary tract with Gram-negative bacteria. *Infection and Immunity* 59, 421–427.

Hedges, S., Svensson, M. and Svanborg, C. (1992). Interleukin-6 response of epithelial cell lines to bacterial stimulation *in vitro*. *Infection and Immunity* 60, 1295–1301.

Hedlund, M., Svensson, M., Nilsson, Å., Duan, R. and Svanborg, C. (1996). Role of the ceramide signalling pathway in cytokine responses to P fimbriated *Escherichia coli*. *Journal of Experimental Medicine* 183, 1–8.

Hedlund, M., Nilsson, Å., Duan, R.D. and Svanborg, C. (1998). Sphingomyelin, glycosphingolipids and ceramide signalling in cells exposed to P fimbriated *Escherichia coli*. *Molecular Microbiology* 29, 1297–1306.

Hedlund, M., Wachtler, C., Johansson, E., Hang, L., Somerville, J.E., Darveau, R.P. and Svanborg, C. (1999). P fimbriae dependent lipopolysaccharide independent activation of epithelial cytokine responses. *Molecular Microbiology* 33, 693–703.

Hedlund, M., Frendéus, B., Wachtler, C., Hang, L., Fischer, H. and Svanborg, C. (2001). Type 1 fimbriae deliver an LPS- and TLR4-dependent activation signal to CD14-negative cells. *Molecular Microbiology* 39, 542–552.

Hull, R., Gill, R., Hsu, P., Minshew, B. and Falkow, S. (1981). Construction and expression of recombinant plasmids encoding type 1 or D- mannose resistant pili from the urinary tract infection *Escherichia coli* isolate. *Infection and Immunity* 33, 933–938.

Johanson, I.M., Plos, K., Marklund, B.I. and Svanborg, C. (1993). Pap, papG and prsG DNA sequences in *Escherichia coli* from the fecal flora and the urinary tract. *Microbial Pathogenesis* 15, 121–129.

Kagnoff, M. and Kiyono, H. (1996). *Mucosal Immunology*. San Diego: Academic Press.

Källenius, G., Möllby, R., Svensson, S.B., Winberg, J., Lundblad, S., Svensson, S. and Cedergren, B. (1980). The Pk antigen as a receptor for the haemagglutination of pyelonephritogenic *E. coli*. *FEMS Microbiology Letters* 7, 297–302.

Kim, M. Y., Linardic, C., Obeid, L. and Hannun, Y. (1991). Identification of sphingomyelin turnover as an effector mechanism for the action of tumor necrosis factor alpha and gamma-interferon. Specific role in cell differentiation. *Journal of Biological Chemistry* 266, 484–489.

Korhonen, T.K., Vaisanen, V., Saxen, H., Hultberg, H. and Svenson, S.B. (1982). P-antigen-recognizing fimbriae from human uropathogenic *Escherichia coli* strains. *Infection and Immunity* 37, 286–291.

Laudanna, C., Mochly-Rosen, D., Liron, T., Constantin, G. and Butcher, E.C. (1998). Evidence of zeta protein kinase C involvement in polymorphonuclear neutrophil integrin-dependent adhesion and chemotaxis. *Journal of Biological Chemistry* 273, 30306–30315.

Leffler, H. and Svanborg-Edén, C. (1980). Chemical identification of a glycosphingolipid receptor for *Escherichia coli* attaching to human urinary tract epithelial cells and agglutinating human erythrocytes. *FEMS Microbiology Letters* 8, 127–134.

Leffler, H. and Svanborg-Edén, C. (1981). Glycolipid receptors for uropathogenic *Escherichia coli* on human erythrocytes and uroepithelial cells. *Infection and Immunity* 34, 920–929.

Lindberg, U. (1975). Asymptomatic bacteriuria in school girls. V. The clinical course and response to treatment. *Acta Paediatrica Scandinavica* 64, 718–724.

Linder, H., Engberg, I., Mattsby Baltzer, I., Jann, K. and Svanborg-Edén, C. (1988). Induction of inflammation by *Escherichia coli* on the mucosal level: requirement for adherence and endotoxin. *Infection and Immunity* 56, 1309–1313.

Linder, H., Engberg, I., Hoschützky, H., Mattsby Baltzer, I. and Svanborg-Edén, C. (1991). Adhesion dependent activation of mucosal IL-6 production. *Infection and Immunity* 59, 4357–4362.

Lindstedt, R., Larson, G., Falk, P., Jodal, U., Leffler, H. and Svanborg-Edén, C. (1991). The receptor repertoire defines the host range for attaching *Escherichia coli* recognizing globo-A. *Infection and Immunity* 59, 1086–1092.

Lomberg, H., Cedergren, B., Leffler, H., Nilsson, B., Carlström, A.-S. and Svanborg-Edén, C. (1986). Influence of blood group on the availability of

243

receptors for attachment of uropathogenic *Escherichia coli. Infection and Immunity* **51**, 919–926.

McCormick, B., Colgan, S., Delp-Archer, C., Miller, S. and Madara, J. (1993). *Salmonella typhimurium* attachment to human intestinal epithelial monolayers: transcellular signalling to subepithelial neutrophils. *Journal of Cellular Biology* **123**, 895–907.

Neish, A.S., Gewirtz, A.T., Zeng, H., Young, A.N., Hobert, M.E., Karmali, V., Rao, A.S. and Madara, J.L. (2000). Prokaryotic regulation of epithelial responses by inhibition of IkappaB-alpha ubiquitination. *Science* **289**, (5484):1560–1563.

Normark, S., Lark, D., Hull, R., Norgren, M., Båga, M., O´Hanley, P., Schoolnik, G. and Falkow, S. (1983). Genetics of digalactose-binding adhesion from a uropathogenic *Escherichia coli. Infection and Immunity* **41**, 942–949.

Öhman, L., Hed, J. and Stendahl, O. (1982). Interaction between human polymorphonuclear leukocytes and two different strains of type 1 fimbriae-bearing *Escherichia coli. Journal of Infectious Diseases* **146**, 751–757.

Orskov, I., Ferencz, A. and Orskov, F. (1980). Tamm–Horsfall protein or uromucoid is the normal urinary slime that traps type 1 fimbriated Escherichia coli. *Lancet* **1**, 887.

Otto, G., Braconier, J., Andreasson, A. and Svanborg, C. (1999). Interleukin-6 and disease severity in patients with bacteremic and nonbacteremic febrile urinary tract infection. *Journal of Infectious Diseases* **179**, 172–179.

Parkos, C., Delp, C., Armin Arnout, M. and Madara, J. (1991). Neutrophil migration across cultured intestinal epithelium. *Journal of Clinical Investigation* **88**, 1605–1612.

Plos, K. (1992). *P-fimbrial Genotype, Phenotype and Clonal Variation in* Escherichia coli *from the Urinary Tract and the Fecal Flora*. Göteborg.

Plos, K., Carter, T., Hull, S., Hull, R. and Svanborg-Edén, C. (1990). Frequency and organization of *pap* homologous DNA in relation to clinical origin of uropathogenic *Escherichia coli. Journal of Infectious Diseases* **161**, 518–524.

Plos, K., Connell, H., Jodal, U., Marklund, B. I., Marild, S., Wettergren, B. and Svanborg, C. (1995). Intestinal carriage of P fimbriated *Escherichia coli* and the susceptibility to urinary tract infection in young children. *Journal of Infectious Diseases* **171**, 625–631.

Poltorak, A., He, X., Smirnova, I., Liu, M.Y., Van Huffel, C., Du, X., Birdwell, D., Alejos, E., Silva, M. and Galanos, C. (1998). Defective LPS signaling in C3H/HeJ and C57BL/10ScCr mice: mutations in Tlr4 gene. *Science* **282**, 2085–2088.

Schütze, S., Potthoff, K., Machleidt, T., Berkovic, C., Wiegman, K. and Krönke, M. (1992). TNF activates NF-κB by phosphatidylcholine-specific phospholipase C-induced 'acidic' sphingomyelin breakdown. *Cell* **71**, 765–776.

Shahin, R.D., Engberg, I., Hagberg, L. and Svanborg-Edén, C. (1987). Neutrophil recruitment and bacterial clearance correlated with LPS responsiveness in local Gram-negative infection. *Journal of Immunology* **138**, 3475–3480.

Smith, H.W. and Linggood, M.A. (1972). Further observations on *Escherichia coli* enterotoxins with particular regard to those produced by atypical piglet strains and by calf and lamb strains: the transmissible nature of these enterotoxins and of a K antigen possessed by calf and lamb strains. *Journal of Medical Microbiology* **5**, 243–250.

Somerville, J.E.J., Cassiano, L., Bainbridge, B., Cunningham, M.D. and Darveau, R.P. (1996). A novel *Escherichia coli* lipid A mutant that produces an anti-inflammatory lipopolysaccharide. *Journal of Clinical Investigation* **97**, 359–365.

Stamey, T. and Sexton, C. (1975). The role of vaginal colonization with Enterobacteriacea in recurrent urinary tract infections. *Journal of Urology* **113**, 214–217.

Svanborg, C., Godaly, G. and Hedlund, M. (1999). Cytokine responses during mucosal infections: role in disease pathogenesis and host defence. *Current Opinion in Microbiology* **2**, 99–105.

Svanborg, C., Frendéus, B., Godaly, G., Hang, L., Hedlund,M. and Wachtler, C. (2001). Mucosal host response to urinary tract infection. *Journal of Infectious Diseases* **183** (Supplement 1) 61–65.

Svanborg-Edén, C. and Hansson, H.A. (1978). *Escherichia coli* pili as possible mediators of attachment to human urinary tract epithelial cells. *Infection and immunity* **21**, 229–237.

Svanborg-Edén, C., Hanson, L.A., Jodal, U., Lindberg, U. and Sohl-Åkerlund, A. (1976). Variable adherence to normal urinary tract epithelial cells of *Escherichia coli* strains associated with various forms of urinary tract infections. *Lancet* **ii**, 490–492.

Svanborg-Edén, C., Eriksson, B., Hanson, L.Å., Jodal, U., Kaijser, B., Lidin Jamson, G., Lindberg, U. and Olling, S. (1978). Adhesion to normal human uroepithelial cells of *Escherichia coli* of children with various forms of urinary tract infections. *Journal of Pediatrics* **93**, 398–403.

Svanborg-Edén, C., Hagberg, L., Hanson, L.A., Korhonen, T., Leffler, H. and Olling, S (1981). Adhesion of *Escherichia coli* in urinary tract infection. In *Adhesion and Microorganism Pathogenicity*, Ciba Foundation Symposium 80, pp. 161–187. London: Pitman Medical Ltd.

Svanborg-Edén, C., Freter, R., Hagberg, L., Hull, R., Hull, S., Leffler, H. and Schoolnik, G. (1982). Inhibition of experimental ascending urinary tract infection by an epithelial cell-surface receptor analogue. *Nature* **298**, 560–562.

Svanborg-Edén, C., Bjursten, L.M., Hull, R., Hull, S., Magnusson, K.E.,

Moldovano, Z. and Leffler, H. (1984). Influence of adhesins on the interaction of *Escherichia coli* with human phagocytes. *Infection and Immunity* **44**, 672–680.

Svanborg-Edén, C., Hagberg, L., Hull, R., Hull, S., Magnusson, K.E. and Ohman, L. (1987). Bacterial virulence versus host resistance in the urinary tracts of mice. *Infection and Immunity* **55**, 1224–1232.

Svensson, M., Lindstedt, R., Radin, N. and Svanborg, C. (1994). Epithelial glycosphingolipid expression as a determinant of bacterial adherence and cytokine production. *Infection and Immunity* **62**, 4404–4410.

Tekamp-Olson, P., Gallegos, C., Bauer, D., McClain, J., Sherry, B., Fabre, M., van Deventer, S. and Cerami, A. (1990). Cloning and characterization of cDNAs for murine macrophage inflammatory protein 2 and its human homologues. *Journal of Experimental Medicine* **172**, 911–919.

Tewari, R., Ikeda, T., Malaviya, R., MacGregor, J.I., Little, J.R., Hultgren, S.J. and Abraham, S.N. (1994). The PapG tip adhesin of P fimbriae protects *Escherichia coli* from neutrophil bactericidal activity. *Infection and Immunity* **62**, 5296–5304.

Ulevitch, R.J. and Tobias, P.S. (1995). Receptor-dependent mechanisms of cell stimulation by bacterial endotoxin. *Annual Reviews of Immunology* **14**, 437–457.

Wold, A., Thorssén, M., Hull, S. and Svanborg-Edén, C. (1988). Attachment of *Escherichia coli* via mannose or Galα1–4Galβ containing receptors to human colonic epithelial cells. *Infection and Immunity* **56**, 2531–2537.

Wu, X.R., Sun, T.T. and Medina, J.J. (1996). In vitro binding of type 1-fimbriated *Escherichia coli* to uroplakins Ia and Ib: relation to urinary tract infections. *Proceedings of the National Academy of Sciences, USA* **93**, 9630–9635.

Wullt, B., Bergsten, G., Connell, H., Röllano, P., Gebratsedik, N., Hang, L. and Svanborg, C. (2000). P fimbriae enhance the early establishment of *Escherichia coli* in the human urinary tract. *Molecular Microbiology* **38**, 456–464.

Wullt, B., Bergsten, G., Connell, H., Rollano, P., Gebratsedik, N., Hang, L. and Svanborg, C. (2001). P-fimbriae trigger mucosal responses to *Escherichia coli* in the human urinary tract. *Cellular Microbiology* **3**, 255–264.

Zobell, C. (1943). The effect of solid surfaces on bacterial activity. *Journal of Bacteriology* **46**, 39–56.

CHAPTER 11

Adhesion of oral spirochaetes to host cells and its cytopathogenic consequences

Richard P. Ellen

11.1 INTRODUCTION

247

Oral spirochaetes colonize the gingival crevice and periodontal pocket adjacent to inflamed gingival tissues as their primary habitat. They are found most often as part of a loosely adherent mass of motile bacteria at the interface of the biofilm known as subgingival dental plaque and the sulcular and junctional epithelium lining the pocket. This location offers the opportunity to co-adhere with other bacteria and to adhere to a variety of host cells and extracellular matrix (ECM) components. Spirochaetes are motile, and they can translocate between epithelial cells, through the basement membrane, and even into the viscous ECM of the lamina propria of the gingiva. *En route*, whole cells or fragments shed from their outer sheath may become attached to epithelial cells, gingival fibroblasts, endothelial cells, and leukocytes, and thereby may initiate cascades of cellular responses that affect the normal functions of host cells and the remodelling of the periodontal tissues. Thus the adhesion of oral spirochaetes to ECM components and cells probably contributes to the exacerbation of periodontal lesions and to the chronicity of lesions by interfering with wound healing in sites where infections are not sufficiently suppressed by host defences or by therapy.

Most oral spirochaetes have not yet been cultivated in the microbiology laboratory (Qui *et al.*, 1994). Until the advent of methodology based on molecular genetics, there was little understanding of the great diversity represented by this group, which was traditionally classified by cell size and by the number of periplasmic flagella originating at the poles of these long, slender bacteria. Only 10 species have been cultivated and described in enough phenotypic detail to be assigned species names according to conventions in nomenclature. All have been classified in the genus *Treponema*. Yet over 50 clusters, or phylotypes, in 10 major groups have been described on the basis of 16 S rRNA sequence diversity (Choi *et al.*, 1996; Dewhirst *et al.*,

2000). Presumably, most humans are colonized early in life by a variety of these phylotypes. It apparently takes the enriched, anaerobic environment of chronically inflamed periodontal pockets to support the emergence of the most diverse variety of spirochaetes at a population density sufficient for detection and discrimination (Choi *et al.*, 1994; Riviere *et al.*, 1995; Moter *et al.*, 1998; Willis *et al.*, 1999).

Treponema denticola is the most frequently isolated species of oral spirochaete and the species most successfully cultivated long term in the laboratory (Chan *et al.*, 1993). Therefore it has become a convenient model organism for the investigation of mechanisms of treponemal adhesion to host ligands and pathways by which host cells respond to oral treponemes. Moreover, routine cultivation of *T. denticola* provided the opportunity for it to serve as a model vector for the first studies of genetic transformation and construction of genetic mutants in treponemes (Chi *et al.*, 1999). The complete genome of the type strain ATCC 35405 is presently being sequenced (see URL http://www.tigr.org/cgi-bin/BlastSearch/blast.cgi?). This chapter will examine the state of knowledge of the mechanisms and cytopathogenic consequences of adhesion of oral treponemes to ECM components and to human cells. By necessity, almost all the information was derived from the literature on *T. denticola* (for reviews, see Thomas, 1996; Fenno and McBride, 1998). New investigations that include some of the other cultivable oral species are just beginning to emerge, and these will appear in the chapter where relevant.

11.2 SURFACE STRUCTURE OF *TREPONEMA DENTICOLA*

Like all spirochaetes, treponemes are spiral-shaped, motile bacteria. External to the cytoplasmic membrane are a periplasm and an outer sheath that are analogous to the periplasm and outer membrane of Gram-negative bacteria. Flagella, varying in number according to species, originate at each pole and extend in the periplasm for more than half the cell length; the flagella from opposite poles often overlap at mid cell. *Treponema denticola* is a small spirochaete that usually has two flagella originating from each pole (Fig. 11.1). The structural proteins of the flagella themselves, the proteins that anchor the flagella, the proteins that function in flagellar export, and the proteins that act as motors for flagellar rotation and thereby control the direction of cellular movement are comparable to those with similar functions in other bacteria (Heinzerling *et al.*, 1997; Li *et al.*, 2000). The outer sheath is composed of a complex lipid-carbohydrate polymer, either lipopolysaccharide or lipo-oligosaccharide, and associated outer membrane protein

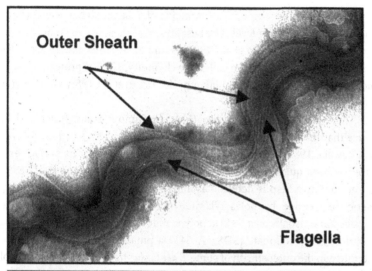

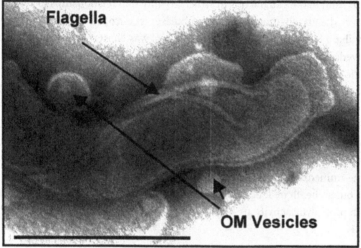

Figure 11.1. Transmission electron photomicrographs of negatively stained cells of *Treponema denticola* ATCC 35405 demonstrating typical ultrastructure of oral treponemes. (*Top*) A *T. denticola* cell in the process of cell division, where daughter cells have begun to segregate but the flagella remain connected. The outer sheath is clearly visible. (*Bottom*) Higher magnification of one pole, showing outer membrane (OM) vesicles and insertion of two periplasmic flagella in basal bodies. Bars = ~0.5 μm. (Photomicrographs by P.F. Yang and R.P. Ellen.)

complexes. Very few outer sheath proteins of *T. denticola* have been characterized at the molecular level. The best known of these are the major surface protein (Msp) (Haapasalo *et al.*, 1992; Fenno et al, 1996) and an associated serine protease that has been called the chymotrypsin-like protease or, more recently, dentilisin by different laboratories (Uitto *et al.*, 1988; Grenier *et al.*, 1990; Ishihara *et al.*, 1996).

There is electron microscopy (EM), immuno-EM and functional evidence that the Msp complex is displayed on the bacterial surface (Masuda and Kawata, 1982; Egli *et al.*, 1993), although the extent of its surface exposure has been questioned in one recent publication (Caimano *et al.*, 1999). Extracted native Msp is detected as a high molecular mass complex, oligomeric polypeptide by polyacrylamide gel electrophoresis (PAGE) when unboiled and usually as a 53 kDa polypeptide when denatured by boiling in sodium dodecyl sulphate (SDS). A 64 kDa putative adhesin has also been identified in the outer sheath of some *T. denticola* strains (Weinberg and Holt, 1991) and may be analogous in structure–function relationships to Msp. The outer sheath chymotrypsin-like protease is encoded by the gene *prtP* (Ishihara *et al.*, 1996). The term 'dentilisin' was chosen because of the conserved deduced amino acid residues that comprise and flank its putative catalytic residues, which resemble those of the subtilisin family of serine proteases. The protease runs in SDS-PAGE as a minimum 91–95 kDa complex; the complex is much larger in unboiled samples (Uitto *et al.*, 1988; Grenier *et al.*, 1990; Rosen *et al.*, 1995). The complex in the type strain ATCC 35405 contains dentilisin, estimated at 72 kDa from its deduced amino acid sequence, and a 43 kDa protein of unknown function that is encoded by a gene immediately upstream from *prtP* (Ishihara *et al.*, 1996). Fragments of the outer sheath of *T. denticola* are known to shed from the surface. Vesicles with proteolytic activity are also thought to derive from the outer sheath (Fig. 11.1).

11.3 ADHESION OF TREPONEMES TO EXTRACELLULAR MATRIX

Oral treponemes encounter ECM components on the surface of host cells, in the intercellular spaces between epithelial cells, in the basement membrane, and in the extracellular compartment of the lamina propria and deeper connective tissues of the gingiva. Knowledge of their binding and adhesion to these components is derived primarily from *in vitro* experimental assays using *T. denticola*. This organism most likely evolved to bind to ECM components both as a mechanism for evading removal by bathing host fluids and as a way to forage for substrates that could be degraded by proteolysis to satisfy its nutritional requirement for peptides. Indeed, some of the ECM proteins to which *T. denticola* binds are known to be degraded by dentilisin, and this activity has been shown to promote the movement of *T. denticola* cells through *in vitro* molecular complexes mimicking the mammalian basement membrane (Grenier *et al.*, 1990). The ECM components bound by *T. denticola* represent a diverse range of molecular structures. These have been identified using assays that detect either the soluble ECM molecules that can be bound by the bacteria, or various polypeptides extracted from the bacteria, or the adhesion of whole *T. denticola* cells (or extracted proteins) to inanimate surfaces that had been preconditioned by the adsorption of ECM components. The distinction is subtle but significant; macromolecules that are bound to inanimate or cell surfaces may undergo conformational changes that affect bacterial adhesion.

Like many wild-type bacteria, the surface of *T. denticola* cells is moderately hydrophobic (Fiehn, 1991); cells in aqueous suspensions of *T. denticola* vary in their net negative surface charge (Cowan *et al.*, 1994). Therefore whole cells of the organism are able to associate transiently or even bind to inanimate surfaces depending on their wettability. *Treponema denticola* cells bind avidly to hydrophobic plastics and non-wettable silicon surfaces conditioned with methyl groups (Ellen *et al.*, 1998). The cells lie flat upon contact. The cells bind less readily to wettable hydrophilic surfaces; those cells that do adhere bind tenuously, mostly in a polar orientation and possibly mediated by ionic interactions. In contrast, *T. denticola* adheres very well to inanimate surfaces that have been preconditioned with human macromolecules that form extracellular matrices, and such coated surfaces favour a polar orientation of the adherent cells (Fig. 11.2). The organism binds selectively to fibronectin (Fn), laminin (Lam), fibrinogen, and hyaluronan-coated surfaces but binds relatively poorly to surfaces conditioned with bovine serum albumin (BSA) and inconsistently to type IV or other collagens (Haapasalo *et al.*, 1991, 1996). Its adhesion to the native ECM conditioning components is usually tip-oriented (Dawson and Ellen, 1990, 1994), similar to the polar adhesion to Fn previously reported for *Treponema pallidum*.

In most cases, the identity of the adhesin(s) and the exact mechanisms of adhesion are unknown. Marked inhibition of adhesion to Lam and other ECM proteins when the bacteria were pretreated with either mixed glycosidases or the sulphydryl reagents *p*-chloromercuribenzoic acid and oxidized glutathione suggests that surface protein SH groups and/or carbohydrate residues may be involved. So far, the only non-proteolytic outer sheath protein known to adhere to the ECM components listed is Msp. Both native and recombinant Msp (rMsp) bind to Lam- and Fn-coated plastic (Haapasalo

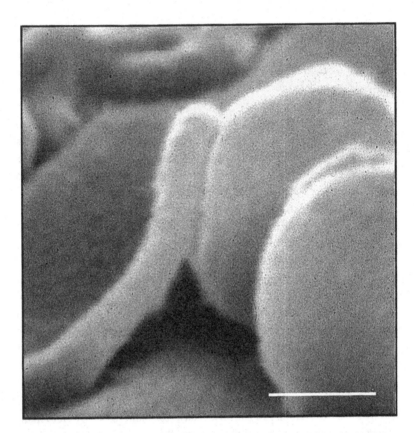

Figure 11.2. Scanning electron photomicrograph of a *T. denticola* cell adhering by one pole to a latex microsphere coated with fibronectin. Bar = ~0.5 μm. (Reprinted from *Trends in Microbiology*, 2, Ellen, R.P., Dawson, J.R. and Yang, P.F.: *Treponema denticola* as a model for polar adhesion and cytopathogenicity of spirochetes, pp. 114–119, Copyright 1994, with permission of Elsevier Science.).

et al., 1991; Fenno *et al.*, 1996). The degree of binding is diminished in the presence of the soluble form of these ECM components. Yet, rMsp does not block the adhesion of *T. denticola* cells, probably due to the presence of other, yet to be identified, outer sheath adhesins that recognize the substrates. Similarly, very little is known about potential domains of ECM components that may serve as the ligand for Msp or other *T. denticola* adhesins. The whole bacterium can bind to an immobilized, elastase-generated 140 kDa α-chain fragment of Lam that contains its eukaryotic cell attachment domain (Haapasalo *et al.*, 1991). Lam and Fn contain a similar cell-binding domain that consists of RGD(Arg-Gly-Asp) residues. *Treponema denticola* cells can adhere in polar orientation to plastic coated with RGDS-containing heptapeptides but not to control peptides. However, the putative target peptides in solution do not inhibit the adhesion of whole cells to Fn (Dawson and Ellen, 1990).

A novel aspect of polar adhesion to Fn-conditioned surfaces is the time-dependent concentration of Fn-binding adhesins at one pole of the bacterium during the adhesion process. *Treponema denticola* cells usually approach Fn-coated surfaces laterally or obliquely and then slowly 'stand up' into a polar orientation. Dawson and Ellen (1994) have shown by ligand–gold electron microscopy that *T. denticola* cells that contact Fn-coated nitrocellulose membranes have a disproportionate labelling of Fn[gold]-binding adhesins at one pole as compared with cells that are distant to, or prevented from contact with, the Fn-coated surface (Fig. 11.3). Cells contacting buffer-treated or Lam-coated surfaces show no redistribution of Fn-binding adhesins. These observations lend support to the contention that treponemal proteins are free to move in the outer sheath of the bacterium (Charon *et al.*, 1981) and imply that a process analogous to 'capping' of specific ligand receptors in eukaryotes occurs in prokaryotic cells. Clustering of adhesins would tend to increase the avidity of the interaction between the bacterium and the specific substrate.

Much of the work on *T. denticola* adhesion to ECM components has concentrated on binding assays in which the ligand is first bound to a surface, probably because native ECM components in tissues associate with cell surface receptors such as integrins and with each other. When extracted proteins of *T. denticola* are first dissociated, electrophoresed by PAGE, transferred to membranes and then exposed to Fn in solution, immunoreactive Fn can be detected bound to many of the polypeptide bands (Umemoto *et al.*, 1993). Few have been characterized, and it is likely that several of the positive bands that migrate differentially actually reflect variously-sized molecular species or degradation products of a single protein. *Treponema denticola* uses ECM proteins as targets for adhesion in order to resist removal by bathing fluids, but it also probably binds soluble substrates to satisfy its nutritional needs. Thus the substrate recognition and binding domains of surface-associated enzymes, for example subtilisin, may serve concomitantly or transiently as adhesins. In addition, several strains of *T. denticola* have a gene for OppA, a 70 kDa putative surface binding protein of an ATP-binding cassette type of transporter (Fenno *et al.*, 2000). *Treponema denticola* OppA evidently binds plasma Fn and soluble plasminogen but binds to neither substrate when it is immobilized on a surface, nor to epithelial cells. OppA is also found in *Treponema vincentii*. *Treponema denticola* also has an operon that encodes MglBAC (Lépine and Ellen, 2000), another putative ATP-binding

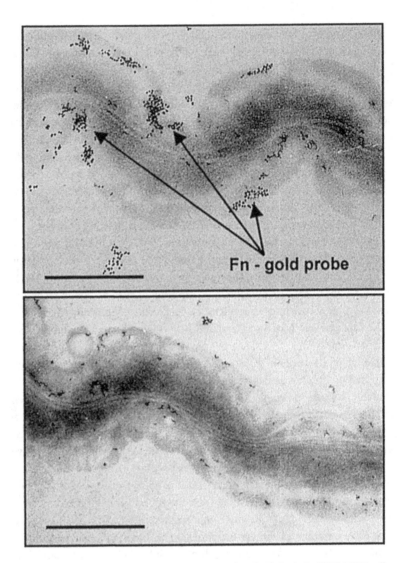

Fn - gold probe

Figure 11.3. Transmission electron photomicrographs of a *T. denticola* ATCC 33520 cell that had migrated through methyl cellulose into contact with a fibronectin-coated filter. After washing, the bacteria were exposed to fibronectin (Fn)-gold probes (10 nm). The probes concentrated towards one pole of the spirochaete (*top*) relative to the other pole (*bottom*). Bars = ~0.5 μm. (Photomicrographs by J.R. Dawson and R.P. Ellen.)

cassette implicated in glucose/galactose binding in enteric bacteria and flagellar export in other bacteria, including *T. pallidum*. MglB and MglA also have 40–55% amino acid similarity to several Spa and Ysc surface proteins of the ATP-binding and type III secretion pathways of enteropathogens.

11.4 ADHESION TO EPITHELIAL CELLS AND DISRUPTION OF EPITHELIAL BARRIER FUNCTION

In their natural habitat, oral treponemes interface directly with the stratified squamous, sulcular and junctional epithelium of the gingival crevice and inflamed periodontal pocket. Their adhesion to a variety of squamous-like epithelial cell lines *in vitro* has been studied by several investigators (Olsen, 1984; Baehni, 1986; Keulers *et al.*, 1993; Walker *et al.*, 1999). Recently, there have been detailed investigations of the cytopathogenic consequences of the adhesion of *T. denticola* to: (i) human KB cells (ATCC CL-17) (DeFilippo *et al.*, 1995), which were originally derived from an epidermoid carcinoma; (ii) HeLa cells (Mathers *et al.*, 1996);, (iii) human Hep-2 cells (Ko *et al.*, 1998b; Ellen *et al.*, 2000), which are of laryngeal origin; and (iv) cell lines established from explants of porcine periodontal ligament epithelial cell rests (Uitto *et al.*, 1995). Together, these studies have clearly demonstrated the potential of *T. denticola* to breach epithelial barriers (Fig. 11.4).

When epithelial monolayers are challenged with *T. denticola* supsensions, the bacteria attach selectively to a subpopulation of cells. Consequently, cytopathic changes are detected in some cells in the monolayer almost immediately, but most cells remain viable even though they may progress through a series of cellular responses affecting their cytoskeleton and intercellular junctional complexes. In time, an increasing proportion of the cells detach from the extracellular substratum. The detached cells are usually dead, while those that remain attached apparently remain viable, even though they may experience profound rearrangement of their cytoskeleton, disassembly of actin filaments, diminished cytokeratin and desmoplakin II expression and demonstrate rather marked abnormalities in volume regulation, which is an actin-dependent function (DeFilippo *et al.*, 1995; Uitto *et al.*, 1995). Although *T. denticola* cells can be detected inside some epithelial cells and endothelial cells during *in vitro* challenge assays (Uitto *et al.*, 1995; Peters *et al.*, 1999), there is no evidence that their uptake represents an evolved survival or virulence factor that is analogous to the survival strategies reported for many other mucosal pathogens. Indeed, intracellular *T. denticola* cells are seen in various stages of degradation inside vesicular organelles of the epithelial cells (Uitto *et al.*, 1995).

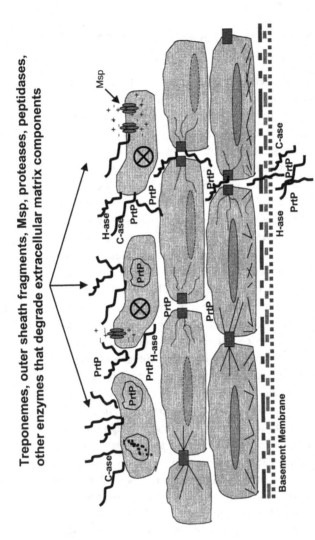

Treponemes, outer sheath fragments, Msp, proteases, peptidases, other enzymes that degrade extracellular matrix components

Figure 11.4. Hypothetical model illustrating the consequences of the adhesion of oral treponemes to stratified squamous epithelium of the gingival crevice, based mostly on research using *T. denticola*. Proteolytic enzymes such as the chymotrypsin-like enzyme (subtilisin, PrtP) and other enzymes that degrade intercellular matrix (chondroitinase (C-ase) and hyaluronidase (H-ase)) disrupt barrier function that is normally maintained by intercellular adhesion molecules and associated junctional proteins. The treponemes migrate between the epithelial cells, degrade matrix, and penetrate through the basement membrane into the lamina propria of the gingiva. Adhesion of treponemes to exposed epithelial cells induces cytoskeletal rearrangement, loss of volume regulation, shrinkage, and diminished expression of junctional proteins, which also contribute to widened intercellular spaces. Msp binds to epithelial cells and may transpose into the plasma membrane to perturb ion fluxes. Msp is also cytotoxic to some cells (⊗). Both whole treponemes and PrtP may be taken up into the cells. The treponemes are evidently degraded in intracellular vesicles.

Transcytosis of *T. denticola* and other treponemes may be an infrequent event, even in cases when spirochaetes are detected in the gingival connective tissues of periodontal lesions. *Treponema denticola* is able to pass through model mucosal barriers *in vitro* but not as readily as some other species of treponemes isolated from periodontal lesions (Riviere *et al.*, 1991). Several species of oral treponemes produce enzymes with hyaluronidase, chondroitinase, and wide-range peptidase activities that would be expected to degrade extracellular matrices in the epithelium and connective tissues (Mikx *et al.*, 1992; Scott *et al.*, 1996b). Moreover, *T. denticola* and outer membrane extracts with proteolytic activity have profound detrimental effects on epithelial junctions *in vitro*, both in typical confluent monolayers and in a multilayer system that mimics stratified epithelium (Uitto *et al.*, 1995; Ko *et al.*, 1998b). Proteolytic protein complexes or vesicles that are shed from the whole bacteria can penetrate between epithelial cells coincident with bacterial abrogation of epithelial barrier function (Uitto *et al.*, 1995). The native chymotrypsin-like serine protease (PrtP; dentilisin), within complexes of outer membrane extracts or in purified preparations, clearly has the ability to eliminate junctional resistance to electrical conductivity and to the diffusion of molecules that are normally excluded (Uitto *et al.*, 1995; Ko *et al.*, 1998b). Insertional inactivation of the *prtP* gene, resulting in a mutant devoid of the prolylphenylalanine peptidase activity typical of serine proteases, prevents the disruptive activity of *T. denticola* outer membrane extracts on epithelial resistance (Ellen *et al.*, 2000). Direct degradation of the intercellular matrix and substratum by dentilisin probably accounts for the time- and concentration-dependent detachment of epithelial cells in culture. Hep-2 cells seeded in the presence of moderate concentrations of an outer membrane extract of wild-type *T. denticola* 35405 never reach confluence, whereas cells cultured with similar extracts from the PrtP-defective mutant grow normally into contact (Ellen *et al.*, 2000). Confluent Hep-2 cells that are exposed to *T. denticola* outer membrane extracts also experience a biphasic response in electrical impedance signals that reflect cellular micromotion. At first, their micromotion increases, which may reflect sublethal actin filament rearrangement, cellular shape changes, and shrinkage due to altered actin assembly or proteolysis of intercellular contacts; then micromotion apparently ceases (Ko *et al.*, 1998b). Thus *T. denticola* affects physiological properties of epithelial cells that would normally maintain exclusive epithelial barriers and migration into confluence that would promote wound healing.

In vitro studies using confluent monolayers emphasize the acute pathogenic effects of *T. denticola* adhesion on epithelial cells, many of which can be attributed to the bacterium's proteolytic activity. Yet, other cell surface molecules such as its non-proteolytic outer sheath proteins may have toxic effects. Prominent among these is Msp, which is cytotoxic for a variety of mammalian cell types (Fenno et al., 1998a). Both native and rMsp have pore-forming properties that have been documented using HeLa cells. Exposure of cultured HeLa cells to Msp induces transient fluctuations in membrane potential that may be caused by the transposition of Msp into the plasma membrane (Mathers et al., 1996) (Fig. 11.4). Msp is known to have the properties of a non-exclusive outer membrane porin when tested in a standard *in vitro* model (Egli et al., 1993; Fenno et al., 1998a). Transient retention of its porin activity when incorporated into the plasma membrane of human epithelial cells would be expected to affect ion fluxes that normally regulate cellular functions. Indeed, the Msp-induced ion-permeable channels found in HeLa cells appear to be inwardly unidirectional and open only intermittently. Some other bacterial porins have been shown to transpose to the target cell plasma membrane and to cause acute elevation of intracellular Ca^{2+} concentration. Msp has recently been found to display analogous effects in gingival fibroblasts, as discussed in section 11.5.

11.5 ADHESION TO FIBROBLASTS AND THE PERTURBATION OF SIGNALLING PATHWAYS FOR ACTIN ASSEMBLY

Fibroblasts are the key cells that maintain homeostasis of non-mineralized collagenous tissues, and they provide a useful model for studying cytopathogenic effects of periodontal pathogens on non-immune cells that are important in inflammatory responses and wound repair in the gingiva. Similar to their adhesion to ECM components and epithelial cells, oral treponemes adhere in a polar orientation to cultured gingival fibroblasts (Fig. 11.5), and they are occasionally engulfed into the intracellular environment. For example, although almost all adherent *T. denticola* cells remain extracellular, intact bacteria have been observed in some fibroblasts by using double label immunofluorescence techniques, electron microscopy and confocal microscopy (Ellen et al., 1994a). The biological implications of this infrequent invasion are unknown, as oral treponemes have not yet been reported to exist intracellularly in fibroblasts in the gingiva.

Many of the fibroblast's key functions in maintaining the homeostasis of the gingival tissues involve its cytoskeleton. The cells migrate along collagen-rich ECM networks using tractional forces generated by cycles of actin-binding focal complex assembly and adhesion of integrins to ECM ligands. Thus their locomotion in response to chemotactic factors would be perturbed by microorganisms that degraded the ECM at the fibroblast

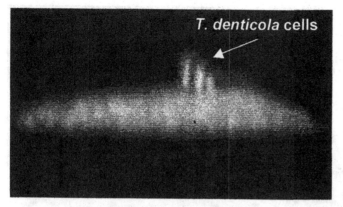

Figure 11.5. *Treponema denticola* ATCC 35405 cells bound in a polar orientation to the dorsal surface of a human gingival fibroblast. The image represents the *x–z* plane derived from a three-dimensional reconstruction from *x–y* optical sections of a fibroblast stained by indirect immunofluorescence with polyclonal rabbit antibodies raised against *T. denticola* ATCC 35405 followed by TRITC-conjugated goat anti-rabbit γ-globulin to label *T. denticola* cells, and by fluorescein isothiocyanate-conjugated concanavalin A to highlight the HGF plasma membrane. (Confocal microscopy image by I. Buivids and R.P. Ellen.)

surface or upset the intracellular signalling pathways that regulate the cytoskeleton. Physiological turnover of collagen in the gingiva is maintained at a controlled rate by the phagocytosis of collagen fibrils, intracellular degradation of the fibrils, and the synthesis of new collagen by fibroblasts. Collagen phagocytosis depends on actin assembly. It should be sensitive to bacterial factors that bind to the cells and that interrupt the normal balance of signalling molecules, such as the phosphoinositides (PIs) and calcium that regulate the activity of actin binding and severing proteins.

Cultured human gingival fibroblasts (HGFs) challenged by suspensions of *T. denticola* undergo cytoskeletal rearrangement, round up, shrink, and eventually detach from the ECM substratum (Baehni et al, 1992; Weinberg and Holt, 1990) (Figs. 11.6 and 11.7). These cytopathic changes occur asynchronously in HGF cultures, beginning with a few cells within minutes and eventually involving most of them. The changes are dependent on time and bacterial cell density, and they are affected by whether the HGFs are challenged as sparsely spaced cells soon after plating or as monolayers that have grown into confluence over a few days. Gross cellular distortion and detachment of HGFs from the substratum may result directly from ECM degradation by bacterial proteases, especially dentilisin, as the serine protease inhibitor phenylmethylsulphonyl fluoride (PMSF) inhibits HGF detachment

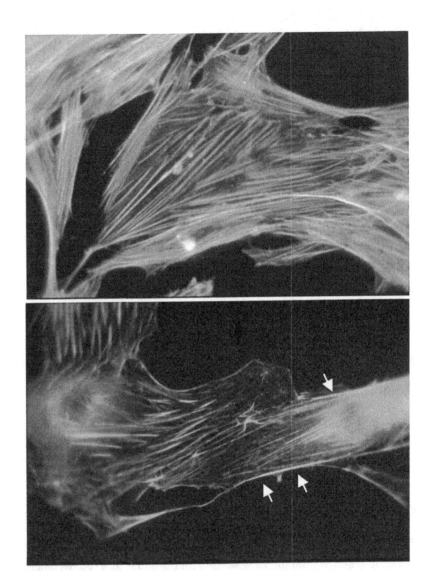

Figure 11.6. Fluorescence microscopy of rhodamine-phalloidin–stained human gingival fibroblasts that had been challenged with an outer membrane extract of *T. denticola* ATCC 35405 (*bottom*) compared with the control that had been exposed to the culture medium vehicle (*top*). Note the disruption of actin filaments in the originally prominent stress fibres and the assembly of actin filaments in a dense perinuclear positon and subjacent to the plasma membrane (arrows). (Photomicrographs by M. Song and R.P. Ellen.)

by *T. denticola* (Baehni *et al.*, 1992). *Treponema denticola* can degrade endogenous fibronectin that decorates the surface of cultured HGFs (Ellen *et al.*, 1994b) and this may account for some of the initial rounding-up that has been observed. Yet inhibition of serine protease activity by PMSF does not reduce the adhesion of *T. denticola* to HGFs (Ellen *et al.*, 1994b) or the disruption of stress fibre networks in HGFs (Yang *et al.*, 1998). Likewise, insertional inactivation of the *prtP* gene does not diminish stress fibre disruption in HGFs by *T. denticola* whole cells (Ellen *et al.*, 2000). It is the stress fibres that normally interface via integrins with the ECM components of the substratum. Many of the cytopathic effects of *T. denticola* whole cells can be mediated by extracts of its outer membrane (OM). While highly proteolytic, these extracts contain other non-proteolytic proteins that may account for some of the impact on intracellular signalling pathways affecting the cytoskeleton and thereby perturb important physiological functions of gingival fibroblasts.

(261)

Treponema denticola OM has profound effects on the actin-dependent phagocytosis of collagen by HGFs, a function crucial for the normal turnover of collagen in the gingiva (Battikhi *et al.*, 1999). Even short-term, transient exposure of HGFs to a concentration of OM that has no detrimental effect on their viablility, cell size and attachment to the substratum can cause a great increase in the uptake of collagen-coated beads *in vitro*. The rapid response, detected within 15 minutes of exposure to the OM, suggests that *T. denticola* OM components exert their effects on pathways that cause rearrangement of actin oligomers near the plasma membrane rather than triggering signalling pathways that would require genetic regulation of protein expression.

Investigations of intracellular signalling pathways affected by binding of *T. denticola* surface components have concentrated on calcium flux and biochemical determination of total inositol phosphates as a downstream indicator of perturbation of membrane PI metabolism (Fig. 11.7). Because the calcium responses of gingival fibroblasts are inherently sensitive to mechanical forces on the plasma membrane, OM extracts have served as the bacterial challenge to avoid confounding effects caused by the motility of tethered treponemes (Ko *et al.*, 1998a). Transients in intracellular calcium concentration ($[Ca^{2+}]_i$) were measured in fura-2-loaded HGFs by ratio fluorimetry. Exposure to *T. denticola* OM causes immediate but short-lived calcium oscillations in resting HGFs. These last for about 30 minutes, followed by a prolonged period of quiescence in which few, if any, calcium fluctuations can be detected. In addition, *T. denticola* OM is a rather potent inhibitor of HGF calcium responses to well-established chemical agonists and to mechanical

T. denticola and its outer sheath proteins

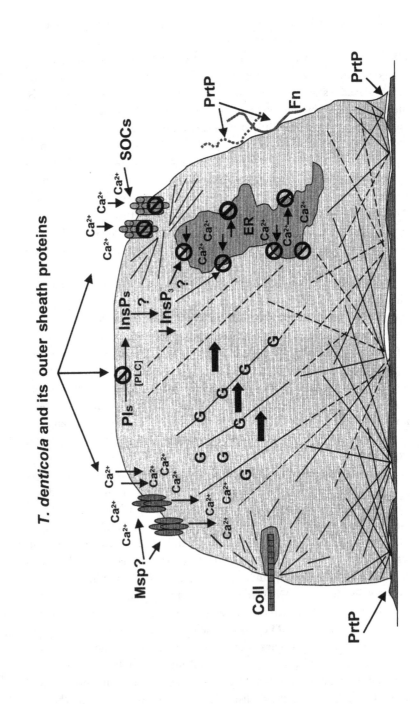

Figure 11.7. Hypothetical model illustrating the pathogenic consequences of the adhesion of *T. denticola*, outer membrane extracts (OM), Msp, and PrtP to human gingival fibroblasts (HGFs) and extracellular matrix proteins. The stop signs (circles with a bar across) indicate that calcium flux through the channels is either blocked or inhibited. Exposure of HGFs to *T. denticola*, OM, or Msp causes disassembly of actin filaments in stress fibres (straight lines→dashed lines) and the reorganization of the actin cytoskeleton associated with increased phagocytosis of collagen (Coll), cell shrinkage, and eventual detachment from the substratum. OM and Msp induce acute near-plasma-membrane and intracellular Ca^{2+} transients that may be due to Ca^{2+} permeation through the plasma membrane and, perhaps in part, to the formation of ion-permeable channels by Msp. The high concentration of Ca^{2+} may activate gelsolin (G) to bind actin filaments (D. Matevsky, D.A. Grove, Q. Wang and R.P. Ellen, unpublished data) and to sever them in a series of enzymatic steps that would yield a pool of actin oligomers that could be reassembled to foster phagocytosis or to uncouple store-operated Ca^{2+} flux. OM and Msp inhibit agonist-induced release of Ca^{2+} from endoplasmic reticulum (ER) stores. One feasible mechanism could be due to diminished inositol phosphates (InsPs) generated from membrane phosphoinositides (PIs) through inhibition of phospholipase C (PLC) or kinases (not shown), with resultant ineffective interactions of $InsP_3$ and its ER receptor leading to incomplete Ca^{2+} depletion–replenishment cycles. OM and Msp also inhibit Ca^{2+} entry through store-operated channels (SOCs), uncoupling them from ER stores, in part by inducing actin filament assembly subjacent to the plasma membrane. The outer sheath serine protease PrtP degrades endogenous HGF matrix proteins such as fibronectin (Fn) and contributes to the subsequent detachment of HGFs from the substratum.

force. It blocks both ATP- and thapsigargin-induced release of Ca^{2+} from endoplasmic reticulum (ER) stores. It inhibits the increase in $[Ca^{2+}]_i$ following physical stretching of the plasma membrane, a rise that is usually caused by a combination of Ca^{2+} influx through stretch-activated channels and store-operated calcium channels (SOCs) that are linked to calcium release from the ER. Indeed, preliminary experiments examining the effect of *T. denticola* OM on store-operated Ca^{2+} flux suggest that the OM inhibits Ca^{2+} entry through putative SOCs. Moreover, both *T. denticola* whole cells and OM cause a diminished inositol phosphate (InsP) response in HGFs to agonists in serum and to ATP (Yang *et al.*, 1998). Inositol 1,4,5-trisphosphate ($InsP_3$) is a key factor in the release and replenishment of Ca^{2+} in the ER stores. The components of the *T. denticola* OM responsible for calcium perturbation in HGFs have not been fully elucidated. The fact that its inhibition of stretch-activated calcium transients is insensitive to PMSF, an inhibitor of serine proteases, but sensitive to heat at 60 °C, a temperature at which dentilisin's peptidase activity is stable, suggests that the OM's calcium perturbing activity is not related to the protease PrtP (Ko *et al.*, 1998a).

The major surface protein, Msp, is apparently one of the OM components that may cause profound effects of *T. denticola* on store-operated calcium flux in HGFs. Recently, Wang and co-workers (Wang *et al.*, 2001) have reported that native Msp complex (i) binds and aggregates on the HGF surface, (ii) induces $[Ca^{2+}]_i$ transients near the plasma membrane that are probably due to persistent Ca^{2+} permeation through the plasma membrane, (iii) blocks ATP- and thapsigargin-induced release of Ca^{2+} from the ER, and (iv) inhibits influx of extracellular Ca^{2+} through SOCs following thapsigargin depletion of ER stores (Fig. 11.7). Msp also induces redistribution of actin filaments, including assembly of filamentous actin subjacent to the plasma membrane, a sign that it may inhibit conformational coupling of SOCs and the ER Ca^{2+} stores (Berridge *et al.*, 2000). These findings are consistent with the novel findings of Mathers and co-workers (1996) that Msp can have profound effects on cation flux in epithelial cells. Fluctuations in cytosolic Ca^{2+} may initiate a cascade of intracellular responses that mediate the cytoskeletal rearrangement that is so common in human cells exposed to *T. denticola*. Yet continuous exposure of HGFs to Msp complex is required for cytotoxic activity resulting in significant cell death (Fenno *et al.*, 1998a; Wang *et al.*, 2001).

11.6 HAEMAGGLUTINATION AND HAEMOLYSIS

Even mildly inflamed periodontal tissues tend to bleed, enriching the environment of the periodontal pocket with erythrocytes and contents of haemolysed erythrocytes. Treponemes are able to bind and agglutinate red blood cells (RBCs). *Treponema denticola* agglutinates RBCs from human blood types A, B and O, as well as rabbit, horse, and bovine sources (Grenier, 1991; Mikx and Keulers, 1992). The haemagglutinating activity is cell associated, influenced by bacterial surface charge, sensitive to heat and proteolysis of the bacteria, and strongest in cells harvested in late exponential and stationary phases of growth (Mikx and Keulers, 1992; Cowan *et al.*, 1994). Although reports of haemagglutination inhibitors differ among laboratories, it is likely that the treponemal haemagglutinin is a protein. It may also be a surface-associated proteolytic enzyme, as soluble protein substrates and inhibitors of trypsin-like and chymotrypsin-like proteases all diminish the haemagglutinating activity of *T. denticola*, *T. vincentii* and *T. socranskii* (Eggert *et al.*, 1995). As arginine can also inhibit haemagglutination of these treponemes, as well as that of some other oral bacteria, Eggert and co-workers (1995) have hypothesized that the bacteria may bind surface proteins of erythrocytes, and likewise other proteins, via the highly-charged guanidinium group of arginine. Sensitivity of haemagglutination by *T. denticola* to periodate (Mikx and Keulers, 1992) suggests that carbohydrate-containing ligands may also be involved, but sugar specificity has not yet been established.

There is evidently a close association between haemagglutination and haemolytic activities of *T. denticola* (Grenier, 1991). Thus oral spirochaetes may have evolved a way to bind erythrocytes in their environment and thereby foster the lysis of cells that would provide enriched exposure to iron-containing compounds that may, in turn, be bound by outer sheath proteins (Scott *et al.*, 1996a). Recently, a cysteine-dependent haemolysin has been described for *T. denticola* (Chu *et al.*, 1999; Kurzban *et al.*, 1999). *Treponema denticola* expresses a 46 kDa cytosolic L-cysteine desulphydrase, named cystalysin, that both chemically oxidizes haemoglobin in the presence of cysteine and degrades RBC membrane proteins, including spectrin, thereby affecting the integrity of the plasma membrane. Chu and co-workers (1999) characterized cystalysin as a putative pyridoxal-5-phosphate-containing enzyme with cystathionase activity, optimum under reduced, H_2S-enriched conditions like those that occur in periodontal pockets. There are no known haemolysins that are structurally homologous. The crystal structure of recombinant cystalysin has recently been resolved and analysed to relate its active site architecture to a putative mechanism for the catabolism of sulphur-containing amino acid substrates (Krupka *et al.*, 2000).

11.7 INTERACTIONS WITH LEUKOCYTES AND MACROPHAGES

The polymorphonuclear leukocyte (PMNL) is a key cell of innate immune responses in the periodontal environment; PMNLs migrate through the junctional epithelium in response to gradients of surface components that shed from bacteria at the surface of the subgingival biofilm (Darveau, 2000). Whole cells of *T. denticola* are readily phagocytosed after contacting PMNLs *in vitro* (Lingaas *et al.*, 1983; Olsen *et al.*, 1984; Boehringer *et al.*, 1986). Yet, without causing disruption of the cells, they can inhibit oxidative bursts of PMNLs (Loesche *et al.*, 1988; Sela *et al.*, 1988) and trigger release and activation of PMNL-derived proteolytic enzymes that would be expected to contribute to connective tissue degradation in periodontal lesions (Ding *et al.*, 1997). LPS, peptidoglycan and a 53 kDa outer membrane protein, probably Msp, cause PMNLs to discharge the matrix metalloproteinases MMP-8 and MMP-9; Msp is particularly active in the release of MMP-8 (Ding *et al.*, 1996). Latent forms of MMPs from both PMNLs and fibroblasts can also be activated by the chymotrypsin-like serine protease (PrtP) of *T. denticola* (Sorsa *et al.*, 1992).

Treponemes express many lipoproteins that are associated with their cytoplasmic and outer membranes. An enriched lipoprotein fraction and the lipo-oligosaccharides of *T. denticola* ATCC 35404 have been shown to have many biological effects on PMNLs and macrophages (Sela *et al.*, 1997; Rosen *et al.*, 1999). The lipoproteins can cause enhanced luminol-dependent chemiluminescence and release of lysozyme from PMNLs and modulate these responses to the experimental agonist *N*-formyl-methionyl-leucyl-phenylalanine (FMLP). They and the lipo-oligosaccharides induce expression of the proinflammatory cytokines tumour necrosis factor α and interleukin 1 from macrophages, which are thought to mediate significant tissue-destructive signalling cascades in periodontal disease. In addition to modulating the innate immunity activities of these key cells, *T. denticola* is also known to suppress monocyte-dependent mitogenic responses in lymphocytes (Shenker *et al.*, 1984). Therefore the binding of whole *T. denticola* cells and fragments shed from their surface to gingival inflammatory cells has the potential to alter host–parasite interactions important in the pathogenesis of periodontitis.

11.8 INTERGENERIC CO-AGGREGATION

The inflamed periodontal pocket is an ideal environment for the emergence of fastidious anaerobes like the treponemes that rely on complex food webs for sustenance. Indeed, the most striking feature of suspended subgingival debris viewed by phase-contrast or darkfield microscopy is the abundance of highly motile spirochaetes surrounding the bacterial biofilm and tethered to it by one pole (Fig. 11.8). Like many bacteria that comprise oral biofilms, *T. denticola* is known to coaggregate with bacteria in other genera. Samples derived from inflamed sites of destructive periodontitis often contain a complex of at least three major proteolytic species: *Porphyromonas gingivalis, Bacteroides forsythus* and *T. denticola*. Together, these species account for very broad substrate degradation, and the individual bacteria in the biofilm probably derive their required peptides through 'community proteolysis'. *Treponema denticola, Treponema medium* and *Treponema socranskii* subsp. *socranskii* coaggregate avidly with *P. gingivalis* over a wide pH range (Grenier, 1992; Onagawa *et al.*, 1994; Umemoto *et al.*, 1999). These and several other oral treponemes co-aggregate with *Fusobacterium* species (Kolenbrander *et al.*, 1995). There has been one report that the co-aggregation reactions of *T. denticola* and *P. gingivalis* are sensitive to heating the *T. denticola* partner (Grenier, 1992); yet, other reports agree that the co-adhesion of *T. denticola* with either *P. gingivalis* or fusobacteria is resistant to heat. Co-aggregation with *P. gingivalis* may be inhibited by D-galactosamine and arginine under some conditions (Grenier, 1992). Co-aggregation of *T. denticola* with fusobacteria is evidently sensitive to galactosamine but not to arginine. The actual surface adhesins responsible for *T. denticola*'s intergeneric co-aggregation reactions are unknown. Inactivation of the *prtP* gene, which eliminates the serine protease activity of dentilisin, concomitantly knocks out the ability of the mutant *T. denticola* strain to coaggregate with fusobacteria (Ishihara *et al.*, 1998). However, PrtP-deficient mutants of *T. denticola* have many other pleiotropic effects, such as lower expression of other surface proteins (e.g. Msp) and diminshed hydrophobicity that may account for the altered co-aggregation (Fenno *et al.*, 1998b; Ishihara *et al.*, 1998).

11.9 TREPONEMES, FASTIDIOUS HANGERS-ON IN A COMPLEX COMMUNITY

Treponemes require a complex assortment of nutrients for energy and growth, and they colonize an environment enriched with a diverse array of solutes and complex molecules. Their need of peptides has probably driven the evolution of their proteases and peptidases. In the case of *T. denticola*, expression of PrtP in its outer sheath probably provides the bacterium an opportunity to adhere with the specificity of an enzyme to protein substrates.

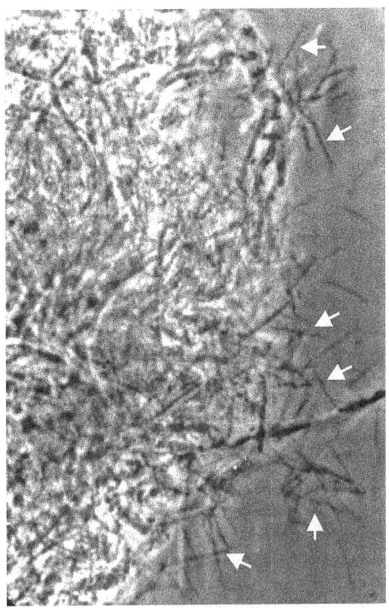

Figure 11.8. Phase-contrast micrograph of subgingival debris taken from a periodontal pocket of a subject with periodontitis. Numerous spirochaetes (arrows) are tethered to the surface of the dense bacterial biofilm (dental plaque) at the interface that would be exposed to the sulcular and junctional epithelium. (Photomicrograph by R.P. Ellen. Adapted from Ten Cate, 1994.)

The use of sulphur as an electron acceptor in treponemes and other anaerobes may have led to *T. denticola*'s ability to bind via sulphydryl groups and
to interact with sulphated macromolecules in the extracellular matrix and
cytokeratins of epithelial cells. The affinity of Msp for matrix proteins and its
close association with PrtP in the outer sheath is an additional example of a
connection between adhesion specificity and nutrient acquisition.
Haemagglutination, haemolysis, and haemin-binding would be another. To
be fastidious in a complex environment like the periodontal pocket has probably driven oral treponemes to adhere to a variety of cell surfaces and extracellular ligands so that they may hang on to a fragile lifestyle in a harsh
climate. This chapter has summarized the pathogenic consequences of the
most obvious adhesion functions that oral treponemes have evolved for their
own survival.

ACKNOWLEDGEMENT

Original research cited in this chapter was supported by grant MT-5619
to R.P.E. from the Canadian Institutes of Health Research.

REFERENCES

Baehni, P. (1986). Interactions between plaque microorganisms and human oral
epithelial cells. In *Borderland Between Caries and Periodontal Disease*, 3rd
European Symposium, ed. T. Lehner and G. Cimasoni, pp. 143–153.
Geneva: Editions Médecine et Hygiène.

Baehni, P.C., Song, M., McCulloch, C.A.G. and Ellen, R.P. (1992). *Treponema denticola* induces actin rearrangement and detachment of human gingival fibroblasts. *Infection and Immunity* **60**, 3360–3368.

Battikhi, T., Lee, W., McCulloch, C.A.G. and Ellen, R.P. (1999). *Treponema denticola* outer membrane enhances the phagocytosis of collagen-coated beads by
gingival fibroblasts. *Infection and Immunity* **67**, 1220–1226.

Berridge, M.J., Lipp, P. and Bootman, M.D. (2000). The calcium entry *pas de deux*.
Science **287**, 1604–1605.

Boehringer, H., Berthold, P.H. and Taichman, N.S. (1986). Studies on the interaction of human neutrophils with plaque spirochetes. *Journal of Periodontal
Research* **21**, 195–209.

Caimano, M.J., Bourell, K.W., Bannister, T.D., Cox, D.L. and Radolf, J.D. (1999). The
Treponema denticola major sheath protein is predominantly periplasmic and
has only limited surface exposure. *Infection and Immunity* **67**, 4072–4083.

Chan, E.C.S., Siboo, R., Keng, T., Psarra, N., Hurley, R., Cheng, S.-L. and Iugovaz,

I. (1993). *Treponema denticola* (ex Brumpt 1925) sp. Nov., nom. Rev., and identification of new spirochete isolates from periodontal pockets. *International Journal of Systematic Bacteriology* **43**, 196–203.

Charon, N.W., Lawrence, C.W. and O'Brien, S. (1981). Movement of antibody-coated latex beads attached to the spirochete *Leptospira interrogans. Proceedings of the National Academy of Sciences, USA* **78**, 7166–7170.

Chi, B., Chauhan, S. and Kuramitsu, H. (1999). Development of a system for expressing heterologous genes in the oral spirochete *Treponema denticola* and its use in expression of the *Treponema pallidum flaA* gene. *Infection and Immunity* **67**, 3653–3656.

Choi, B.K., Paster, B.J., Dewhirst, F.E. and Gobel, U.B. (1994). Diversity of cultivable and uncultivable oral spirochetes from a patient with severe destructive periodontitis. *Infection and Immunity* **62**, 1889–1895.

Choi, B.K., Wyss, C. and Göbel, U.B. (1996). Phylogenetic analysis of pathogen related oral spirochetes. *Journal of Clinical Microbiology* **34**, 1922–1925.

Chu, L., Ebersole, J.L., Kurzban, G.P. and Holt, S.C. (1999). Cystalysin, a 46-kDa L-cysteine desulfhydrase from *Treponema denticola*: biochemical and biophysical characterization. *Clinical Infectious Diseases* **28**, 442–450.

Cowan, M.M., Mikx, F.H.M. and Busscher, H.J. (1994). Electrophoretic mobility and haemagglutination of *Treponema denticola* ATCC 33520. *Colloids and Surfaces B: Biointerfaces* **2**, 407–410.

Darveau, R. (2000). Oral innate host defense response: interactions with microbial communities and their role in the development of disease. In *Oral Bacterial Ecology, the Molecular Basis*, ed. H.K. Kuramitsu and R.P. Ellen, pp. 169–218. Wymondham, Norfolk: Horizon Scientific Press.

Dawson, J.R. and Ellen, R.P. (1990). Tip-oriented adherence of *Treponema denticola* to fibronectin. *Infection and Immunity* **58**, 3924–3928.

Dawson, J.R. and Ellen, R.P. (1994). Clustering of fibronectin adhesins toward *Treponema denticola* tips upon contact with immobilized fibronectin. *Infection and Immunity* **62**, 2214–2221.

DeFilippo, A.B., Ellen, R.P. and McCulloch, C.A.G. (1995). Induction of cytoskeletal rearrangments and loss of volume regulation in epithleial cells by *Treponema denticola. Archives of Oral Biology* **40**, 199–207.

Dewhirst, F.E., Tamer, M.A., Ericson, R.E., Lau, C.N., Levanos, V.A., Boches, S.K., Galvin, J.L. and Paster, B.J. (2000). The diversity of periodontal spirochetes by 16S rRNA analysis. *Oral Microbiology and Immunology* **15**, 196–202.

Ding, Y., Uitto, V.-J., Haapasalo, M, Lounatmaa, K., Konttinen, Y.T., Salo, T., Grenier, D. and Sorsa, T. (1996). Membrane components of *Treponema denticola* trigger proteinase release from human polymorphonuclear leukocytes. *Journal of Dental Research* **75**, 1986–1993.

Ding, Y., Haapasalo, M., Kerosuo, E., Lounatmaa, K., Kotiranta, A. and Sorsa, T. (1997). Release and activation of human neutrophil matrix metallo- and serine proteinases during phagocytosis of *Fusobacterium nucleatum*, *Porphyromonas gingivalis*, and *Treponema denticola*. *Journal of Clinical Periodontology* **24**, 237–248.

Eggert, F.-M., Chan, E.C.S., Klitorinos, A. and Flowerdew, G. (1995). Arginine is a common ligand for haemagglutination and protein binding by organisms inhabiting mucosal surfaces. *Microbial Ecology in Oral Health and Disease* **8**, 267–280.

Egli, C., Leung, W.K., Müller, K.H., Hancock, R.E.W. and McBride, B.C. (1993). Pore-forming properties of the major 53 kilodalton surface antigen from the outer sheath of *Treponema denticola*. *Infection and Immunity* **61**, 1694–1699.

Ellen, R.P., Dawson, J.R. and Yang, P.F. (1994a). *Treponema denticola* as a model for polar adhesion and cytopathogenicity of spirochetes. *Trends in Microbiology* **2**, 114–119.

Ellen, R.P., Song, M. and McCulloch, C.A.G. (1994b). Degradation of endogenous plasma membrane fibronectin concomitant with *Treponema denticola* 35405 adhesion to gingival fibroblasts. *Infection and Immunity* **62**, 3033–3037.

Ellen, R.P., Wikström, M., Grove, D.A., Song, M. and Elwing, H. (1998). Polar adhesion of *Treponema denticola* on wettability gradient surfaces. *Colloids and Surfaces. B: Biointerfaces* **11**, 177–186.

Ellen, R.P., Ko, K.S.-C., Lo, C.-M., Grove, D.L. and Ishihara, K. (2000). Insertional inactivation of the prtP gene of *Treponema denticola* confirms dentilisin's disruption of epithelial junctions. *Journal of Molecular Microbiology and Biotechnology* **2**, 581–586.

Fenno, J.C. and McBride, B.C. (1998). Virulence factors of oral treponemes. *Anaerobe* **4**, 1–17.

Fenno, J.C., Müller, K.-H. and McBride, B.C. (1996). Sequence analysis, expression, and binding activity of recombinant major outer sheath protein (Msp) of *Treponema denticola*. *Journal of Bacteriology* **178**, 2489–2497.

Fenno, J.C., Hannam, P.M., Leung, W.K., Tamura, M., Uitto, V.-J. and McBride, B.C. (1998a). Cytopathic effects of the major surface protein and the chymotrypsin-like protease of *Treponema denticola*. *Infection and Immunity* **66**, 1869–1877.

Fenno, J.C., Wong, G.W., Hannam, P.M. and McBride, B.C. (1998b). Mutagenesis of outer membrane virulence determinants of the oral spirochete *Treponema denticola*. *FEMS Microbiology Letters* **163**, 209–215.

Fenno, J.S., Tamura, M., Hannam, P.M., Wong, G.W., Chan, R.A. and McBride, B.C. (2000). Identification of a *Treponema denticola* OppA homologue that binds host proteins present in the subgingival environment. *Infection and Immunity* **68**, 1884–1892.

Fiehn, N.-E. (1991). Surface hydrophobicity of small oral spirochetes. *Acta Odontologica Scandinavica* **49**, 1–6.

Grenier, D. (1991). Characteristics of hemolytic and hemagglutinating activities of *Treponema denticola*. *Oral Microbiology and Immunology* **6**, 246–249.

Grenier, D. (1992). Demonstration of a bimodal coaggregation reaction between *Porphyromonas gingivalis* and *Treponema denticola*. *Oral Microbiology and Immunology* **7**, 280–284.

Grenier, D., Uitto, V.-J. and McBride, B.C. (1990). Cellular location of a *Treponema denticola* chymotrypsin like protease and importance of the protease in migration through the basement membrane. *Infection and Immunity* **58**, 347–351.

Haapasalo, M., Singh, U., McBride, B.C. and Uitto, V.-J. (1991). Sulfhydryl-dependent attachment of *Treponema denticola* to laminin and other proteins. *Infection and Immunity* **59**, 4230–4237.

Haapasalo, M., Müller, K.-H., Uitto, B.-J., Leung, W.K. and McBride, B.C. (1992). Characterization, cloning, and binding properties of the major 53-kilodalton *Treponema denticola* surface antigen. *Infection and Immunity* **60**, 2058–2065.

Haapasalo, M., Hannam, P., McBride, B.C. and Uitto, V.-J. (1996). Hyaluronan, a possible ligand mediating *Treponema denticola* binding to periodontal tissue. *Oral Microbiology and Immunology* **11**, 156–160.

Heinzerling, H.F., Olivares, M. and Burne, R.A. (1997). Genetic and transcriptional analysis of *flgB* flagellar operon constituents in the oral spirochete *Treponema denticola* and their heterologous expression in enteric bacteria. *Infection and Immunity* **65**, 2041–2051.

Ishihara, K., Miura, T., Kuramitsu, H.K. and Okuda, K. (1996). Characterization of the *Treponema denticola prtP* gene encoding a prolyl-phenylalanine-specific protease (dentilisin). *Infection and Immunity* **64**, 5178–5186.

Ishihara, K., Kuramitsu, H.K., Miura, T. and Okuda, K. (1998). Dentilisin activity affects the organization of the outer sheath of *Treponema denticola*. *Journal of Bacteriology* **180**, 3837–3844.

Keulers, R.A., Maltha, J.C., Mikx, F.H.M. and Wolters-Lutgerhorst, J.M. (1993). Attachment of *Treponema denticola* strains to monolayers of epithelial cells of different origin. *Oral Microbiology and Immunology* **8**, 84–88.

Ko, K.S.-C., Glogauer, M., McCulloch, C.A.G. and Ellen, R.P. (1998a). *Treponema denticola* outer membrane inhibits calcium flux in gingival fibroblasts. *Infection and Immunity* **66**, 703–709.

Ko, S.-C., Lo, C.-M., Ferrier, J., Hannam, P., Tamura, M., McBride, B.C. and Ellen, R.P. (1998b). Cell–substrate impedance analysis of epithelial cell shape and micromotion upon challenge with bacterial proteins that perturb extracellular matrix and cytoskeleton. *Journal of Microbiological Methods* 34, 125–132.

Kolenbrander, P.E., Parrish, K.D., Andersen, R.N. and Greenberg, E.P. (1995). Intergeneric coaggregation of oral *Treponema* spp. with *Fusobacterium* spp. and intrageneric coaggregation among *Fusobacterium* spp. *Infection and Immunity* 63, 4584–4588.

Krupka, H.I., Huber, R., Holt, S.C. and Clausen, T. (2000). Crystal structure of cystalysin from *Treponema denticola*: a pyridoxal 5'-phosphate-dependent protein acting as a haemolytic enzyme. *EMBO Journal* 19, 3168–3178.

Kurzban, G.P., Chu, L., Ebersole, J.L. and Holt, S.C. (1999). Sulfhemoglobin formation in human erythrocytes by cystalysin, an L-cysteine desulfhydrase from *Treponema denticola*. *Oral Microbiology and Immunology* 14, 153–164.

Lépine G. and Ellen, R.P. (2000). *mglA* and *mglB* of *Treponema denticola*; similarity to ABC transport and *spa* genes. *DNA Sequence* 11, 419–431.

Li, C., Motaleb, M.A., Sal, M., Goldstein, F. and Charon, N.W. (2000). Spirochete periplasmic flagella and motility. *Journal of Molecular Microbiology and Biotechnology* 2, 345–354.

Lingaas, E., Olsen, I, Midtvedt, T. and Hurlen, B. (1983). Demonstration of the in vitro phagocytosis of *Treponema denticola* by human polymorphonuclear neutrophils. *Acta Pathologica, Microbiologica et Immunologica Scandinavica [B]* 91, 333–337.

Loesche, W.J., Robinson, J.P., Flynn, M., Hudson, J.L. and Duque, R.E. (1988). Reduced oxidative function in gingival crevicular neutrophils in periodontal disease. *Infection and Immunity* 56, 156–160.

Masuda, K. and Kawata, T. (1982). Isolation, properties, and reassembly of outer sheath carrying a polygonal array from an oral treponeme. *Journal of Bacteriology* 150, 1405–1413.

Mathers, D.A., Leung, W.K., Fenno, J.C., Hong, Y. and McBride, B.C. (1996). The major surface protein complex of *Treponema denticola* depolarizes and induces ion channels in HeLa cell membranes. *Infection and Immunity* 64, 2904–2910.

Mikx, F.H.M. and Keulers, R.A. (1992). Haemagglutination activity of *Treponema denticola* grown in serum-free medium in continuous culture. *Infection and Immunity* 60, 1761–1766.

Mikx, F.H.M., Jacobs, F. and Satumalay, C. (1992). Cell-bound peptidase activities of *Treponema denticola* ATCC 33520 in continuous culture. *Journal of General Microbiology* 138, 1837–1842.

Moter, A.C., Hoenig, B.K., Choi, B.K. and Göbel, U.B. (1998). Molecular epidemiology of oral treponemes associated with periodontal disease. *Journal of Clinical Microbiology* 36, 1399–1403.

Olsen, I. (1984). Attachment of *Treponema denticola* to cultured human epithelial cells. *Scandinavian Journal of Dental Research* **92**, 55–63.

Olsen, I., Ligaas, E., Hurlen, B. and Midtvedt, T. (1984). Scanning and transmission electron microscopy of the phagocytosis of *Treponema denticola* and *Escherichia coli* by human neutrophils in vitro. *Scandinavian Journal of Dental Research* **92**, 282–293.

Onagawa, M., Ishihara, K. and Okuda, K. (1994). Coaggregation between *Porphyromonas gingivalis* and *Treponema denticola. Bulletin of Tokyo Dental College* **35**, 171–181.

Peters, S.R., Valdez, M., Riviere, G. and Thomas, D.D. (1999). Adherence to and penetration through endothelial cells by oral treponemes. *Oral Microbiology and Immunology* **14**, 379–383.

Qui, Y.-S., Klitorinos, A., Rahal, M.D., Siboo, R. and Chan, E.C.S. (1994). Enumeration of viable oral spirochetes from periodontal pockets. *Oral Microbiology and Immunology* **9**, 301–304.

Riviere, G.R., Weisz, K.S., Adams, D.F. and Thomas, D.D. (1991). Pathogen-related oral spirochetes from dental plaque are invasive. *Infection and Immunity* **59**, 3377–3380.

Riviere, G.R., Smith, K.S., Carranza, N., Jr, Tzagaroulaki, E., Kay, S.L. and Dock, M. (1995). Subgingival distribution of *Treponema denticola, Treponema socranskii,* and pathogen-related oral spirochetes: prevalence and relationship to periodontal status of sample sites. *Journal of Periodontology* **66**, 829–837.

Rosen, G., Naor, R., Rahamim, E., Yishai, R. and Sela, M.N. (1995). Proteases of *Treponema denticola* outer sheath and extracellular vesicles. *Infection and Immunity* **63**, 3973–3979.

Rosen, G., Sela, M.N., Naor, R., Halabi, A., Barak, V. and Shapira, L. (1999). Activation of murine macrophages by lipoprotein and lipooligosaccharide of *Treponema denticola. Infection and Immunity* **67**, 1180–1186.

Scott, D., Chan, E.C. and Siboo, R. (1996a). Iron acquisition by oral hemolytic spirochetes: isolation of a hemin-binding protein and identification of iron reductase activity. *Canadian Journal of Microbiology* **42**, 1072–1079.

Scott, D., Siboo, R., Chan, E.C.S. and Siboo, R. (1996b). An extracellular enzyme with hyaluronidase and chondroitinase activities from some oral anaerobic spirochetes. *Microbiology* **142**, 2567–2576.

Sela, M.N., Weinberg, A., Borinsky, R., Holt, S.C. and Dishon, T. (1988). Inhibition of superoxide production in human polymorphonuclear leukocytes by treponemal factors. *Infection and Immunity* **56**, 589–594.

Sela, M.N., Bolotin, A., Naor, R., Weinberg, A. and Rosen, G. (1997). Lipoproteins of *Treponema denticola*: their effect on human polymorphonuclear neutrophils. *Journal of Periodontal Research* **32**, 455–466.

Shenker, B.J., Listgartedn M.A. and Taichman, N.S. (1984). Suppression of human lymphocytic responses by oral spirochetes: a monocyte-dependent phenomenon. *Journal of Immunology* **132**, 2039–2045.

Sorsa, T. Ingman, T., Suomalainen, K., Haapasalo, M., Konttinen, Y.T., Lindy, O., Saari, H. and Uitto, V.-J. (1992). Identification of proteases from periodonto-pathogenic bacteria as activators of latent human neutrophil and fibroblast-type interstitial collagenases. *Infection and Immunity* **60**, 4491–4495.

Ten Cate, A.R. (ed.) (1994). *Oral Histology: Development, Structure and Function*, 4th edn. St Louis: Mosby.

Thomas, D.D. (1996). Aspects of adherence of oral spirochetes. *Critical Reviews of Oral Biology and Medicine* **7**, 4–11.

Uitto, V.-J., Grenier, D., Chan, E.C. and McBride, B.C. (1988). Isolation of a chymotrypsin like enzyme from *Treponema denticola*. *Infection and Immunity* **56**, 2717–2722.

Uitto, V.-J., Pan, Y.-M., Leung, W.K., Larjava, H., Ellen, R.P., Finlay, B.B. and McBride, B.C. (1995). Cytopathic effects of *Treponema denticola* chymotrypsin-like proteinase on migrating and stratified epithelial cells. *Infection and Immunity* **63**, 3401–3410.

Umemoto, T., Nakatani, Y., Nakamura, Y. and Namikawa, I. (1993). Fibronectin-binding proteins of a human oral spirochete *Treponema denticola*. *Microbiology and Immunology* **37**, 75–78.

Umemoto, T., Yoshimura, F., Kureshiro, H., Hayashi, J., Noguchi, T. and Ogawa, T. (1999). Fimbria-mediated coaggregation between human oral anaerobes *Treponema medium* and *Porphyromonas gigivalis*. *Microbiology and Immunology* **43**, 837–845.

Walker, S.G., Ebersole, J.L. and Holt, S.C. (1999). Studies on the binding of *Treponema pectinovorum* to HEp-2 epithelial cells. *Oral Microbiology and Immunology* **14**, 165–171.

Wang, Q., Ko, S.C., Kapus, A., McCulloch, C.A.G. and Ellen, R.P. (2001). A spirochete surface protein uncouples store-operated calcium channels in fibroblasts. A novel cytotoxic mechanism. *Journal of Biological Chemistry* **276**, 23056–23064

Weinberg, A. and Holt, S.C. (1990). Interaction of *Treponema denticola* TD-4, GM-1, and MS25 with human gingival fibroblasts. *Infection and Immunity* **58**, 1720–1729.

Weinberg, A. and Holt, S.C. (1991). Chemical and biological activities of a 64-kilodalton outer sheath protein from *Treponema denticola* strains. *Journal of Bacteriology* **173**, 6935–6947.

Willis, S.G., Smith, K.S., Dunn, V.L., Gapter, L.A., Riviere, K.H. and Riviere, G.R. (1999). Identification of seven *Treponema* species in health- and disease-associated dental plaque by nested PCR. *Journal of Clinical Microbiology* **37**, 867–869.

Yang, P.F., Song, M., Grove, D.A. and Ellen, R.P. (1998). Filamentous actin disruption and diminished inositol phosphate response in gingival fibroblasts caused by *Treponema denticola*. *Infection and Immunity* **66**, 696–702.

CHAPTER 12

Interactions between enteropathogenic *Escherichia coli* and epithelial cells

Elizabeth L. Hartland, Gad Frankel and Stuart Knutton

12.1 INTRODUCTION

Each year diarrhoeal diseases contribute to the deaths of more than 2 million people in developing countries, most of whom are children. Enteropathogenic *Escherichia coli* (EPEC) is an important cause of diarrhoea in young children and makes a major contribution to infant morbidity and mortality in the developing world. A striking feature of EPEC diarrhoea is the age-dependent susceptibility of patients. Infections occur primarily in children less than 2 years of age and symptoms are usually acute but may be very severe and protracted (Nataro and Kaper, 1998). Patients routinely experience profuse watery diarrhoea, but vomiting and low grade fever are also common symptoms. EPEC rarely causes diarrhoea in adults and in volunteers only at very high doses (Donnenberg *et al.*, 1993; Tacket *et al.*, 2000)

EPEC is principally a pathogen of the small bowel and one of several gastrointestinal pathogens of humans and animals able to cause distinctive lesions in the gut, termed attaching and effacing (A/E) lesions. This group of A/E pathogens includes the closely related human pathogen enterohaemorrhagic *E. coli* (EHEC) (see Chapter 9), and the animal pathogens, rabbit enteropathogenic *E. coli* (REPEC) and *Citrobacter rodentium* (Nataro and Kaper, 1998). The remarkable histopathology of A/E lesions can be observed in intestinal biopsies from patients infected with EPEC and other host species. The lesions are characterized by localized destruction of intestinal microvilli, intimate attachment of the bacteria to the host cell surface and the formation of pedestal-like structures underneath tightly adherent bacteria (Fig. 12.1) (Frankel *et al.*, 1998; Vallance and Finlay, 2000).

A/E lesion formation is a dynamic process involving substantial rearrangement and reassembly of host cytoskeletal proteins. Lesions may be reproduced *in vitro* using cultured epithelial cells (Moon *et al.*, 1983). In this system, pedestal-like structures can extend up to 10 μm away from the apical

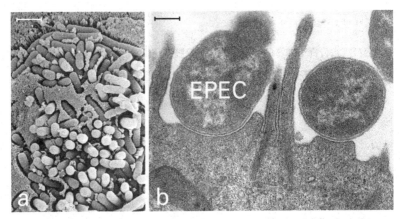

Figure 12.1. (a) Scanning and (b) transmission electron micrographs, showing EPEC infected human intestinal mucosa and illustrating key features of A/E histopathology including localized destruction of brush border microvilli, intimate bacterial attachment and pedestal formation. Bars: (a) 0.2 μm; (b) 0.5 μm).

surface of the epithelial cell and video microscopy has revealed that the pedestals can bend and change length while remaining attached to the cell surface (Sanger *et al.*, 1996). The body of the pedestal beneath adherent bacteria is rich in cytoskeletal proteins and contains an abundance of polymerized, filamentous actin (F-actin) (Knutton *et al.*, 1989). In epithelial cells, the characteristic accumulation of filamentous actin beneath adherent EPEC can be visualized with fluorescently labelled phalloidin, a compound that binds specifically to filamentous actin. This fluorescent actin-staining (FAS) test is widely used as an *in vitro* assessment of A/E lesion formation in cultured epithelial cells infected with EPEC (Knutton *et al.*, 1989). Other cytoskeletal proteins present in the pedestal include α-actinin, ezrin and talin, each of which plays a role in the cross-linking of actin microfilaments (Finlay *et al.*, 1992).

12.2 THE LOCUS FOR ENTEROCYTE EFFACEMENT AND PROTEIN SECRETION

The bacterial genes required for A/E lesion formation by EPEC are present in a pathogenicity island termed the locus for enterocyte effacement (LEE) (Fig. 12.2) (McDaniel *et al.*, 1995). Transfer of the LEE pathogenicity island from EPEC to *E. coli* K12 confers on the latter the ability to form A/E lesions on cultured epithelial cells (McDaniel and Kaper, 1997). DNA sequence analysis of LEE has revealed that the G + C base composition of this region is only 38.4% while the G + C content of the *E. coli* chromosome is

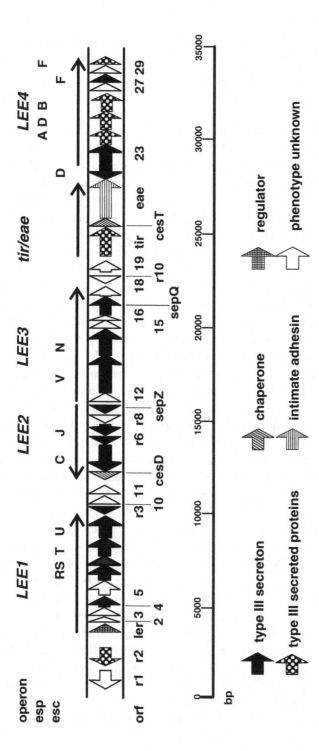

Figure 12.2. The locus of enterocyte effacement (LEE) of *E. coli* E2348/69 showing the arrangement of polycistronic operons (thin arrows), the 41 open reading frames (ORFs) (large arrows) and their nomenclature as *esp*, *esc*, *ces*, *tir* and *eae*; remaining orfs are designated as *orf* or *rof* depending on the direction of transcription relative to *eae*.

significantly higher (50.8%). This divergence implies that EPEC acquired LEE by horizontal gene transfer from a non-*E. coli* source (Elliott *et al.*, 1998). Molecular evolutionary studies speculate that this event took place several times throughout the development of clonal groups of A/E pathogens (Reid *et al.*, 2000).

The LEE region contains 41 open reading frames organized into several polycistronic operons (Fig. 12.2) (Elliott *et al.*, 1998). Most of the genes in LEE code for a type III secretion system that directs the secretion of several other LEE-encoded proteins, the EPEC secreted proteins or Esps. These include EspA, D, B, F, G and Tir/EspE (Elliott *et al.*, 1998). The LEE region also contains the *eae* gene, which codes for an outer membrane protein adhesin, intimin (Jerse *et al.*, 1990). Type III secretion systems are found in a number of human, animal and plant pathogens including species of *Bordetella*, *Burkholderia*, *Chlamydia*, *Erwinia*, *Pseudomonas*, *Salmonella*, *Shigella*, *Xanthomonas* and *Yersinia* (Hueck, 1998). These systems are designed to secrete and translocate virulence proteins across the bacterial cell envelope and the plasma membrane of the target eukaryotic cell into the cell cytoplasm. Recent electron microscopical studies of the *Shigella* and *Salmonella* (SP1) type III secretion systems have revealed the existence of a macromolecular complex that spans both bacterial membranes and consists of a basal structure with two upper and two lower rings connected by a cylindrical structure, and a needle-like projection that extends out from the bacterial cell surface (Kubori *et al.*, 1998; Blocker *et al.*, 1999; Tamano *et al.*, 2000). Many of the proteins that are believed to form part of the LEE-encoded type III secretion apparatus exhibit sequence homology with the prototype type III system of *Yersinia* and thus their role in secretion is assumed from sequence similarity (see Chapter 8). At this stage, the function of very few of these proteins (designated Esc) has been established experimentally.

Like other type III secretion systems, export of some, if not all, Esps involves a system of unique chaperones. Two chaperones have been identified to date: CesD, which is involved in the secretion of EspD; and CesT, which is required for the secretion of Tir (Hueck, 1998; Wainwright and Kaper, 1998; Abe *et al.*, 1999; Elliott *et al.*, 1999). Both chaperones bind directly to their cognate proteins and stabilize the protein prior to secretion. CesT is a small, acidic protein (15 kDa, predicted pI 4.3) and thus has the features of a classic type III chaperone (Abe *et al.*, 1999; Elliott *et al.*, 1999). By contrast, the 17.5 kDa CesD protein is more basic (predicted pI 7.1). As well as being deficient in the secretion of EspD, *cesD* mutants are also unable to efficiently secrete EspB. However, CesD does not appear to interact with EspB and so the reason for this secretion deficiency is unknown (Wainwright and Kaper, 1998). Genes encoding the Esps and type III secretion components are organized into four polycistronic operons designated *LEE1–4*. A fifth operon comprises *tir, cesT* and *eae* (Fig. 12.2) (Friedberg *et al.*, 1999). At least three factors regulate the expression of these transcriptional units, the plasmid-encoded regulator (Per), the LEE-encoded regulator (Ler) and the integration host factor (IHF). The Per regulon is present on a large plasmid (the EAF plasmid), which also carries the genes coding for a type IV pilus known as the bundle-forming pilus or BFP (Nataro *et al.*, 1987; see section 12.7). The Per regulon consists of three components, PerA, B and C, of which PerA is an AraC-like protein and PerB shows homology to eukaryotic DNA binding proteins (Gomez-Duarte and Kaper, 1995). No homologues of PerC have been identified so far. Unlike other AraC homologues, PerA acts in concert with PerB and PerC to induce full transcriptional activation of target genes (Gomez-Duarte and Kaper, 1995). The Per locus acts primarily on *LEE1*, and in particular on the first open reading frame of *LEE1*, Ler (Mellies *et al.*, 1999). Ler is similar to the H-NS (histone-like non-structural protein) binding proteins that together form a family of DNA-binding proteins (Elliott *et al.*, 2000). Following activation by Per, Ler then initiates transcription of *LEE2, 3, 4* and the *tir* operon. A recent study showed that site-directed mutagenesis of hydrophobic residues in the amphipathic region of Ler abolished the ability of Ler to activate *LEE2* and bind to *LEE2* regulatory region (Sperandio *et al.*, 2000). Ler is also regulated by the DNA-binding protein, IHF (Friedberg *et al.*, 1999). In addition, Per, Ler and LEE2 are controlled globally by quorum sensing (Sperandio *et al.*, 1999). Taken as a whole, expression of the LEE operons is modulated at several levels by different controlling elements that together form a kind of regulatory cascade.

12.3 INTIMIN

The *eae* gene was the first locus to be associated with A/E lesion formation. *eae* was identified following the screening of Tn*phoA* mutants by FAS (Jerse *et al.*, 1990). The product of *eae*, intimin, is a 94 kDa outer membrane protein adhesin (Jerse and Kaper, 1991). Experimental infection of human volunteers with an *eae* mutant of EPEC and data from several animal models have shown that intimin is essential for colonization of the host but not for the elaboration of diarrhoea (Donnenberg *et al.*, 1993). Overall, the intimins from different A/E pathogens show a high degree of similarity and also exhibit homology to the invasin protein of *Yersinia* (Isberg *et al.*, 1987). The N-terminal region comprising the first 700 amino acid residues is highly conserved and exhibits around 94% identity. Greatest sequence divergence is found in the C-terminal region comprising the last 280 residues which shows only 49% identity (Frankel *et al.*, 1994). Based on sequence variation in the C-terminal 280 residues, we now recognize a family of intimins that includes at least five antigenic groups, α, β, γ, δ and ε (Adu-Bobie *et al.*, 1998). Different intimins are associated with different clonal groups of A/E pathogens; for example, the prototype O127:H7 EPEC strain expresses intimin α, while the closely related A/E pathogen, O157:H7 EHEC, expresses intimin γ (Adu-Bobie *et al.*, 1998).

The cell-binding activity of intimin is contained within the C-terminal 280 amino acid residues (Int280) (Frankel *et al.*, 1995). Although this domain exhibits most variation, the intimins share a number of conserved residues. Like invasin, cell-binding activity depends on the formation of a disulphide link between two conserved cysteine residues, Cys-860 and Cys-937. Disruption of this 77 amino acid residue loop by site-directed mutagenesis of Cys-937 to alanine or serine (Int280C/S) abrogates cell-binding function (Frankel *et al.*, 1995). Purified Int280 but not Int280C/S is capable of binding to the surface of epithelial cells and inducing ultrastructural changes in the cell, although cell binding by intimin alone is not sufficient to cause A/E lesions (Frankel *et al.*, 1995). *In vitro*, Int280 induces elongated protrusions of the HEp-2 epithelial cell membrane in the early stages of infection. These abnormal extensions of the host cell membrane have been termed microvillus-like processes (MLPs) and their number and length reduce as A/E lesion formation progresses (Phillips *et al.*, 2000). In this particular study, latex beads coated with Int280 were also capable of generating MLPs, demonstrating that intimin stimulates substantial remodelling of the host cell surface.

Infection of mice with *C. rodentium* expressing intimin β or intimin α results in gross thickening of the colonic epithelium (Frankel *et al.*, 1996b). This marked pathology coincides with a strong T helper cell 1 (T_H1) mucosal immune response and closely resembles mouse models of inflammatory bowel disease (Higgins *et al.*, 1999b). Dead bacteria expressing intimin mediate severe colonic epithelial cell hyperplasia if given intrarectally to mice and drive a mucosal T_H1 response. The mechanism behind the immunomodulatory properties of intimin are unknown, as intimin does not appear to be directly mitogenic for T-lymphocytes of the lamina propria (Higgins *et al.*, 1999a), although intimin can bind directly to T-cells through β_1-integrins (Frankel *et al.*, 1996a).

Intimin also potentially contributes to tissue tropism and, to some extent, host specificity. Infection of paediatric small intestinal human *in vitro* organ cultures (IVOCs) show that O127:H6 EPEC, which expresses intimin α, will colonize the entire surface of the small intestinal mucosa, including the proximal and distal small intestine and Peyer's patches, while O157:H7 EHEC, which expresses intimin γ, is restricted to the follicle-associated epithelium of Peyer's patches. Intimin exchange studies have shown that this tissue tropism can be attributed to the intimin type (Phillips and Frankel, 2000). Similar studies have been performed in animal models of infection (Tzipori *et al.*, 1995; Hartland *et al.*, 2000b). An *eae* null mutant of *C. rodentium* that normally expresses intimin β was engineered to express intimin γ from O157:H7 EHEC and intimin α from O127:H6 EHEC. *Citrobacter rodentium* expressing intimin α colonized mice as efficiently as the wild type whereas *C. rodentium* expressing intimin γ was unable to colonize mice (Hartland *et al.*, 2000b). These data suggest that different intimins have different binding specificities and/or affinities.

Recently, the three-dimensional structure of Int280 was determined by multidimensional nuclear magnetic resonance and X-ray crystallography (Kelly *et al.*, 1999; Luo *et al.*, 2000). The derived structure showed that the Int280 molecule comprises three discrete domains, two immunoglobulin-like domains and a C-type, lectin-like module that contains the 77 amino acid residue disulphide loop (Kelly *et al.*, 1999). While the presence of a lectin-like structure at the tip of intimin suggests this part of the protein may interact with a carbohydrate moiety, the absence of a critical calcium-binding motif does not support a role for a protein–carbohydrate interaction. Nevertheless, although the molecular mechanism by which intimin mediates ultrastructural changes on the cell surface is unknown, intimin presumably interacts with a surface carbohydrate or protein receptor encoded by the host cell.

12.4 INTIMATE ADHERENCE

Intimate adherence of EPEC to the host epithelial cell surface is a defining feature of A/E lesions. As well as mediating a direct effect on the host cell through an as yet uncharacterized interaction, intimin also binds to the LEE-encoded protein, Tir (the translocated intimin receptor), also termed EspE (Kenny *et al.*, 1997; Deibel *et al.*, 1998). Tir is a 78 kDa bacterial protein that is translocated into the host epithelial cell where it localizes to the cell membrane (Kenny *et al.*, 1997). Tir contains two putative transmembrane domains and is predicted to have a hairpin loop topology in the host cell membrane, with the middle region exposed extracellularly and the N- and C-terminal regions located intracellularly (Hartland *et al.*, 1999; Kenny, 1999). Protein–protein interactions occur between the most distal, lectin-like domain of intimin and the middle, extracellularly exposed region of Tir (termed TirM) (Hartland *et al.*, 1999; Liu *et al.*, 1999; Batchelor *et al.*, 2000). Intimate adherence of the bacterium to the host membrane depends on a specific protein–protein interaction between intimin and Tir. Recently, the interaction between Int280 and TirM was resolved structurally by crystallography. The structure revealed that the intimin-binding domain of Tir exists in a dimeric form, which is stabilized by an antiparallel, four-helix bundle (Luo *et al.*, 2000). Overall, the entire intimin/Tir complex consists of two Tir molecules and two intimin molecules. Extensive interactions occur between the TirM and the lectin-like module of intimin, creating a largely hydrophobic binding interface, which is presumably responsible for intimate adherence of the bacteria to the host cell membrane containing Tir (Luo *et al.*, 2000).

Upon translocation into the host cell membrane, Tyr-474 in the C-terminal region of Tir becomes phosphorylated (Kenny, 1999). Indeed Tir was first identified as a tyrosine phosphorylated host protein, Hp90, which was later found to be of bacterial origin (Kenny *et al.*, 1997). In EPEC, tyrosine phosphorylation of Tir at this residue is essential for A/E lesion formation (Kenny, 1999). Site-directed mutagenesis of this residue from tyrosine to serine abrogated the ability of EPEC to induce A/E lesions. In O157 EHEC, however, Tir has a serine residue at position 474 and tyrosine phosphorylation of this site is not necessary for the development of A/E lesions (DeVinney *et al.*, 1999). At this stage, the significance of tyrosine phosphorylation of Tir to A/E lesion formation is uncertain.

12.5 DELIVERY OF TIR TO THE HOST CELL MEMBRANE

The translocation of Tir to the host cell membrane depends on at least three other LEE-encoded proteins, EspA, B and D (Knutton *et al.*, 1998; Wolff *et al.*, 1998). Like Tir, all three proteins depend on the type III secretion system of LEE for their export. While the precise mechanism of protein translocation is still unclear, we are slowly beginning to learn more about the function of the secreted proteins. EspA is a major structural component of a large (~ 50 nm diameter), transiently expressed extracellular, filamentous organelle that forms a direct link between the bacterial cell surface and the host cell membrane (Fig. 12.3) (Knutton *et al.*, 1998). Ultrastructural analysis suggests that the EspA filaments are made up of bundles of smaller fibrils, that together form a rigid, hollow, cylindrical rod (Knutton *et al.*, 1998). Recent studies have shown that EspA exists as multimeric isoforms in culture supernatants of EPEC (Delahay *et al.*, 1999). The C-terminal end

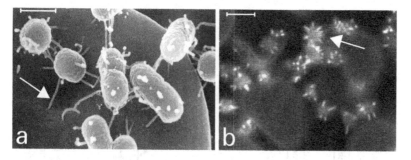

Figure 12.3. (a) Scanning electron micrograph and (b) EspA immunofluorescence showing that EspA filaments form a direct link between the bacterium and the host cell membrane, in this example between the bacterium and human red blood cells (arrows). Bars: (a) 0.5 μm; (b) 1 μm.

of EspA is strongly predicted to contain an alpha-helical region with a propensity to form a coiled-coil interaction. Non-conservative site-directed mutagenesis of hydrophobic amino acid residues predicted to be essential for coiled-coil formation abrogated the ability of EspA to form multimers, assemble filaments and induce A/E lesions (Delahay et al., 1999). This suggests that, similar to the well-characterized coiled-coil region of flagellins (Hyman and Trachtenberg, 1991), coiled-coil interactions are important for EspA filament assembly and/or stability. Interestingly, espD mutants secrete only low levels of EspA and are unable to produce mature EspA filaments, instead producing short, vestigial filaments (Knutton et al., 1998; Kresse et al., 1999). Although this suggests that EspD has a structural role in filament assembly, EspD is not required for EspA multimerization and, at this stage, no staining of the filaments has been demonstrated with antisera to EspD and no protein–protein associations between EspA and EspD have been observed. Nevertheless, it is possible that EspD forms a minor component of the filament.

espB mutants produce normal EspA filaments and antibodies to EspB do not stain the filament, indicating that EspB does not play a integral role in filament construction (Knutton et al., 1998). EspB does, however, show a specific and close association with the filament and can be co-purified with EspA filaments by immunoprecipitation (Hartland et al., 2000a). Although EspB is essential for protein translocation, this protein is not required for attachment of EspA filaments to the host cell surface (Hartland et al., 2000a). Thus the association between the EspA filament and EspB may represent the filament's transition from an adhesive function to a role in protein translocation.

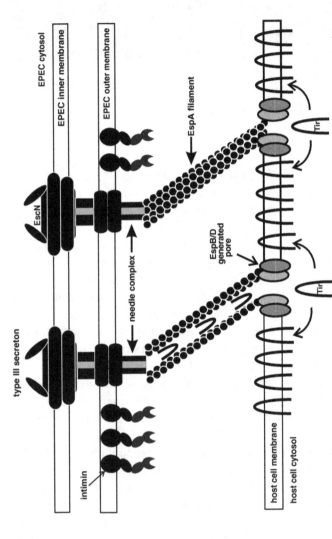

Figure 12.4. Model of EPEC protein translocation. The proposed protein translocation apparatus (translocon) consists of pores in the bacterial envelope (type III secreton) and in the host cell membrane (EspB/D-generated pore) connected by a hollow EspA filament, thereby providing a continuous channel from the bacterial to the host cell cytosol. The translocon is used to translocate Tir into the host cell, where it becomes inserted into the host cell membrane. Subsequent retraction/elimination of EspA filaments allows intimin-Tir interaction and pedestal formation.

Although the molecular basis of protein translocation has not yet been elucidated for any type III secretion system, each model predicts the presence of pore-forming proteins of bacterial origin in both the bacterial outer membrane and the host plasma membrane, which are connected by a needle-like structure (Fig. 12.4) (Frankel et al., 1998). EspD and EspB show weak homology with YopB and YopD, respectively, of Yersinia (Håkansson et al., 1993). Like EspD and EspB, YopB and YopD are hydrophobic, type III secreted proteins that are inserted into the host cell membrane during infection (Neyt and Cornelis, 1999). YopB and YopD are pore-forming proteins that together form a channel in the host cell membrane through which other effector proteins enter the cytoplasm of the target cell (Tardy et al., 1999). Pore formation by the type III secreted proteins of Yersinia spp. and Shigella spp. has been correlated with the ability of the bacteria to mediate contact-dependent haemolysis of red blood cells (RBCs) (Håkansson et al., 1996; Blocker et al., 1999). EspD and EspB are both essential for protein translocation and mediate EPEC-induced haemolysis of RBCs, suggesting that they may also be pore-forming proteins (Warawa et al., 1999). Recently, Shaw et al. (2001) showed that EspD was the only bacterial protein detected in the membrane of RBCs following infection with EPEC. By contrast, EspB did not appear to be associated with the RBC membrane, suggesting that EspD plays a dominant role in pore formation. An investigation of protein–protein interactions found that EspD interacts with itself, suggesting that the pore may comprise EspD multimers (S.J. Daniell et al, unpublished data). Interestingly, interactions between EspD and other secreted proteins have not been observed thus far. Given its absence from the RBC membrane and apparent lack of interaction with EspD, the precise role of EspB in the formation of a membrane pore, and indeed protein translocation, remains unclear.

12.6 HOST CELL CYTOSKELETAL AND SIGNALLING CHANGES

A number of bacterial pathogens mediate a direct effect on the host cell cytoskeleton during infection. The development of an A/E lesion following infection of the epithelium with EPEC reflects gross ultrastructural changes in the host cell underneath, and around, adherent bacteria. Cytoskeletal reorganization involves the recruitment of several cytoskeletal proteins to the body of the pedestal, including actin, α-actinin, talin and ezrin (Finlay et al., 1992). Although the precise cellular processes that govern cytoskeletal rearrangement during infection with EPEC are unknown, recently the N-terminal region of Tir was shown to interact directly with α-actinin (Goosney *et al.*, 2000). α-Actinin is one of the cytoskeletal proteins recruited to the pedestal during A/E lesion formation. The N-terminal region of Tir corresponds to an area of the protein that is located inside the host cell (Hartland *et al.*, 1999), and provides the first demonstration of a direct link between an EPEC-derived factor and a host cell target protein. This interaction occurred independently of tyrosine phosphorylation of Tir, unlike the recruitment of other cytoskeletal proteins (actin, vasodilator-stimulated phosphoprotein (VASP) and N-Wiscott–Aldrich syndrome protein (N-WASP) where tyrosine phosphorylation of Tir appears to be an essential step (Goosney *et al.*, 2000).

As well as its putative role in the formation of a membrane pore, EspB has been suggested to have an independent effect on the cytoskeleton of the target cell. EspB is translocated to both the host cell membrane and cytosolic fractions, and transfection of EspB into the cytosol of epithelial cells induces the reorganization of actin stress fibres (Taylor *et al.*, 1999). This suggests that EspB may harbour some cytoskeletal-altering activity of its own, although no host cell targets have been identified to date. EPEC has also been shown to induce a number of signalling changes in the host cell that may indirectly lead to cytoskeletal alterations. Attachment of the bacteria to the epithelial cell surface leads to inositol phosphate fluxes and activation of protein kinase C (PKC) and phospholipase C-γ (PLC-γ) (Crane and Oh, 1997). These cellular responses occur only if the bacteria express an intact type III secretion system and intimin. Hence, intimate attachment is required to induce the signalling changes.

12.7 LOCALIZED ADHERENCE AND MICROCOLONY FORMATION

In addition to A/E lesions, EPEC adheres to the surface of epithelial cells in characteristic microcolonies. This pattern of attachment is termed localized adherence (LA) and denotes the adhesion of bacteria to host cells in aggregates or clusters. The bacterial microcolonies are believed to be essential for efficient colonization of the host and subsequent dispersal of the pathogen. Localized adherence and microcolony formation by EPEC depends on the expression of a type IV pilus called the bundle-forming pilus, BFP (Giron *et al.*, 1991; Donnenberg *et al.*, 1992). While the role and timing of microcolony formation *in vivo* remains uncertain, recent studies with IVOCs suggest that BFP mediates microcolony formation subsequent to intimate attachment and the development of A/E lesions (Hicks *et al.*, 1998), rather than initiating attachment of the bacteria to the host epithelium prior to A/E lesion formation.

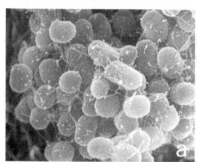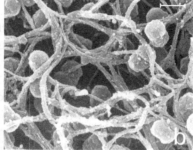

Figure 12.5. Scanning electron micrographs illustrating bundle-forming pilus (BFP)-mediated microcolony formation and dispersal. (a) Thin BFP filaments initially promote bacteria–bacteria interaction and microcolony formation; (b) rearrangement to a network of much thicker BFP filaments promotes release of bacteria and dispersal of the microcolony. Bar, 0.5 μm.

Production and assembly of BFP depends on a cluster of 14 genes, which are present on the large, 50–70 MDa EAF plasmids (Donnenberg *et al.*, 1992; Stone *et al.*, 1996). The major pilin subunit (BFP pilin or bundlin) is encoded by the first gene of the cluster, *bfpA*. BFP pilin is produced as a pre-protein that undergoes proteolytic cleavage by the *bfp*P-encoded peptidase into a mature form (Zhang *et al.*, 1994). BFP production is regulated by the plasmid-encoded Per (also known as BfpTWV) (Tobe *et al.*, 1996). Human volunteer studies have shown that BFP is essential for colonization of the host intestine (Bieber *et al.*, 1998); *bfpA* mutants caused significantly less diarrhoea than wild-type EPEC in experimentally infected adults. In addition, a *bfpF* mutant that lacks the ability to disperse from microcolonies was 200 times less virulent than the wild-type strain in the same infection model, suggesting that microcolony dispersal is an important virulence trait. *In vitro* studies have shown that BFP undergoes marked changes in quaternary structure as the infection proceeds. The pilus changes from a thin to a thick bundle structure that results in disruption of the interactions between bacteria and the subsequent release of bacteria from the microcolony (Fig. 12.5) (Knutton *et al.*, 1999). This transformation relies on BfpF, a putative nucleotide-binding factor (Anantha *et al.*, 1998), which is either directly or indirectly involved in the shift from thin to thick bundles. BfpF is not required for thin filament production but may act as an energizer for pilus retraction in a manner analogous to PilT of *Pseudomonas aeruginosa* (Knutton *et al.*, 1999). Thus, through the structural transformation of the pilus, BFP is able to promote initial bacterial aggregation and formation of the microcolony and then dispersal of bacteria already primed to form new A/E lesions and microcolonies at fresh sites in the gut.

12.8 CONCLUSIONS

Although our understanding of the mechanisms employed by EPEC to interact with the host intestinal epithelium has improved dramatically in recent years, the precise mechanism of EPEC-induced diarrhoea remains obscure. Features of EPEC diarrhoea suggest the existence of a toxin but to date no toxigenic factors have been isolated. Extensive effacement of absorptive microvilli due to A/E lesions may contribute to fluid loss but the short incubation period between ingestion and onset of diarrhoea (less than 4 hours) suggests active fluid secretion by the intestine (Nataro and Kaper, 1998). One possible mechanism involves EPEC-induced alterations in transepithelial electric resistance that result from disruption of epithelial tight junctions due to phosphorylation of myosin light chain (Philpott *et al.*, 1996; Yuhan *et al.*, 1997). Adherence by EPEC also stimulates some secretion of chloride ions by epithelial cells (Collington *et al.*, 1998). These modes of interference with the intestinal barrier may lead to increased permeability of the gut and contribute to EPEC-induced diarrhoea, although host factors such as the release of inflammatory mediators may also play a role.

A/E lesion formation and intimate adherence of EPEC to the host epithelium is a complex, multistage process. The LEE-encoded secreted proteins and intimin represent a potent set of virulence determinants that act to undermine the structure of the host cell cytoskeleton and effect tight adherence of the bacteria to the intestinal epithelium. Gradually, through the use of animal models and powerful tools in molecular biology, cell biology, microscopy and structural biology, we are beginning to unravel the individual and combined roles of the LEE-encoded proteins and other virulence factors in pathogenesis. A more complete understanding of the molecular basis of protein translocation and subsequent bacterial reorganization of the host cell cytoskeleton would aid the design of strategies to intervene in the disease process.

ACKNOWLEDGEMENTS

Research in the authors' laboratories is supported by the Wellcome Trust. E.L.H. is the recipient of a Royal Society/NHMRC Howard Florey fellowship.

REFERENCES

Abe, A., de Grado, M., Pfuetzner, R.A., Sanchez-Sanmartin, C., Devinney, R., Puente, J.L., Strynadka, N.C. and Finlay, B.B. (1999). Enteropathogenic

Escherichia coli translocated intimin receptor, Tir, requires a specific chaperone for stable secretion. *Molecular Microbiology* 33, 1162–1175.

Adu-Bobie, J., Frankel, G., Bain, C., Goncalves, A.G., Trabulsi, L.R., Douce, G., Knutton, S. and Dougan, G. (1998). Detection of intimins alpha, beta, gamma, and delta, four intimin derivatives expressed by attaching and effacing microbial pathogens. *Journal of Clinical Microbiology* 36, 662–668.

Anantha, R.P., Stone, K.D. and Donnenberg, M.S. (1998). Role of BfpF, a member of the PilT family of putative nucleotide-binding proteins, in type IV pilus biogenesis and in interactions between enteropathogenic *Escherichia coli* and host cells. *Infection and Immunity* 66, 122–131.

Batchelor, M., Prasannan, S., Daniell, S., Reece, S., Connerton, I., Bloomberg, G., Dougan, G., Frankel, G. and Matthews, S. (2000). Structural basis for recognition of the translocated intimin receptor (Tir) by intimin from enteropathogenic *Escherichia coli*. *EMBO Journal* 19, 2452–2464.

Bieber, D., Ramer, S.W., Wu, C.Y., Murray, W.J., Tobe, T., Fernandez, R. and Schoolnik, G.K. (1998). Type IV pili, transient bacterial aggregates, and virulence of enteropathogenic *Escherichia coli*. *Science* 280, 2114–2118.

Blocker, A., Gounon, P., Larquet, E., Niebuhr, K., Cabiaux, V., Parsot, C. and Sansonetti, P. (1999). The tripartite type III secreton of *Shigella flexneri* inserts IpaB and IpaC into host membranes. *Journal of Cell Biology* 147, 683–693.

Collington, G.K., Booth, I.W. and Knutton, S. (1998). Rapid modulation of electrolyte transport in Caco-2 cell monolayers by enteropathogenic *Escherichia coli* (EPEC) infection. *Gut* 42, 200–207.

Crane, J.K. and Oh, J.S. (1997). Activation of host cell protein kinase C by enteropathogenic *Escherichia coli*. *Infection and Immunity* 65, 3277–3285.

Deibel, C., Kramer, S., Chakraborty, T. and Ebel, F. (1998). EspE, a novel secreted protein of attaching and effacing bacteria, is directly translocated into infected host cells, where it appears as a tyrosine-phosphorylated 90 kDa protein. *Molecular Microbiology* 28, 463–474.

Delahay, R.M., Knutton, S., Shaw, R.K., Hartland, E.L., Pallen, M.J. and Frankel, G. (1999). The coiled-coil domain of EspA is essential for the assembly of the type III secretion translocon on the surface of enteropathogenic *Escherichia coli*. *Journal of Biological Chemistry* 274, 35969–35974.

DeVinney, R., Stein, M., Reinscheid, D., Abe, A., Ruschkowski, S. and Finlay, B.B. (1999). Enterohemorrhagic *Escherichia coli* O157:H7 produces Tir, which is translocated to the host cell membrane but is not tyrosine phosphorylated. *Infection and Immunity* 67, 2389–2398.

Donnenberg, M.S., Giron, J.A., Nataro, J.P. and Kaper, J.B. (1992). A plasmid-encoded type IV fimbrial gene of enteropathogenic *Escherichia coli* associated with localized adherence. *Molecular Microbiology* 6, 3427–3437.

Donnenberg, M.S., Tacket, C.O., James, S.P., Losonsky, G., Nataro, J.P., Wasserman, S.S., Kaper, J.B. and Levine, M.M. (1993). Role of the *eaeA* gene in experimental enteropathogenic *Escherichia coli* infection. *Journal of Clinical Investigation* **92**, 1412–1417.

Elliott, S.J., Wainwright, L.A., McDaniel, T.K., Jarvis, K.G., Deng, Y.K., Lai, L.C., McNamara, B.P., Donnenberg, M.S. and Kaper, J.B. (1998). The complete sequence of the locus of enterocyte effacement (LEE) from enteropathogenic *Escherichia coli* E2348/69. *Molecular Microbiology* **28**, 1–4.

Elliott, S.J., Hutcheson, S.W., Dubois, M.S., Mellies, J.L., Wainwright, L.A., Batchelor, M., Frankel, G., Knutton, S. and Kaper, J.B. (1999). Identification of CesT, a chaperone for the type III secretion of Tir in enteropathogenic *Escherichia coli*. *Molecular Microbiology* **33**, 1176–1189.

Elliott, S.J., Sperandio, V., Giron, J.A., Shin, S., Mellies, J.L., Wainwright, L., Hutcheson, S.W., McDaniel, T.K. and Kaper, J.B. (2000). The locus of enterocyte effacement (LEE)-encoded regulator controls expression of both LEE- and non-LEE-encoded virulence factors in enteropathogenic and enterohemorrhagic *Escherichia coli*. *Infection and Immunity* **68**, 6115–6126.

Finlay, B.B., Rosenshine, I., Donnenberg, M.S. and Kaper, J.B. (1992). Cytoskeletal composition of attaching and effacing lesions associated with enteropathogenic *Escherichia coli* adherence to HeLa cells. *Infection and Immunity* **60**, 2541–2543.

Frankel, G., Candy, D.C., Everest, P. and Dougan, G. (1994). Characterization of the C-terminal domains of intimin-like proteins of enteropathogenic and enterohemorrhagic *Escherichia coli*, *Citrobacter freundii*, and *Hafnia alvei*. *Infection and Immunity* **62**, 1835–1842.

Frankel, G., Candy, D.C., Fabiani, E., Adu-Bobie, J., Gil, S., Novakova, M., Phillips, A.D. and Dougan, G. (1995). Molecular characterization of a carboxy-terminal eukaryotic-cell-binding domain of intimin from enteropathogenic *Escherichia coli*. *Infection and Immunity* **63**, 4323–4328.

Frankel, G., Lider, O., Hershkoviz, R., Mould, A.P., Kachalsky, S.G., Candy, D.C.A., Cahalon, L., Humphries, M.J. and Dougan, G. (1996a). The cell-binding domain of intimin from enteropathogenic *Escherichia coli* binds to beta1 integrins. *Journal of Biological Chemistry* **271**, 20359–20364.

Frankel, G., Phillips, A.D., Novakova, M., Field, H., Candy, D.C., Schauer, D.B., Douce, G. and Dougan, G. (1996b). Intimin from enteropathogenic *Escherichia coli* restores murine virulence to a *Citrobacter rodentium eaeA* mutant: induction of an immunoglobulin A response to intimin and EspB. *Infection and Immunity* **64**, 5315–5325.

Frankel, G., Phillips, A.D., Rosenshine, I., Dougan, G., Kaper, J.B. and Knutton,

S. (1998). Enteropathogenic and enterohaemorrhagic *Escherichia coli*: more subversive elements. *Molecular Microbiology* **30**, 911–921.

Friedberg, D., Umanski, T., Fang, Y. and Rosenshine, I. (1999). Hierarchy in the expression of the locus of enterocyte effacement genes of enteropathogenic *Escherichia coli*. *Molecular Microbiology* **34**, 941–952.

Giron, J.A., Ho, A.S. and Schoolnik, G.K. (1991). An inducible bundle-forming pilus of enteropathogenic *Escherichia coli*. *Science* **254**, 710–713.

Gomez-Duarte, O.G. and Kaper, J.B. (1995). A plasmid-encoded regulatory region activates chromosomal *eaeA* expression in enteropathogenic *Escherichia coli*. *Infection and Immunity* **63**, 1767–1776.

Goosney, D.L., DeVinney, R., Pfuetzner, R.A., Frey, E.A., Strynadka, N.C. and Finlay, B.B. (2000). Enteropathogenic *Escherichia coli* translocated intimin receptor, Tir, interacts directly with alpha-actinin. *Current Biology* **10**, 735–738.

Håkansson, S., Bergman, T., Vanooteghem, J.C., Cornelis, G. and Wolf-Watz, H. (1993). YopB and YopD constitute a novel class of *Yersinia* Yop proteins. *Infection and Immunity* **61**, 71–80.

Håkansson, S., Schesser, K., Persson, C., Galyov, E.E., Rosqvist, R., Homble, F. and Wolf-Watz, H. (1996). The YopB protein of *Yersinia pseudotuberculosis* is essential for the translocation of Yop effector proteins across the target cell plasma membrane and displays a contact-dependent membrane disrupting activity. *EMBO Journal* **15**, 5812–5823.

Hartland, E.L., Batchelor, M., Delahay, R.M., Hale, C., Matthews, S., Dougan, G., Knutton, S., Connerton, I. and Frankel, G. (1999). Binding of intimin from enteropathogenic *Escherichia coli* to Tir and to host cells. *Molecular Microbiology* **32**, 151–158.

Hartland, E.L., Daniell, S.J., Delahay, R.M., Neves, B.C., Wallis, T., Shaw, R.K., Hale, C., Knutton, S. and Frankel, G. (2000a). The type III protein translocation system of enteropathogenic *Escherichia coli* involves EspA–EspB protein interactions. *Molecular Microbiology* **35**, 1483–1492.

Hartland, E.L., Huter, V., Higgins, L.M., Goncalves, N.S., Dougan, G., Phillips, A.D., MacDonald, T.T. and Frankel, G. (2000b). Expression of intimin gamma from enterohemorrhagic *Escherichia coli* in *Citrobacter rodentium*. *Infection and Immunity* **68**, 4637–4646.

Hicks, S., Frankel, G., Kaper, J.B., Dougan, G. and Phillips, A.D. (1998). Role of intimin and bundle-forming pili in enteropathogenic *Escherichia coli* adhesion to pediatric intestinal tissue *in vitro*. *Infection and Immunity* **66**, 1570–1578.

Higgins, L.M., Frankel, G., Connerton, I., Goncalves, N.S., Dougan, G. and MacDonald, T.T. (1999a). Role of bacterial intimin in colonic hyperplasia and inflammation. *Science* **285**, 588–591.

Higgins, L.M., Frankel, G., Douce, G., Dougan, G. and MacDonald, T.T. (1999b). *Citrobacter rodentium* infection in mice elicits a mucosal Th1 cytokine response and lesions similar to those in murine inflammatory bowel disease. *Infection and Immunity* **67**, 3031–3039.

Hueck, C.J. (1998). Type III protein secretion systems in bacterial pathogens of animals and plants. *Microbiology and Molecular Biology Reviews* **62**, 379–433.

Hyman, H.C. and Trachtenberg, S. (1991). Point mutations that lock *Salmonella typhimurium* flagellar filaments in the straight right-handed and left-handed forms and their relation to filament superhelicity. *Journal of Molecular Biology* **220**, 79–88.

Isberg, R.R., Voorhis, D.L. and Falkow, S. (1987). Identification of invasin: a protein that allows enteric bacteria to penetrate cultured mammalian cells. *Cell* **50**, 769–778.

Jerse, A.E. and Kaper, J.B. (1991). The *eae* gene of enteropathogenic *Escherichia coli* encodes a 94– kilodalton membrane protein, the expression of which is influenced by the EAF plasmid. *Infection and Immunity* **59**, 4302–4309.

Jerse, A.E., Yu, J., Tall, B.D. and Kaper, J.B. (1990). A genetic locus of enteropathogenic *Escherichia coli* necessary for the production of attaching and effacing lesions on tissue culture cells. *Proceedings of the National Academy of Sciences, USA* **87**, 7839–7843.

Kelly, G., Prasannan, S., Daniell, S., Fleming, K., Frankel, G., Dougan, G., Connerton, I. and Matthews, S. (1999). Structure of the cell-adhesion fragment of intimin from enteropathogenic *Escherichia coli*. *Nature Structural Biology* **6**, 313–318.

Kenny, B. (1999). Phosphorylation of tyrosine 474 of the enteropathogenic *Escherichia coli* (EPEC) Tir receptor molecule is essential for actin nucleating activity and is preceded by additional host modifications. *Molecular Microbiology* **31**, 1229–1241.

Kenny, B., DeVinney, R., Stein, M., Reinscheid, D.J., Frey, E.A. and Finlay, B.B. (1997). Enteropathogenic *Escherichia coli* (EPEC) transfers its receptor for intimate adherence into mammalian cells. *Cell* **91**, 511–520.

Knutton, S., Baldwin, T., Williams, P.H. and McNeish, A.S. (1989). Actin accumulation at sites of bacterial adhesion to tissue culture cells: basis of a new diagnostic test for enteropathogenic and enterohemorrhagic *Escherichia coli*. *Infection and Immunity* **57**, 1290–1298.

Knutton, S., Rosenshine, I., Pallen, M.J., Nisan, I., Neves, B.C., Bain, C., Wolff, C., Dougan, G. and Frankel, G. (1998). A novel EspA-associated surface organelle of enteropathogenic *Escherichia coli* involved in protein translocation into epithelial cells. *EMBO Journal* **17**, 2166–2176.

Knutton, S., Shaw, R.K., Anantha, R.P., Donnenberg, M.S. and Zorgani, A.A.

(1999). The type IV bundle-forming pilus of enteropathogenic *Escherichia coli* undergoes dramatic alterations in structure associated with bacterial adherence, aggregation and dispersal. *Molecular Microbiology* **33**, 499–509.

Kresse, A.U., Rohde, M. and Guzmán, C.A. (1999). The EspD protein of entero-hemorrhagic *Escherichia coli* is required for the formation of bacterial surface appendages and is incorporated in the cytoplasmic membranes of target cells. *Infection and Immunity* **67**, 4834–4842.

Kubori, T., Matsushima, Y., Nakamura, D., Uralil, J., Lara-Tejero, M., Sukhan, A., Galan, J.E. and Aizawa, S.I. (1998). Supramolecular structure of the *Salmonella typhimurium* type III protein secretion system. *Science* **280**, 602–625.

Liu, H., Magoun, L., Luperchio, S., Schauer, D.B. and Leong, J.M. (1999). The Tir-binding region of enterohaemorrhagic *Escherichia coli* intimin is sufficient to trigger actin condensation after bacterial-induced host cell signalling. *Molecular Microbiology* **34**, 67–81.

Luo, Y., Frey, E.A., Pfuetzner, R.A., Creagh, A.L., Knoechel, D.G., Haynes, C.A., Finlay, B.B. and Strynadka, N.C. (2000). Crystal structure of enteropathogenic *Escherichia coli* intimin-receptor complex. *Nature* **405**, 1073–1077.

McDaniel, T.K. and Kaper, J.B. (1997). A cloned pathogenicity island from enteropathogenic *Escherichia coli* confers the attaching and effacing phenotype on E. coli K-12. *Molecular Microbiology* **23**, 399–407.

McDaniel, T.K., Jarvis, K.G., Donnenberg, M.S. and Kaper, J.B. (1995). A genetic locus of enterocyte effacement conserved among diverse enterobacterial pathogens. *Proceedings of the National Academy of Sciences, USA* **92**, 1664–1668.

Mellies, J.L., Elliott, S.J., Sperandio, V., Donnenberg, M.S. and Kaper, J.B. (1999). The Per regulon of enteropathogenic *Escherichia coli*: identification of a regulatory cascade and a novel transcriptional activator, the locus of enterocyte effacement (LEE)-encoded regulator (Ler). *Molecular Microbiology* **33**, 296–306.

Moon, H.W., Whipp, S.C., Argenzio, R.A., Levine, M.M. and Giannella, R.A. (1983). Attaching and effacing activities of rabbit and human enteropathogenic *Escherichia coli* in pig and rabbit intestines. *Infection and Immunity* **41**, 1340–1351.

Nataro, J.P. and Kaper, J.B. (1998). Diarrheagenic *Escherichia coli*. *Clinical Microbiology Reviews* **11**, 142–201.

Nataro, J.P., Maher, K.O., Mackie, P. and Kaper, J.B. (1987). Characterization of plasmids encoding the adherence factor of enteropathogenic *Escherichia coli*. *Infection and Immunity* **55**, 2370–2377.

Neyt, C. and Cornelis, G.R. (1999). Insertion of a Yop translocation pore into the macrophage plasma membrane by *Yersinia enterocolitica*: requirement for translocators YopB and YopD, but not LcrG. *Molecular Microbiology* **33**, 971–981.

Phillips, A.D. and Frankel, G. (2000). Intimin-mediated tissue specificity in enteropathogenic *Escherichia coli* interaction with human intestinal organ cultures. *Journal of Infectious Diseases* **181**, 1496–500.

Phillips, A.D., Giron, J., Hicks, S., Dougan, G. and Frankel, G. (2000). Intimin from enteropathogenic *Escherichia coli* mediates remodelling of the eukaryotic cell surface. *Microbiology* **146**, 1333–1344.

Philpott, D.J., McKay, D.M., Sherman, P.M. and Perdue, M.H. (1996). Infection of T84 cells with enteropathogenic *Escherichia coli* alters barrier and transport functions. *American Journal of Physiology* **270**, G634–G645.

Reid, S.D., Herbelin, C.J., Bumbaugh, A.C., Selander, R.K. and Whittam, T.S. (2000). Parallel evolution of virulence in pathogenic *Escherichia coli*. *Nature* **406**, 64–67.

Sanger, J.M., Chang, R., Ashton, F., Kaper, J.B. and Sager, J.W. (1996). Novel form of actin motility transports bacteria on the surface of infected cells. *Cell Motility and the Cytoskeleton* **34**, 279–287.

Shaw, R.K., Daniell, S., Ebel, F., Frankel, G. and Knutton, S. (2001). EspA filament-mediated protein translocation into red blood cells. *Cellular Microbiology* **3**, 213–222.

Sperandio, V., Mellies, J.L., Nguyen, W., Shin, S. and Kaper, J.B. (1999). Quorum sensing controls expression of the type III secretion gene transcription and protein secretion in enterohemorrhagic and enteropathogenic *Escherichia coli*. *Proceedings of the National Academy of Sciences, USA* **96**, 15196–15201.

Sperandio, V., Mellies, J.L., Delahay, R.M., Frankel, G., Crawford, J.A., Nguyen, W. and Kaper, J.B. (2000). Activation of enteropathogenic *E. coli* (EPEC) LEE2 and LEE3 operons by Ler. *Molecular Microbiology* **38**, 781–793.

Stone, K.D., Zhang, H.Z., Carlson, L.K. and Donnenberg, M.S. (1996). A cluster of fourteen genes from enteropathogenic *Escherichia coli* is sufficient for the biogenesis of a type IV pilus. *Molecular Microbiology* **20**, 325–337.

Tacket, C.O., Sztein, M.B., Losonsky, G., Abe, A., Finlay, B.B., McNamara, B.P., Fantry, G.T., James, S.P., Nataro, J.P., Levine, M.M. and Donnenberg, M.S. (2000). Role of EspB in experimental human enteropathogenic *Escherichia coli* infection. *Infection and Immunity* **68**, 3689–3695.

Tamano, K., Aizawa, S. I., Katayama, E., Nonaka, T., Imajoh-Ohmi, S., Kuwae, A., Nagai, S. and Sasakawa, C. (2000). Supramolecular structure of the *Shigella* type III secretion machinery: the needle part is changeable in length and essential for delivery of effectors. *EMBO Journal* **19**, 3876–3887.

Tardy, F., Homble, F., Neyt, C., Wattiez, R., Cornelis, G.R., Ruysschaert, J.M. and

Cabiaux, V. (1999). *Yersinia enterocolitica* type III secretion-translocation system: channel formation by secreted Yops. *EMBO Journal* **18**, 6793–6799.

Taylor, K.A., Luther, P.W. and Donnenberg, M.S. (1999). Expression of the EspB protein of enteropathogenic *Escherichia coli* within HeLa cells affects stress fibers and cellular morphology. *Infection and Immunity* **67**, 120–125.

Tobe, T., Schoolnik, G.K., Sohel, I., Bustamante, V.H. and Puente, J.L. (1996). Cloning and characterization of *bfpTVW*, genes required for the transcriptional activation of bfpA in enteropathogenic *Escherichia coli*. *Molecular Microbiology* **21**, 963–975.

Tzipori, S., Gunzer, F., Donnenberg, M.S., de Montigny, L., Kaper, J.B. and Donohue-Rolfe, A. (1995). The role of the *eaeA* gene in diarrhoea and neurological complications in a gnotobiotic piglet model of enterohemorrhagic *Escherichia coli* infection. *Infection and Immunity* **63**, 3621–3627.

Vallance, B.A. and Finlay, B.B. (2000). Exploitation of host cells by enteropathogenic *Escherichia coli*. *Proceedings of the National Academy of Sciences, USA* **97**, 8799–8806.

Wainwright, L.A. and Kaper, J.B. (1998). EspB and EspD require a specific chaperone for proper secretion from enteropathogenic *Escherichia coli*. *Molecular Microbiology* **27**, 1247–1260.

Warawa, J., Finlay, B.B. and Kenny, B. (1999). Type III secretion-dependent hemolytic activity of enteropathogenic *Escherichia coli*. *Infection and Immunity* **67**, 5538–5540.

Wolff, C., Nisan, I., Hanski, E., Frankel, G. and Rosenshine, I. (1998). Protein translocation into host epithelial cells by infecting enteropathogenic *Escherichia coli*. *Molecular Microbiology* **28**, 143–155.

Yuhan, R., Koutsouris, A., Savkovic, S.D. and Hecht, G. (1997). Enteropathogenic *Escherichia coli*-induced myosin light chain phosphorylation alters intestinal epithelial permeability. *Gastroenterology* **113**, 1873–1882.

Zhang, H.Z., Lory, S. and Donnenberg, M.S. (1994). A plasmid-encoded prepilin peptidase gene from enteropathogenic *Escherichia coli*. *Journal of Bacteriology* **176**, 6885–6891.

CHAPTER 13

Host cell responses to *Porphyromonas gingivalis* and *Actinobacillus actinomycetemcomitans*

Richard J. Lamont

13.1 INTRODUCTION

The area of contact between the teeth and the gums (gingiva) is an ana-tomically unique region that comprises mineralized tissue embedded in epi-thelium and exposed to a microbially abundant environment. The small (1–4 mm deep) gap between the surfaces of the tooth and the gingiva is known as the gingival sulcus or crevice. The gingiva is highly vascularized and the crevice is lined with sulcular epithelial cells that differ from oral epi-thelial cells by exhibiting less keratinization. Apically, sulcular epithelium becomes junctional epithelium that is characterized by a lack of keratiniza-tion, limited differentiation and a relatively permeable structure. It is this junctional epithelium that directly interposes between the gingiva and the tooth surface (Fig. 13.1). In destructive periodontal disease there is migra-tion of the junctional epithelium resulting in enlargement of the crevice into a deeper periodontal pocket that contains inflammatory cells such as neu-trophils and T-cells. The gingiva itself also contains immune cells including B-cells, T-cells and dendritic cells. The microbiota of the gingival area in both health and disease is complex, with at least 500 species of bacteria present in the gingival crevice. Although many of these have pathogenic potential, the strongest causal associations have been demonstrated between *Porphyromonas gingivalis* and severe adult periodontitis, and between *Actinobacillus actinomycetemcomitans* and localized juvenile periodontitis.

Many factors contribute to the maintenance or disruption of the ecolog-ical balance in the subgingival area. The immunological status of the host, the relative and absolute numbers of specific organisms or groups of organ-isms, and environmental parameters such as tobacco use, all play a role in determining gingival health or disease (Socransky and Haffajee, 1992). Periodontal pathogens such as *P. gingivalis* and *A. actinomycetemcomitans* can, therefore, be present in the absence of disease (Greenstein and

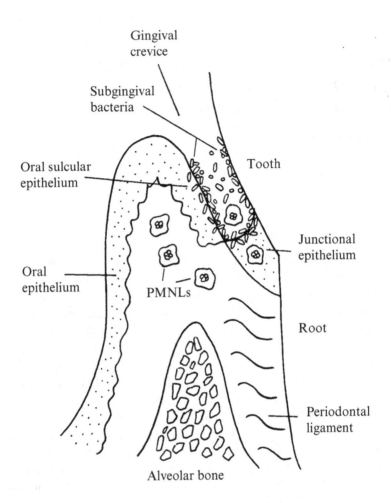

Figure 13.1. Schematic representation (not to scale) of the gingival crevice and surrounding periodontal tissues. PMNLs, polymorphonuclear leukocytes.

Lamster, 1997). A complex interplay between bacteria and host is thus apparent, an interplay that begins upon initial colonization by the organisms.

In order to colonize the gingival crevice, bacteria adhere to available surfaces that include gingival epithelial cells. In the case of *P. gingivalis* and *A. actinomycetemcomitans*, this adherence is not a passive 'hand-holding' event. Rather, evidence is accumulating that adhesion involves sensing and response reactions whereby the bacteria and the gingival epithelium form an interactive interface. This involves an elaborate communications network that generates and transmits signals among bacteria, epithelial cells and the underlying cells in the periodontal tissues. Bacteria possess a variety of adhesins, with differing receptor specificities and affinities, that can potentially impinge to varying degrees upon diverse receptor-dependent host cell biochemical pathways. The consequences of this adhesion process include uptake of bacteria within the host cells and modification of phenotypic properties that are important for tissue integrity and maintenance of the local innate host defence mechanisms that serve to control the microbial challenge.

13.2 ADHESINS DISPLAYED BY *P. GINGIVALIS* AND *A. ACTINOMYCETEMCOMITANS*

As adherence is an important early step in defining the nature of host cell responses to *P. gingivalis* and *A. actinomycetemcomitans*, we will begin our story with a description of the adhesins of these organisms and their mechanisms of action.

13.2.1 *Porphyromonas gingivalis*

Porphyromonas gingivalis can bind to host cells including epithelial cells, endothelial cells, fibroblasts, and erythrocytes, and to components of the extracellular matrix namely laminin, elastin, fibronectin, type I collagen, thrombospondin and vitronectin (Kontani *et al.*, 1997; Lamont and Jenkinson, 1998; Nakamura *et al.*, 1999; Sojar *et al.*, 1999; Dorn *et al.*, 2000). Adhesion is multimodal, involving fimbriae, outer membrane proteins and proteinases, molecules that are inextricably linked at both the transcriptional and post-translational levels.

The major fimbriae of *P. gingivalis* constitute a unique class of these Gram-negative organelles and comprised an ∼43 kDa fimbrillin (FimA) monomer. Fimbrillin possesses a number of binding domains for individual substrate recognition. The functional domain of FimA for epithelial cells spans amino acid residues 49–90 whereas the domains for fibronectin-binding are located within amino acid residues 126–146 along with 318–337, and involve a conserved VXXXA sequence (Sojar *et al.*, 1995, 1999). Although the *fimA* gene is monocistronic, immediately downstream are four genes whose products may be associated with the mature fimbriae (Watanabe *et al.*, 1996). The *fimA* upstream region contains functionally active σ^{70}-like promoter consensus sequences along with a potential UP element (Xie and Lamont, 1999). AT-rich sequences upstream from the RNA polymerase-binding sites are involved in positive regulation of transcriptional activity. Environmental cues to which *fimA* responds include temperature, haemin concentration and salivary molecules (Xie *et al.*, 1997), parameters with relevance to conditions in the oral cavity. The *fimA* gene can also be positively autoregulated by the FimA protein (Xie *et al.*, 2000).

Although the primary function of proteinases secreted by the asaccharolytic *P. gingivalis* is the provision of nutrients, proteinases are also involved both directly and indirectly in adhesion. At least eight distinct proteinases are produced by *P. gingivalis* (for reviews, see Potempa *et al.*, 1995; Kuramitsu, 1998; Curtis *et al.*, 1999) and direct enzyme–substrate interactions can effect adhesion, albeit short lived. Possibly more importantly, the C-terminal coding regions of the Arg-X- and Lys-X-specific proteases RgpA and Kgp contain extensive amino acid repeat blocks that have up to 90% identity with sequences that are also found in the HagA, D and E haemagglutinin proteins (discussed below) (Barkocy-Gallagher *et al.*, 1996). These regions may thus allow direct binding to human cell surface receptors. An additional adherence-related activity of proteinases is the partial degradation of substrates that subsequently exposes epitopes for adhesin recognition. Hydrolysis of fibronectin or other matrix proteins by the Arg-X-specific proteases RgpA and RgpB displays C-terminal Arg residues that mediate fimbriae-dependent binding (Kontani *et al.*, 1996). RgpA and RgpB also contribute to the fimbriae-mediated adhesive process by processing the leader peptide from the fimbrillin precursor (Nakayama *et al.*, 1996), and by up-regulating transcription of the *fimA* gene (Tokuda *et al.*, 1996; Xie *et al.*, 2000).

In addition to the haemagglutinin-associated activities of the RgpA and Kgp proteinases, five *hag* genes encoding haemagglutinins have now been sequenced. The *hagA*, *hagD* and *hagE* genes encode polypeptides with 73–93% identical amino acid sequences, while *hagB* and *hagC* genes are at distinct chromosomal loci and encode 39 kDa polypeptides that are 98.6% identical (Progulske-Fox *et al.*, 1995; Lépine and Progulske-Fox, 1996). A minimal peptide motif PVQNLT has recently been shown to be associated with haemagglutinating activity and is found within the proteinase–haemagglutinin sequences (Shibita *et al.*, 1999) and at multiple chromosomal sites (Barokocy-Gallagher *et al.*, 1996).

13.2.2 *Actinobacillus actinomycetemcomitans*

Actinobacillus actinomycetemcomitans can also adhere to host cells and to matrix components such as fibronectin, collagen and laminin (Meyer and

Fives-Taylor 1994; Alugupalli *et al.*, 1996; Mintz and Fives-Taylor, 1999). Adhesion is associated with fimbriae, outer membrane proteins, and extracellular vesicles and amorphous material (Fives-Taylor *et al.*, 1999). Fimbriae are found only on recent clinical isolates; laboratory subculture results in loss of fimbrial expression (Scannapieco *et al.*, 1987; Rosan *et al.*, 1988). Although the mechanism of this transition is unknown, progress is being made on the genetics of fimbrial structure and assembly. The fimbrial subunit appears to be a 6.5 kDa protein (Flp) that bears homology to the *Neisseria gonorrhoeae* type IV fimbriae, which are synthesized via the general protein secretion pathway for fimbrial assembly (Hultgren *et al.*, 1996; Inoue *et al.*, 1998). The *flp* gene comprises part of an operon that includes genes that may be involved in protein secretion and fimbrial assembly (Haase *et al.*, 1999). Whether the fimbrial structural component per se mediates adhesion to host cells, or whether fimbria-associated proteins (Ishihara *et al.*, 1997) act as adhesins, remains to be determined.

Adhesion to epithelial cells is also associated both with extracellular vesicles that bud from outer membrane extrusions and with extracellular amorphous material that may be predominantly protein or glycoprotein. These components may also enhance binding of other adherence-deficient strains (Meyer and Fives-Taylor, 1993, 1994). The functional adhesin has yet to be identified; however, an ~34 kDa OmpA-like outer membrane protein can bind laminin (Alugupalli *et al.*, 1996).

13.3 UPTAKE OF BACTERIA BY HOST CELLS

Perhaps the most dramatic outcome of the interaction between *P. gingivalis* or *A. actinomycetemcomitans* and host cells is the internalization of bacterial cells. *Porphyromonas gingivalis* can invade epithelial cells, endothelial cells and dendritic cells (for a review, see Lamont and Jenkinson, 1998). Interestingly, although the overall mechanistic basis is similar in these cell systems, the signal transduction pathways activated by the organism and the intracellular locations of the bacteria differ according to cell type. Invasion by *A. actinomycetemcomitans* has been demonstrated in primary and transformed epithelial cells (for a review, see Fives-Taylor *et al.*, 1999). For both organisms, viable bacterial cells are required for this active bacterially driven process. The ability to induce self-uptake by non-professional phagocytic cells is a property of a number of important pathogens including *Salmonella*, *Shigella*, *Listeria* and *Yersinia*, and is considered to be an important virulence determinant. An intracellular location may benefit bacteria by providing a nutritionally rich environment that is largely sheltered from the host immune response. Moreover, whether inadvertently or by design, the perturbation of host cell information flow that occurs during the invasion process can also compromise the normal phenotypic properties of the cell. The molecular basis of the invasive mechanisms of *P. gingivalis* and *A. actinomycetemcomitans* is under investigation in a number of laboratories and has the potential to provide insights into not only bacterial pathogenicity but also eukaryotic cell biology.

13.3.1 *Porphyromonas gingivalis*

13.3.1.1 Epithelial cells

Studies of *P. gingivalis* invasion of epithelial cells have, in the main, utilized two models: primary gingival epithelial cells and transformed oral epithelial lines such as KB cells. Invasion of *P. gingivalis* was first demonstrated in primary gingival epithelial cells (Lamont *et al.*, 1992). These cells are cultured from basal epithelial cells extracted from gingival explants and can be maintained in culture for several generations. Immunohistochemical staining has shown that the cells are non-differentiated and non-cornified, features of junctional epithelium (Oda and Whatsin, 1990). Thus, although not derived from junctional epithelium, these epithelial cells demonstrate similar properties and provide a relevant *ex vivo* model for the events that occur at the base of the gingival crevice.

When in contact with primary gingival epithelial cells, *P. gingivalis* is induced to secrete a novel set of extracellular proteins (Park and Lamont, 1998). Such contact-dependent protein secretion is generally indicative of the presence of a type III protein secretion apparatus. Type III secretion machines are utilized by Gram-negative bacteria to secrete effector proteins directly into the cytoplasm of the host cell (Cheng and Schneewind, 2000). The effector molecules can exhibit, for example, phosphatase or kinase activity that allows the pathogen to interfere directly with host signalling pathways (DeVinney *et al.*, 2000). However, a search of the *P. gingivalis* genome database (http://www.tigr.org) reveals that *P. gingivalis* does not possess obvious structural equivalents of components of the type III secretion system. None the less, some functional equivalence is implied by the finding that one of the contact-dependent secreted proteins of *P. gingivalis* bears homology to bacterial phosphoserine phosphatases (Laidig *et al.*, 2001). The extent to which this protein is translocated into the host cell and is functional therein remains to be determined.

Following proximate association with the epithelial cells, adhesion is realized. As mentioned above, *P. gingivalis* binding to epithelial cells is multimodal. However, studies with fimbriae-deficient mutants and manipulations of *fimA* expression suggest that the FimA-mediated component of adhesion is necessary (although probably not sufficient) for subsequent invasion (Weinberg *et al.*, 1997; Xie *et al.*, 1997). The *P. gingivalis* cysteine proteases (RpgA and Kgp) contribute to optimal invasion, possibly by exposing cryptitopes in epithelial cell receptors for fimbrial recognition (Park and Lamont, 1998). Sensitivity of the invasion process to the inhibitors cytochalasin D and nocodazole provides indirect evidence that both actin microfilament and microtubule rearrangements are required for the membrane invaginations that bring the bacteria into the host cell (Lamont *et al.*, 1995). The whole process is remarkably rapid and efficient. Invasion is complete within 15 minutes (Belton *et al.*, 1999) and conventional antibiotic protection assays reveal that laboratory strains can invade to around 10% of the initial inoculum, with clinical isolates exhibiting over 20% invasion (Lamont *et al.*, 1995). Moreover, direct fluorescent image analysis indicates that these percentages may be an underestimate of the number of internalized bacteria (Belton *et al.*, 1999). Once inside the cells, the bacteria are not confined to a membrane-bound vacuole and congregate in the perinuclear region (Belton *et al.*, 1999) (Plate 13.1). The ultimate metabolic fate of bacteria and epithelial cells *in vivo* is uncertain. The bacteria, however, remain viable and are capable of intracellular replication (Lamont *et al.*, 1995). The epithelial cells do not undergo necrotic or apoptotic cell death, although the cells contract and there is condensation of the actin cytoskeleton after prolonged cohabitation with *P. gingivalis* (Belton *et al.*, 1999). The invasion process is represented diagrammatically in Fig. 13.2.

Invasion of transformed cells such as KB cells by *P. gingivalis* is somewhat less efficient, with values less than 0.1% of the initial inoculum generally reported (Duncan *et al.*, 1993; Sandros *et al.*, 1993, 1994; Njoroge *et al.*, 1997). This may be a consequence of the alterations in signal transduction pathways and surface protein expression that accompany transformation (Hynes *et al.*, 1978; Cantley *et al.*, 1991). Initial adherence is FimA mediated (Njoroge *et al.*, 1997); however, engulfment of bacteria then occurs by classic receptor-mediated endocytosis (Sandros *et al.*, 1996) and bacteria can be found both free in the cytoplasm and contained within membrane-bound vacuoles (Sandros *et al.*, 1993; Njoroge *et al.*, 1997). Bacteria do remain viable and can replicate within the KB cells (Madianos *et al.*, 1996).

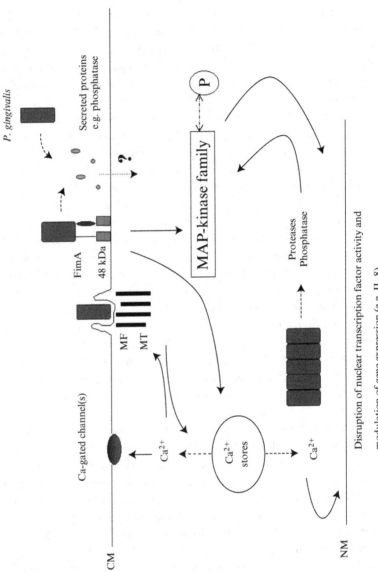

P. gingivalis

Secreted proteins
e.g. phosphatase

FimA

48 kDa

MF
MT

Ca-gated channel(s)

CM

MAP-kinase family

P

Proteases
Phosphatase

Ca^{2+}

Ca^{2+}
stores

Ca^{2+}

Disruption of nuclear transcription factor activity and
modulation of gene expression (e.g. IL-8)

NM

Figure 13.2. Model of currently understood *P. gingivalis* interactions with primary gingival epithelial cells. *Porphyromonas gingivalis* cells bind through adhesins on fimbriae (e.g. FimA) to surface receptors on gingival cells (e.g. the 48 kDa receptor for FimA). Microtubules and microfilaments are rearranged to facilitate invagination of the membrane that results in the engulfment of bacterial cells. Bacteria rapidly locate in the perinuclear area where they replicate. Calcium ions are released from intracellular stores and other signalling molecules such as the MAP-kinase family can be phosphorylated/dephosphorylated or degraded. Gene expression in the epithelial cells is ultimately affected. Abbreviations: Ca, calcium; CM, cytoplasmic membrane; IL-8, interleukin 8; MF, actin microfilaments; MT, tubulin microtubules; NM, nuclear membrane; P, phosphate; → pathway with potential intermediate steps; ----->, translocation; ···--->, release; <·····>, reversible association. (Reproduced in modified form, Lamont and Jenkinson, 1998.)

13.3.1.2 Endothelial cells

Invasion of bovine and human heart and aortic endothelial cells by *P. gingivalis* has been established (Deshpande *et al.*, 1998; Dorn *et al.*, 1999). Initial attachment requires FimA (Deshpande *et al.*, 1998); however, the presence of FimA is not sufficient to direct maximal invasion (Dorn *et al.*, 2000), indicating the need for additional adhesins and/or invasins. Invasion requires microfilament and microtubule remodelling (Deshpande *et al.*, 1998) and, once inside the cells, the bacteria are present in multimembranous vacuoles that resemble autophagosomes (Dorn *et al.*, 1999). *Porphyromonas gingivalis* can thus potentially gain access to cells of vascular walls, where the induction of autophagocytic pathways may alter the properties of the cells. Invasion of vascular endothelial cells may also provide a portal for bacterial entry into the bloodstream and subsequent systemic spread.

13.3.1.3 Dendritic cells

Dendritic cells are antigen-presenting cells that can activate lymphocytes including, distinctively, naive T-cells (Banchereau and Steinman, 1998). An immunological role for dendritic cells in chronic periodontal disease was postulated by DiFranco *et al.* (1985), and later Saglie *et al.* (1987) reported an increased number of dendritic cells in sites of diseased oral epithelium containing intragingival bacteria. These perceptive and provocative observations were largely neglected until a recent revival of the concept (Cutler *et al.*, 1999). Evidence has also been presented that *P. gingivalis* can internalize within cultured dendritic cells and that invasion is associated with sensitization and activation of the dendritic cells (Cutler *et al.*, 1999). This process has parallels with contact hypersensitivity responses that may therefore play a role in periodontal diseases.

13.3.2 *Actinobacillus actinomycetemcomitans*

Actinobacillus actinomycetemcomitans invasion has been studied most extensively in the KB cell model where internalization occurs by a dynamic multistep process (Fives-Taylor *et al.*, 1999). Initial attachment to the transferrin receptor appears to be the primary stimulus for invasion (Meyer *et al.*, 1997), although binding to integrins may constitute a secondary entry pathway (Meyer *et al.*, 1997; Fives-Taylor *et al.*, 1999). Adhesion induces effacement of the microvilli and the bacteria enter through ruffled apertures in the cell membrane (Meyer *et al.*, 1996). Entry of most strains requires restructuring of actin microfilaments that translocate from the periphery of the cell to a focus surrounding bacterial cells (Fives-Taylor *et al.*, 1995). Other strains internalize through actin-independent receptor-mediated endocytosis (Brissette and Fives-Taylor, 1999). Internal bacteria are initially constrained within a host-derived membrane vacuole, but this membrane is soon broken down and the bacteria are present in the cytoplasm where they can replicate (Sreenivasan *et al.*, 1993). The molecule(s) required for lysis of the vacuole membrane are unknown; however, *A. actinomycetemcomitans* produces a phospholipase C, an enzyme with potential membrane lytic abilities (Fives-Taylor *et al.*, 1999). Leukotoxin, a member of the RTX family of pore-forming cytolysins (Hritz *et al.*, 1996) could also be involved, although a study of the invasive ability of clinical isolates of *A. actinomycetemcomitans* indicated that leukotoxin production is more associated with non-invasive than with invasive strains (Lépine *et al.*, 1998). Interestingly, once in the cytoplasm, intracellular replication is more rapid than division in laboratory growth medium (Meyer *et al.*, 1996). The events that follow *A. actinomycetemcomitans* entry provide one of the most dramatic examples of bacterial orchestration of host cell function. *Actinobacillus actinomycetemcomitans* induces the formation of surface membrane protrusions through which the organism can migrate and enter into adjacent cells (Plate 13.2). The formation of these protrusions is consequent to bacterial interaction with the plusends of microtubules and movement through them may be driven by bacterial cell division (Fives-Taylor *et al.*, 1999; Meyer *et al.*, 1999). This unique process will facilitate penetration of the organism through the gingival tissues. The invasion processes are represented diagrammatically in Fig. 13.3.

13.3.3 Host cell signalling and phenotypic consequences associated with bacterial entry

What has become apparent from the study of a variety of invasive pathogens is that the process of bacterial entry into non-phagocytic host cells is dependent upon the bacteria seizing control of host intracellular communication pathways. Although it is difficult to discriminate signalling processes that are strictly required for invasion from those that are a response to the stress of a microbial onslaught, any alteration of normal information flow may have consequences for the status of the host cell.

In primary gingival epithelial cells, invasive *P. gingivalis* induces a transient increase in cytosolic [Ca^{2+}], as a result of release of Ca^{2+} from intracellular stores (Izutsu *et al.*, 1996). Such calcium ion fluxes are likely to be important in many signalling events and may converge on calcium-gated ion

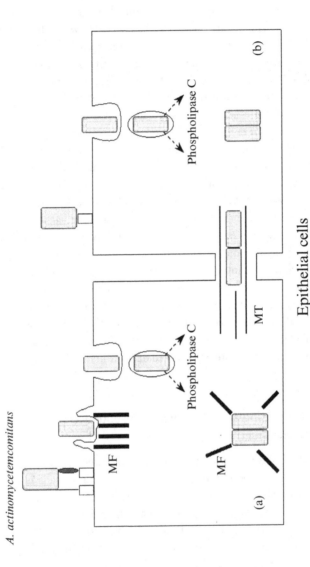

Figure 13.3. Model of currently understood *A. actinomycetemcomitans* interactions with epithelial cells. *Actinobacillus actinomycetemcomitans* cells bind through adhesins such as fimbriae, outer membrane proteins, and extracellular vesicles and amorphous material, to surface receptors on gingival cells such as the transferrin receptor. The epithelial cell membrane ruffles and effaces, and invaginations engulf the bacteria, which then become internalized within a membrane vesicle. Invasion can be (a) actin dependent, resulting in an actin focus around the bacteria, or (b) actin independent. The bacterial cells destroy the membrane vesicles (possibly by secretion of phospholipase C) releasing the bacteria into the cytoplasm, where they grow and divide rapidly. Bacteria become localized at membrane protrusions, through which they enter adjoining epithelial cells in a microtubule-dependent process.

Abbreviations: MF, actin microfilaments; MT, tubulin microtubules; ---→, release.

channels in the cytoplasmic membrane, the cytoskeletal apparatus, or nuclear transcription factors. Of similar importance is the observation that invasive *P. gingivalis* causes tyrosine phosphorylation of a eukaroytic 43 kDa protein, possibly an extracellular signal-regulated protein kinase (ERK) (Sandros *et al.*, 1996). ERKs are members of the mitogen-activated protein (MAP) kinase family that are involved in multiple intracellular signalling pathways. The collective results of the subversion of epithelial cell intracellular pathways by *P. gingivalis* can have phenotypic effects with immediate relevance to the disease process. Transcription and secretion of interleukin (IL)-8 (a potent neutrophil chemokine) by gingival epithelial cells is inhibited following *P. gingivalis* invasion. *Porphyromonas gingivalis* is even able to antagonize IL-8 secretion following stimulation of epithelial cells by common plaque commensals (Darveau *et al.*, 1998). Reduced expression of epithelial cell intercellular adhesion molecule (ICAM)-1 may also contribute to down-regulation of the innate host response (Madianos *et al.*, 1997). Regulation of matrix metalloproteinase (MMP) production by gingival epithelial cells is disrupted following contact with *P. gingivalis* (Fravalo *et al.*, 1996), thus interfering with extracellular matrix repair and reorganization. These activities are distinct from, but probably complementary to, the direct action of proteolytic enzymes that will be delivered in close proximity to their substrates during the adhesion and entry process. *Porphyromonas gingivalis* proteinases can also activate and up-regulate the transcription of MMP enzymes (DeCarlo *et al.*, 1997, 1998). Furthermore, *P. gingivalis* proteinases can degrade IL-8 and other cytokines (Fletcher *et al.*, 1997; Darveau *et al.*, 1998; Yun *et al.*, 1999; Zhang *et al.*, 1999) along with occludin, cadherins and integrins, proteins important in maintaining the barrier function of the epithelium (Katz *et al.*, 2000).

Less is known concerning the outcome of the invasive interaction between *A. actinomycetemcomitans* and epithelial cells. Recent studies show that gingival epithelial cells secrete IL-8 and up-regulate ICAM-1 in response to *A. actinomycetemcomitans* invasion (Huang *et al.*, 1998a,b). Thus the initial encounter with the innate immune response differs between *A. actinomycetemcomitans* and *P. gingivalis*. Whereas *P. gingivalis* has stealth-like properties that avoid immune recognition, *A. actinomycetemcomitans* is a more overt pathogen, readily recognized by the sentinels of mucosal immunity. These distinguishing features may have relevance to the differing clinical presentation of periodontal diseases associated with the two organisms, as discussed below.

13.3.4 Relevance to health and disease

Invasion of gingival tissues by periodontal organisms such as *P. gingivalis* and *A. actinomycetemcomitans* has been recognized, although not universally appreciated, for some time. Early immunofluorescence studies revealed the presence of both organisms within cells from biopsy tissue (Saglie *et al.*, 1988). In addition, *A. actinomycetemcomitans* was cultivated in large numbers from similar gingival biopsy samples (Christersson *et al.*, 1987). Comparison of healthy and diseased sites showed that bacterial penetration is a characteristic of diseased tissue (Pekovic and Fillery, 1984; Saglie *et al.*, 1986). More recent immunological studies have also provided supportive evidence for the ability of *P. gingivalis* to localize within the cells of the gingiva (Noiri *et al.*, 1997; Culter *et al.*, 1999). On an elementary level then, invasion, be it inter- or intracellular, may be an extension of subgingival colonization that exacerbates tissue destruction or serves as an occult bacterial reservoir for reactivation of disease episodes following cell death and release of bacteria. While this may be true to some extent, the significance of intracellular invasion is likely to be more complex. In the first instance, epithelial cells invaded with either *P. gingivalis* or *A. actinomycetemcomitans* do not show any immediate signs of cell death (Madianos *et al.*, 1996; Fives-Taylor *et al.*, 1999; Katz *et al.*, 2000). Even primary gingival epithelial cells that have a finite life span in culture can tolerate high levels of intracellular *P. gingivalis* without loss of viability over at least a 24 hour period (Belton *et al.*, 1999). Given that gingival epithelial cells turn over about every 5–7 days *in vivo*, a significant period of their existence could be in association with internalized bacteria. Indeed, one could advance an alternative hypothesis that bacterial entry is a means by which epithelial cells sequester pathogenic organisms that are then disposed of following normal cell cycling. However, there are several observations that suggest a role for invasion in the pathogenic process rather than in the maintenance of gingival health. Invasive *P. gingivalis* induce down-regulation of IL-8 production by epithelial cells and also antagonize IL-8 secretion following stimulation by common plaque constituents such as *Fusobacterium nucleatum* (Darveau *et al.*, 1998). In clinically healthy tissue, IL-8 forms a gradient of expression that is highest at the bacteria–epithelia interface and decreases deeper in the gingiva (Tonetti *et al.*, 1994). This gradient directs neutrophils to sites of bacterial accumulation thereby protecting the tissues from the bacteria and from neutrophil-mediated damage. Low-level expression of these inflammatory mediators is thus important in ensuring gingival health. Inhibition of IL-8 accumulation by *P. gingivalis* at sites of bacterial invasion could have a debilitating effect on innate host

R. J. LAMONT

defense in the periodontium, where bacterial exposure is constant. The host would no longer be able to detect the presence of bacteria and direct leukocytes for their removal. The ensuing overgrowth of bacteria would then contribute to a burst of disease activity. Furthermore, a delay in polymorphonuclear leukocyte (PMNL) recruitment from the vasculature could cause premature release of lytic enzymes and contribute to tissue destruction (Van Dyke, 1984). None the less, host PMNLs and other defence mechanisms do eventually become mobilized, as evidenced by the pyogenic nature of *P. gingivalis*-associated periodontal diseases. The overgrowth of subgingival plaque bacteria, or of *P. gingivalis* itself, that ensues after initial immune suppression may trigger reactivation of the immune response. Alternatively, or concomitantly, the encounter with different host cells as the infection progresses may result in a more vigorous immune response. For example, *P. gingivalis* invasion of dendritic cells results in maturation, increased co-stimulatory molecule expression and stimulatory activity for T-cells. The migration and proliferation of *P. gingivalis*-specific effector T-cells could be one means by which the immune system gears up in periodontal disease (Cutler *et al.*, 1999). In the case of *A. actinomycetemcomitans*, the initial interaction between invasive bacteria and epithelial cells is accompanied by an increase in levels of innate response effectors, including IL-8. However, *A. actinomycetemcomitans* may be able to avoid the encounter with the immune system by remaining intracellular and spreading from cell to cell through the interconnecting intercellular protrusions. The immune response may then burn out leaving a situation, as observed clinically, whereby some *A. actinomycetemcomitans* infections do not involve significant inflammation.

Although tissue destruction in periodontal diseases is limited to the supporting structures of the teeth, epidemiological evidence has emerged recently for an association between periodontal infections and serious systemic diseases including coronary artery disease (Scannapieco and Genco, 1999). Several observations provide a credible, though as yet very preliminary, basis for a causal link between infections with periodontal organisms such as *P. gingivalis* and heart disease. *Porphyromonas gingivalis* has been detected in carotid and coronary atheromas (Chiu, 1999), and the organism can induce platelet aggregation, which is associated with thrombus formation (Herzberg *et al.*, 1994). While common dental procedures, even vigorous tooth brushing, can lead to the presence of oral bacteria in the bloodstream, it is also possible that tissue and cell invasion by *P. gingivalis* or *A. actinomycetemcomitans* in the highly vascularized gingiva may be a means by which these bacteria can gain access to the circulating blood and establish infections at remote sites. Once located at sites such as the heart vessel walls, the invasion of endothelial cells (Deshpande *et al.*, 1998; Dorn *et al.*, 1999) could constitute a chronic insult to arterial walls. Injured, or activated, endothelial cells may demonstrate a variety of artherogenic properties, including increased pro-coagulant activity, secretion of vasoactive and inflammatory mediators, and expression of adhesion molecules (Deshpande *et al.*, 1998–9). If verified by additional experimentation, these bacterial-endothelial cell interactions could contribute to the pathology of cardiovascular disease.

It is likely that we are only beginning to uncover the full range of consequences of the interactions between invasive oral bacteria and host cells. In the future, high throughput techniques such as DNA array analysis and proteomics can be expected to reveal a greater range of responses of both bacteria and host cells to their coexistence. An understanding of these sensing-response mechanisms and their biological implications will make a significant contribution to the aetiology of periodontal diseases and possibly provide insight into serious systemic diseases.

ACKNOWLEDGEMENTS

Thanks go to Gwyneth Lamont for art work.

REFERENCES

Alugupalli, K.R., Kalfas, S. and Forsgren, A. (1996). Laminin binding to a heat-modifiable outer membrane protein of *Actinobacillus actinomycetemcomitans*. *Oral Microbiology and Immunology* **11**, 326–331.

Banchereau, J. and Steinman, R.M. (1998). Dendritic cells and the control of immunity. *Nature* **392**, 245–252.

Barkocy-Gallagher, G.A., Han, N., Patti, J.M., Whitlock, J., Progulske-Fox, A. and Lantz, M.S. (1996). Analysis of the *prtP* gene encoding porphypain, a cysteine proteinase of *Porphyromonas gingivalis*. *Journal of Bacteriology* **178**, 2734–2741.

Belton, C.M., Izutsu, K.T., Goodwin, P.C., Park, Y. and Lamont, R.J. (1999). Fluorescence image analysis of the association between *Porphyromonas gingivalis* and gingival epithelial cells. *Cellular Microbiology* **1**, 215–224.

Brissette, C.A. and Fives-Taylor, P.M. (1999). *Actinobacillus actinomycetemcomitans* may utilize either actin-dependent or actin-independent mechanisms of invasion. *Oral Microbiology and Immunology* **14**, 137–142.

Cantley, L.C., Auger, K.R., Carpenter, C., Duckworth, B., Graziani, A., Kapeller,

R. and Soltoff, S. (1991). Oncogenes and signal transduction. *Cell* **64**, 281–302.

Cheng, L.W. and Schneewind, O. (2000). Type III machines of Gram-negative bacteria: delivering the goods. *Trends in Microbiology* **8**, 214–220.

Chiu, B. (1999). Multiple infections in carotid atherosclerotic plaques. *American Heart Journal* **138**, S534–S536.

Christersson, L.A., Wikesjo, U.M., Albini, B., Zambon, J.J. and Genco, R.J. (1987). Tissue localization of *Actinobacillus actinomycetemcomitans* in human periodontitis. II. Correlation between immunofluorescence and culture techniques. *Journal of Periodontology* **58**, 540–545.

Curtis, M.A., Kuramitsu, H.K., Lantz, M., Macrina, F.L., Nakayama, K., Potempa, J., Reynolds, E.C. and Aduse-Opoku, J. (1999). Molecular genetics and nomenclature of proteases of *Porphyromonas gingivalis*. *Journal of Periodontal Research* **34**, 464–472.

Cutler C.W., Jotwani, R., Palucka, K.A., Davoust, J., Bell, D. and Banchereau, J. (1999). Evidence and a novel hypothesis for the role of dendritic cells and *Porphyromonas gingivalis* in adult periodontitis. *Journal of Periodontal Research* **34**, 406–412.

Darveau, R.P., Belton, C.M., Reife, R.A. and Lamont, R.J. (1998). Local chemokine paralysis: a novel pathogenic mechanism of *Porphyromonas gingivalis*. *Infection and Immunity* **66**, 1660–1665.

DeCarlo, A.A., Windsor, L.J., Bodden, M.K., Harber, G.J., Birkedal-Hansen, B. and Birkedal-Hansen, H. (1997). Activation and novel processing of matrix metalloproteinases by thiol-proteinase from the oral anaerobe *Porphyromonas gingivalis*. *Journal of Dental Research* **76**, 1260–1270.

DeCarlo, A.A., Grenett, H.E., Harber, G.J., Windsor, L.J., Bodden, M.K., Birkedal-Hansen, B. and Birkedal-Hansen, H. (1998). Induction of matrix metalloproteinases and a collagen-degrading phenotype in fibroblasts and epithelial cells by a secreted *Porphyromonas gingivalis* proteinase. *Journal of Periodontal Research* **33**, 408–420.

Deshpande, R.G., Khan, M.B. and Genco, C.A. (1998). Invasion of aortic and heart endothelial cells by *Porphyromonas gingivalis*. *Infection and Immunity* **66**, 5337–5343.

Deshpande, R.G., Khan, M. and Genco, C.A. (1998–99). Invasion strategies of the oral pathogen *Porphyromonas gingivalis*: implications for cardiovascular disease. *Invasion and Metastasis* **18**, 57–69.

DeVinney, I., Steele-Mortimer, I. and Finlay, B.B. (2000). Phosphatases and kinases delivered to the host cell by bacterial pathogens. *Trends in Microbiology* **8**, 29–33.

DiFranco, C.F., Toto, P.D., Rowden, G., Gargiulo, A.W., Keene, J.J. and Connelly,

E. (1985). Identification of Langerhans cells in human gingival epithelium. *Journal of Periodontology* **56**, 48–54.

Dorn, B.R., Dunn, W.A. and Progulske-Fox, A. (1999). Invasion of human coronary artery cells by periodontal pathogens. *Infection and Immunity* **67**, 5792–5798.

Dorn B.R., Burks, J.N., Seifert, K.N. and Progulske-Fox, A. (2000). Invasion of endothelial and epithelial cells by strains of *Porphyromonas gingivalis*. FEMS *Microbiology Letters* **187**, 139–144.

Duncan, M.J., Nakao, S., Skobe, Z. and Xie, H. (1993). Interactions of *Porphyromonas gingivalis* with epithelial cells. *Infection and Immunity* **61**, 2260–2265.

Fives-Taylor, P.M., Meyer, D.H. and Mintz, K.P. (1995). Characteristics of *Actinobacillus actinomycetemcomitans* invasion of and adhesion to cultured epithelial cells. Advances in Dental Research **9**, 55–62.

Fives-Taylor, P.M., Meyer, D.H., Mintz, K.P. and Brissette, C. (1999). Virulence factors of *Actinobacillus actinomycetemcomitans*. *Periodontology 2000* **20**, 136–167.

Fletcher, J., Reddi, K., Poole, S., Nair, S., Henderson, B., Tabona, P. and Wilson, M. (1997). Interactions between periodontopathogenic bacteria and cytokines. *Journal of Periodontal Research* **32**, 200–205.

Fravalo, P., Menard, C. and Bonnaure-Mallet, M. (1996). Effect of *Porphyromonas gingivalis* on epithelial cell MMP-9 type IV collagenase production. *Infection and Immunity* **64**, 4940–4945.

Greenstein, G. and Lamster, I. (1997). Bacterial transmission in periodontal diseases: a critical review. *Journal of Periodontology* **68**, 421–431.

Haase, E.M., Zmuda, J.L. and Scannapieco, F.A. (1999). Identification and molecular analysis of rough-colony-specific outer membrane proteins of *Actinobacillus actinomycetemcomitans*. *Infection and Immunity* **67**, 2901–2908.

Herzberg, M.C., MacFarlane, G.D., Liu, P. and Erickson, P.R. (1994). The platelet as an inflammatory cell in periodontal diseases: the interactions with *Porphyromonas gingivalis*. In *Molecular Pathogenesis of Periodontal Disease*, ed. R. Genco, S. Hamada, T. Lehner, J. McGhee and S. Mergenhagen, pp. 247–256, Washington, DC: ASM Press.

Hritz, M., Fisher, E. and Demuth, D.R. (1996). Differential regulation of the leukotoxin operon in highly leukotoxic and minimally leukotoxic strains of *Actinobacillus actinomycetemcomitans*. *Infection and Immunity* **64**, 2724–2729.

Huang, G.T., Haake, S.K., Kim, J.W. and Park, N. (1998a). Differential expression of interleukin-8 and intercellular adhesion molecule-1 by human gingival epithelial cells in response to *Actinobacillus actinomycetemcomitans* or *Porphyromonas gingivalis* infection. *Oral Microbiology and Immunology* **13**, 301–309.

Huang, G.T., Haake, S.K. and Park, N. (1998b). Gingival epithelial cells increase interleukin-8 secretion in response to *Actinobacillus actinomycetemcomitans* challenge. *Journal of Periodontology* **69**, 1105–1110.

Hultgren, S.J., Jones, C.H. and Normark, S. (1996). Bacterial adhesins and their assembly. In Escherichia coli *and* Salmonella, ed. F.C. Neidhardt, 2nd edn, pp. 2730–2756, Washington, DC: ASM Press.

Hynes, R., Ali, I., Destree, A., Mautner, V., Perkins, M., Senger, D., Wagner, D. and Smith, K. (1978). A large glycoprotein lost from the surfaces of transformed cells. *Annals of the New York Academy of Sciences* **312**, 317–343.

Inoue, T., Tanimoto, I., Ohta, H., Kato, K., Murayama, Y. and Fukui, K. (1998). Molecular characterization of low-molecular-weight component protein, Flp, in *Actinobacillus actinomycetemcomitans* fimbriae. *Microbiology and Immunology* **42**, 253–258.

Ishihara, K., Honma, K., Miura, T., Kato, T. and Okuda, K. (1997). Cloning and sequence analysis of the fimbriae associated protein (*fap*) gene from *Actinobacillus actinomycetemcomitans*. *Microbial Pathogenesis* **23**, 63–69.

Izutsu, K.T., Belton, C.M., Chan, A., Fatherazi, S., Kanter, J.P., Park, Y. and Lamont, R.J. (1996). Involvement of calcium in interactions between gingival epithelial cells and *Porphyromonas gingivalis*. *FEMS Microbiology Letters* **144**, 145–150.

Katz, J., Sambandam, V., Wu, J.H., Michalek, S.M. and Balkovetz, D.F. (2000). Characterization of *Porphyromonas gingivalis*-induced degradation of epithelial cell junctional complexes. *Infection and Immunity* **68**, 1441–1449.

Kontani, M., Ono, H., Shibata, H., Okamura, Y., Tanaka, T., Fujiwara, T., Kimura, S. and Hamada, S. (1996). Cysteine protease of *Porphyromonas gingivalis* 381 enhances binding of fimbriae to cultured human fibroblasts and matrix proteins. *Infection and Immunity* **64**, 756–762.

Kontani, M., Kimura, S., Nakagawa, I. and Hamada, S. (1997). Adherence of *Porphyromonas gingivalis* to matrix proteins via a fimbrial cryptic receptor exposed by its own arginine-specific protease. *Molecular Microbiology* **24**, 1179–1187.

Kuramitsu, H.K. (1998). Proteases of *Porphyromonas gingivalis*: what don't they do? *Oral Microbiology and Immunology* **13**, 263–270.

Laidig, K.E., Chen, W., Park, Y., Park, K., Yates, J.R., Lamont, R.J. and Hackett, M. (2001). Searching the *Porphyromonas gingivalis* genome with peptide fragmentation mass spectra. *The Analyst* **126**, 52–57.

Lamont, R.J. and Jenkinson, H.F. (1998). Life below the gum line: pathogenic mechanisms of *Porphyromonas gingivalis. Microbiology and Molecular Biology Reviews* **62**, 1244–1263.

Lamont, R.J., Oda, D., Persson, R.E. and Persson, G.R. (1992). Interaction of *Porphyromonas gingivalis* with gingival epithelial cells maintained in culture. *Oral Microbiology and Immunology* **7**, 364–367.

Lamont, R.J., Chan, A., Belton, C.M., Izutsu, K.T., Vasel, D. and Weinberg, A. (1995). *Porphyromonas gingivalis* invasion of gingival epithelial cells. *Infection and Immunity* **63**, 3878–3885.

Lépine, G., and Progulske-Fox, A. (1996). Duplication and differential expression of hemagglutinin genes in *Porphyromonas gingivalis. Oral Microbiology and Immunology* **11**, 65–78.

Lépine, G., Caudry, S., DiRienzo, J.M. and Ellen, R.P. (1998). Epithelial cell invasion by *Actinobacillus actinomycetemcomitans* strains from restriction fragment-length polymorphism groups associated with juvenile periodontitis or carrier status. *Oral Microbiology and Immunology* **13**, 341–347.

Madianos, P.N., Papapanou, P.N., Nannmark, U., Dahlen, G. and Sandros, J. (1996). *Porphyromonas gingivalis* FDC381 multiplies and persists within human oral epithelial cells *in vitro. Infection and Immunity* **64**, 660–664.

Madianos, P.N., Papapanou, P.N. and Sandros, J. (1997). *Porphyromonas gingivalis* infection of oral epithelium inhibits neutrophil transepithelial migration. *Infection and Immunity* **65**, 3983–3990.

Meyer, D.H. and Fives-Taylor, P.M. (1993). Evidence that extracellular components function in adherence of *Actinobacillus actinomycetemcomitans* to epithelial cells. *Infection and Immunity* **61**, 4933–4936.

Meyer, D.H. and Fives-Taylor, P.M. (1994). Characteristics of adherence of *Actinobacillus actinomycetemcomitans* to epithelial cells. *Infection and Immunity* **62**, 928–935.

Meyer, D.H., Lippmann, J.E. and Fives-Taylor, P.M. (1996). Invasion of epithelial cells by *Actinobacillus actinomycetemcomitans*: a dynamic, multistep process. *Infection and Immunity* **64**, 2988–2997.

Meyer, D.H., Mintz, K.P. and Fives-Taylor, P.M. (1997). Models of invasion of enteric and periodontal pathogens into epithelial cells: a comparative analysis. *Critical Reviews of Oral Biology and Medicine* **8**, 389–409.

Meyer, D.H., Rose, J.E., Lippmann, J.E. and Fives-Taylor, P.M. (1999). Microtubules are associated with intracellular movement and spread of the periodontopathogen *Actinobacillus actinomycetemcomitans. Infection and Immunity* **67**, 6518–6525.

Mintz, K.P, and Fives-Taylor, P.M. (1999). Binding of the periodontal pathogen *Actinobacillus actinomycetemcomitans* to extracellular matrix proteins. *Oral Microbiology and Immunology* **14**, 109–116.

Nakamura, T., Amano, A., Nakagawa, I. and Hamada, S. (1999). Specific interactions between *Porphyromonas gingivalis* fimbriae and human extracellular matrix proteins. FEMS *Microbiology Letters* **175**, 267–272.

Nakayama, K., Yoshimura, F., Kadowaki, T. and Yamamoto, K. (1996). Involvement of arginine-specific cysteine proteinase (Arg-gingipain) in fimbriation of *Porphyromonas gingivalis*. *Journal of Bacteriology* **178**, 2818–2824.

Njoroge, T., Genco, R.J., Sojar, H.T. and Genco, C.A. (1997). A role for fimbriae in *Porphyromanas gingivalis* invasion of oral epithelial cells. *Infection and Immunity* **65**, 1980–1984.

Noiri, Y., Ozaki, K., Nakae, H., Matsuo, T. and Ebisu, S. (1997). An immunohistochemical study on the localization of *Porphyromonas gingivalis*, *Campylobacter rectus* and *Actinomyces viscosus* in human periodontal pockets. *Journal of Periodontal Research* **32**, 598–607.

Oda, D. and Whatsin, E. (1990). Human oral epithelial cell culture. I. Improved conditions for reproducible culture in serum-free medium. *In Vitro Cell Developmental Biology* **26**, 589–595.

Park, Y. and Lamont, R.J. (1998). Contact-dependent protein secretion in *Porphyromonas gingivalis*. *Infection and Immunity* **66**, 4777–4782.

Pekovic, D.D. and Fillery, E.D. (1984). Identification of bacteria in immunopathological mechanisms of human periodontal diseases. *Journal of Periodontal Research* **19**, 329–351.

Potempa, J., Pavloff, N. and Travis, J. (1995). *Porphyromonas gingivalis*: a proteinase/gene accounting audit. *Trends in Microbiology* **3**, 430–434.

Progulske-Fox, A., Tumwasorn, S., Lepine, G., Whitlock, J., Savett, D., Ferretti, J.J. and Banas, J.A. (1995). The cloning, expression and sequence analysis of second *Porphyromonas gingivalis* gene that encodes for a protein involved in haemagglutination. *Oral Microbiology and Immunology* **10**, 311–318.

Rosan, B., Slots, J., Lamont, R.J., Listgarten, M.A. and Nelson, G.M. (1988). *Actinobacillus actinomycetemcomitans* fimbriae. *Oral Microbiology and Immunology* **3**, 58–63.

Saglie, F.R., Smith, C.T., Newman, M.G., Carranza, F.A., Pertuiset, J.H., Cheng, L., Auil, E. and Nisengard, R.J. (1986). The presence of bacteria in oral epithelium in periodontal disease. II. Immunohistochemical identification of bacteria. *Journal of Periodontology* **57**, 492–500.

Saglie, F.R., Pertuiset, J.H., Smith, C.T., Nestor, M.G., Carranza, F.A., Newman, M.G., Rezende, M.T. and Nisengard, R. (1987). The presence of bacteria in the oral epithelium in periodontal disease. III. Correlation with Langerhans cells. *Journal of Periodontology* **58**, 417–422.

Saglie, F.R., Marfany, A. and Camargo, P. (1988). Intragingival occurrence of

Actinobacillus actinomycetemcomitans and *Bacteroides gingivalis* in active destructive periodontal lesions. *Journal of Periodontology* **59**, 259–265.

Sandros, J., Papapanou, P.N. and Dahlen, G. (1993). *Porphyromonas gingivalis* invades oral epithelial cells *in vitro*. *Journal of Periodontal Research* **28**, 219–226.

Sandros, J., Papapanou, P.N., Nannmark, U. and Dahlen, G. (1994). *Porphyromonas gingivalis* invades human pocket epithelium *in vitro*. *Journal of Periodontal Research* **29**, 62–69.

Sandros, J., Madianos, P.N. and Papapanou, P.N. (1996). Cellular events concurrent with *Porphyromonas gingivalis* invasion of oral epithelium in vitro. *European Journal of Oral Sciences* **104**, 363–371.

Scannapieco, F.A. and Genco, R.J. (1999). Association of periodontal infections with atherosclerotic and pulmonary diseases. *Journal of Periodontal Research* **34**, 340–345.

Scannapieco, F.A, Millar, S.J., Reynolds, H.S., Zambon, J.J. and Levine, M.J. (1987). Effect of anaerobiosis on the surface ultrastructure and surface proteins of *Actinobacillus actinomycetemcomitans* (*Haemophilus actinomycetemcomitans*). *Infection and Immunity* **55**, 2320–2323.

Shibita Y., Hayakawa, M., Takiguchi, H., Shiroza, T. and Abiko, Y. (1999). Determination and characterization of the hemagglutinin-associated short motifs found in *Porphyromonas gingivalis* multiple gene products. *Journal of Biological Chemistry* **274**, 5012–5020.

Socransky S.S. and Haffajee, A.D. (1992). The bacterial etiology of destructive periodontal disease: current concepts. *Journal of Periodontology* **63**, 322–331.

Sojar, H.T., Lee, J-Y. and Genco, R.J. (1995). Fibronectin binding domain of *P. gingivalis* fimbriae. *Biochemical and Biophysical Research Communications* **216**, 785–792.

Sojar, H.T., Han, Y., Hamada, N., Sharma, A. and Genco, R.J. (1999). Role of the amino-terminal region of *Porphyromonas gingivalis* fimbriae in adherence to epithelial cells. *Infection and Immunity* **67**, 6173–6176.

Sreenivasan, P.K., Meyer, D.H. and Fives-Taylor, P.M. (1993). Requirements for invasion of epithelial cells by *Actinobacillus actinomycetemcomitans*. *Infection and Immunity* **61**, 1239–1245.

Tokuda, M., Duncan, M., Cho, M.I. and Kuramitsu, H.K. (1996). Role of *Porphyromonas gingivalis* protease activity in colonization of oral surfaces. *Infection and Immunity* **64**, 4067–4073.

Tonetti, M.S., Imboden, M.A., Gerber, L., Lang, N.P., Laissue, J. and Mueller, C. (1994). Localized expression of mRNA for phagocyte-specific chemotactic cytokines in human periodontal infections. *Infection and Immunity* **62**, 4005–4014.

Van Dyke, T.E. (1984). Neutrophil receptor modulation in the pathogenesis of periodontal diseases. *Journal of Dental Research* **63**, 452–454.

Watanabe, K., Onoe, T., Ozeki, M., Shimizu, Y., Sakayori, T., Nakamura, H. and Yoshimura, F. (1996). Sequence and product analyses of the four genes downstream from the fimbrillin gene (*fimA*) of the oral anaerobe *Porphyromonas gingivalis. Microbiology and Immunobiology* **40**, 725–734.

Weinberg, A., Belton, C.A., Park, Y. and Lamont, R.J. (1997). Role of fimbriae in *Porphyromonas gingivalis* invasion of gingival epithelial cells. *Infection and Immunity* **65**, 313–316.

Xie, H. and Lamont, R.J. (1999). Promoter architecture of the *Porphyromonas gingivalis* fimbrillin gene. *Infection and Immunity* **67**, 3227–3235.

Xie, H., Cai, S. and Lamont, R.J. (1997). Environmental regulation of fimbrial gene expression in *Porphyromonas gingivalis. Infection and Immunity* **65**, 2265–2271.

Xie, H., Chung, W., Park, Y. and Lamont, R.J. (2000). Regulation of the *Porphyromonas gingivalis fimA* (fimbrillin) gene. *Infection and Immunity* **68**, 6574–6579.

Yun, P.L., DeCarlo, A.A. and Hunter, N. (1999). Modulation of major histocompatibility complex protein expression by human gamma interferon mediated by cysteine proteinase-adhesin polyproteins of *Porphyromonas gingivalis. Infection and Immunity* **67**, 2986–2995.

Zhang, J., Dong, H., Kashket, S. and Duncan, M.J. (1999). IL-8 degradation by *Porphyromonas gingivalis* proteases. *Microbial Pathogenesis* **26**, 275–280.

Index

INDEX

INDEX